Introduction to
ATOMIC AND
MOLECULAR
COLLISIONS

Introduction to
ATOMIC AND MOLECULAR COLLISIONS

R. E. Johnson

University of Virginia
Charlottesville, Virginia

Illustrated by **T. J. Johnson**

PLENUM PRESS • NEW YORK AND LONDON

Library of Congress Cataloging in Publication Data

Johnson, R. E., 1939–
    Introduction to atomic and molecular collisions.

    Includes bibliographies and index.
    1. Collisions (Nuclear physics). 2. Atoms. 3. Molecules. 4. Quantum chemistry.
I. Title.
QC794.6.C6.J63   539.7′54   82-3776
ISBN 0-306-40787-6   AACR2

© 1982 Plenum Press, New York
A Division of Plenum Publishing Corporation
233 Spring Street, New York, N.Y. 10013

Printed in the United States of America

This book is dedicated to
Barbara, Amanda, and Sarah

Preface

In working with graduate students in engineering physics at the University of Virginia on research problems in gas kinetics, radiation biology, ion–materials interactions, and upper-atmosphere chemistry, it became quite apparent that there was no satisfactory text available to these students on atomic and molecular collisions. For graduate students in physics and quantum chemistry and researchers in atomic and molecular interactions there are a large number of excellent advanced texts. However, for students in applied science, who require some knowledge and understanding of collision phenomena, such texts are of little use. These students often have some background in modern physics and/or chemistry but lack graduate-level course work in quantum mechanics. Such students, however, tend to have a good intuitive grasp of classical mechanics and have been exposed to wave phenomena in some form (e.g., electricity and magnetism, acoustics, etc.). Further, their requirements in using collision processes and employing models do not generally include the use of formal scattering theory, a large fraction of the content of many advanced texts. In fact, most researchers who work in the area of atomic and molecular collisions tend to pride themselves on their ability to describe results using simple theoretical models based on classical and semiclassical methods.

This book was written in order to allow a student to develop an understanding of atomic and molecular collision phenomena based on those classical and semiclassical methods and approximations employed frequently in the literature. The book is aimed at the scientist or engineer in such fields as engineering physics, materials science, chemistry, astronomy, aerospace engineering, nuclear engineering, electrical engineering, radiation biology, atmospheric and solar physics, and plasma science, and who requires an understanding of atomic and molecular interactions and collisions to better understand, for example, chemical kinetics, gas dynamics, laser processes, interactions of radiation with materials, etc. However, the book should also be useful to physics and quantum chemistry students actually working on atomic collisions and molecular interaction problems. Having been such a student myself, I found it took a remarkable amount of time to develop the intuitive grasp that was used by workers already in the field in discussing collision-related problems. It was even more difficult to obtain a clear view of the usefulness of the measurements and calculations produced. Therefore, in this book some time is spent discussing, first qualitatively and later in a simple quantitiative way, applications of cross section and reaction rate results to a few macroscopic phenomena controlled by molecular-level events. The problem areas chosen for consideration are ones of which I had at least peripheral knowledge, and I apologize in advance for not including particularly obvious areas dear to some readers' hearts. However, a final purpose of this book is to make clear the commonality of methods and approximations used in dealing with effects controlled by electrons, ions, atoms, and molecules.

This book is written as a graduate-level text for a one-semester course and includes problems and a general bibliography at the end of each chapter. However, it is also hoped that, with the material presented in the appendixes, this book will be a useful reference for classical and semi-classical formulas that are often employed in the literature. Some of the problems and the appendixes are, simply, derivations of results presented in order to keep the text short and, hopefully, keep the interest and involvement of the reader high. In the first chapter, a rationale and a perspective are given for understanding the collisions of atoms and molecules. A number of collision-related macroscopic phenomena are discussed qualitatively with the role of atomic and molecular interactions delineated. In this chapter the type of information required regarding collisions is determined. In the second chapter, the notion of a cross section is developed and then related to collision quantities discussed above. This is followed by a classical treatment of collisions. The pace in this chapter is slow at first, with words used where in many texts mathematical expressions would suffice. However, I felt that if the definition of a cross section was quite clear at the outset even readers unfamiliar with collision concepts could follow the subsequent, much more terse presentations. In the rest of the text I continuously relate back to the ideas developed in Chapter 2.

Background material on wave mechanics and atomic and molecular notation are given in Chapter 3 in order that students who have had only basic instruction in this area can easily follow the discussion of the wave mechanical description of collisions. The description of molecular interactions is completed in Chapter 4, where interaction potentials (which are required to determine the motion of the colliding particles) and transition probabilities (which describe inelastic effects) are calculated. Students specializing in collision research, having a strong background in quantum mechanics, should be able to read this portion of the text and work out the problems on their own in a short period of time as, for instance, preparation for starting dissertation work. The latter part of Chapter 4, on the other hand, may prove to be rather terse for readers unfamiliar with the methods of modern physics. However, I feel the support material is available in earlier chapters and in the appendix, and this material, with patience, is understandable and clearly worthwhile. Up to this point in the presentation, little in the way of experimental data is presented since such results cannot be described clearly without an understanding of both the calculation of the cross sections and the nature of the potentials and transition probabilities. Therefore, in Chapter 5, I present a summary of experimental results for cross sections and rate constants and relate these to the cross-section models and potentials discussed earlier. This is followed, in Chapter 6, by a return to the material presented in Chapter 1. Now, however, estimates of the collision quantities are made in order to describe, in a simple, quantitative way, certain phenomena determined by molecular collisions.

In writing this book I would first like to acknowledge the benefit of a sabbatical leave of absence from the University of Virginia spent at Harvard University. I would also like to thank R. G. Cooks (Purdue University) and M. Inokuti (Argonne National Laboratory) for a careful reading of the manuscript and many helpful comments and criticisms. The following students, who read a draft of the manuscript as part of a course in engineering physics, were also very helpful: G. Cooper, R. Evatt, P. Wantuck, and E. Sieveka. A number of colleagues also read and criticized various sections of the book: J. R. Scott and J. W. Boring (University of Virginia), T. A. Green (Sandia Corporation), G. Victor and W. A. Traub (Harvard-Smithsonian Astrophysical Observatory), W. L. Brown (Bell Laboratory), S. Hamasaki (Jaycor Corporation), J. B. Delos (College of William and Mary), R. B. Bernstein (Columbia University), P. Sigmund (Odense University), and U. Fano (University of Chicago). Finally, I would like to thank my father, T. J. Johnson, for producing many of the illustrations used and my family for their support and encouragement while writing this text.

R. E. Johnson

Contents

4. Interaction Potentials and Transition Probabilities

5. Cross Sections and Rate Constants: Results

6. Application of Results

Appendixes

1

Application Areas

Introduction

The need to understand the behavior of colliding atoms and molecules is self-evident as we live in a world constructed from atomic building blocks. This is not a static construction of the type envisaged early in our history; rather, it is a dynamic construction of moving particles constrained by a few fundamental forces. It is also a world with large differences in density, temperature, and types of material. The combination of mobile, interactive atoms that are distributed nonuniformly provides the basis for the rich variety of phenomena observed in our universe, from exploding stars to the evolution of life.

One would assume that an understanding of macroscopic phenomena occurring in such a universe should require a corresponding understanding of the atomic level phenomena. In many cases this assumption, although true, is not of much importance. For instance, the forces between large objects are clearly the residuals of the forces between the atoms with which they are made, but it is seldom useful to pursue this. However, there are many macroscopic phenomena for which the atomic behavior does play a fundamental, controlling role and an understanding of the atomic level dynamics is important. In passing, we note that the study of the atomic and molecular basis of macroscopic phenomena has always been strongly linked to notions of scientific reductionism by the nonscientific community. This was a special concern because atoms and molecules are the building blocks for biology and hence life. Much of the general resistance to the idea of trying to understand our world by reducing it to its building blocks is justified, as there often has been an accompanying lack of sensitivity toward larger issues. However, part of the opposition is a matter of style. It is based on the mistaken notion that the description of the behavior of atoms is not an art form, when in fact it clearly is. Although this text will obviously place a heavy emphasis on the quantitative aspects of the description, I hope the reader will also obtain a feeling for the qualitative aspects, the envisaging of "unseen" phenomena that is the art of physics.

In this chapter some problem areas that require an understanding of atomic and molecular collisions will be considered. The discussion will be primarily descriptive and we will return to treat some of these same phenomena quantitatively in the last chapter. This discussion also should clarify which problems will and will not be treated in the text. Put simply, the primary emphasis will be on the collisions and interactions of ions, atoms and molecules, and the heavy particles. Attention will be paid to the result of these interactions, like the production of photons and electrons from excitations and ionization of the atoms. Similarities and differences between heavy-particle collisions and collisions involving incident electrons will be considered and examples involving incident electrons will be used at certain points in the text. I will ignore primary events initiated by photons and changes in the nuclear structure of the atoms although secondary events involving the ions or electrons produced will be treated. Finally, nonrelativistic velocities only will be considered. Although the following presentation is necessarily simplified, it is hoped that the important collisional information required will be clear and the similarities between disparate research areas will be evident.

Radiation Cascades

When radiations (particles or photons) from an outside source impinge on and penetrate into a material, various observable phenomena occur that are a direct result of a chain of events referred to as a radiation cascade. For instance, when the sun shines on the atmosphere, the concentrations of chemical species change and an ionized region appears that reflects radio waves. Likewise, a source of fast particles, like a nuclear reactor, can cause embrittlement of the container or general damage to the surrounding materials, including the inactivation of biological materials. In both cases the initial radiation loses intensity as it passes through the material and initiates new radiation within the material.

In the first example, the constituents of the atmosphere are ionized by ultraviolet light, creating a plasma of ions and fast electrons. These photoelectrons are in turn slowed, transferring their energy to the atmospheric constituents and, in the process, cause further excitations and ionizations. The excited atoms and molecules eventually emit photons of lower frequency, hence the initial photons are responsible for a cascade of electrons and lower-energy photons. The ions created in the process also initiate a series of events, referred to as chemical reactions. These processes will be considered shortly.

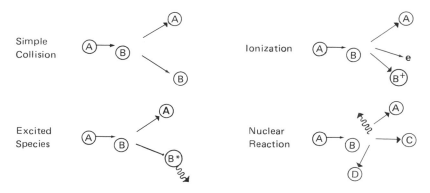

Figure 1.1. Events that occur frequently when atoms collide.

Changes in the structure of materials subjected to fast heavy-particle radiation, like alpha particles, protons, or neutrons, can be understood as a series of collision events of the incident particles with the atoms of the target material. The fast moving, incident particles in these collisions transfer kinetic energy to a number of target atoms and, in addition, may ionize or excite these atoms and even initiate a nuclear reaction. Therefore, a single collision event can produce fast heavy particles, referred to as secondaries, electrons, or photons, as shown in Figure 1.1. If a secondary atom or ion is excited in a very dense material (e.g., a solid), it may lose its internal energy in a subsequent collision with another target atom. Very high energy photons, x-rays and γ-rays, are often produced in materials when the tightly bonded, inner-shell electrons of heavy atoms are removed or when nuclear reactions are initiated. These high-energy photons may also produce additional ionization and, hence, electrons. Therefore, the incident particles initiate two separate cascades—secondary electrons and fast heavy particles. The embrittlement of material, which is one result of such a cascade, is caused primarily by displacement damage, the knocking out of place of target particles. On the other hand, the damage to biological materials is mostly due to the cascade of ionization events set up by the secondary electrons and high-energy photons. Radiation cascades are also used to modify the nature of a material, by implanting incident atoms, and to remove layers of a material, by "knocking off" or sputtering the surface atoms.

The biological case is worth pursuing as a means of understanding the relationship between initial radiation events and final macroscopic effects. In considering the effect of high-energy radiation on cells (or the biological constituents of the cell: DNA, RNA, and various enzymes), the simplest question one asks is: What happens to a cell after it receives a given quantity of radiation?

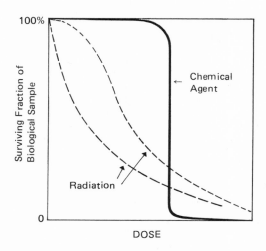

Figure 1.2. Diagram illustrating typical dose–effect curves for the action of chemical agents and radiation [cf. K. G. Zimmer, *Studies on Quantitative Radiation Biology*, Oliver & Boyd, Edinburgh and London (1961)].

Dose–response curves of the type shown in Figure 1.2 clearly indicate the causal nature of radiation damage, which is different in nature from damage induced by administering chemical agents for which there is a clear threshold. Simply relating damage to dose, although useful as an initial attempt at radiation protection, is limited by the distinctiveness and distribution of the initial energy-absorbing events between, for instance, processes initiated by γ-rays and α-particles. To simplify the problem, it is customary to subdivide the set of events initiated by the radiation into various stages, associating approximate characteristic times with the events occurring at each stage, as shown in Table 1.1.

The characteristic times shown in Table 1.1 indicate that, although a chemical reaction can occur, in a probabilistic sense, during the absorption of the radiation, the relative slowness of these reactions means it is highly unlikely. The reactions occur in response to the background milieu of broken bonds and ionizations and are not dependent *directly* on the type of radiation. The effect of the type and energy of the radiation will determine only the initial state of the material. In the first two stages indicated, the material is in a very nonequilibrium state. When thermal equilibrium of moving particles is established locally, chemical reaction kinetics involving reactive radicals dominates the material transformations. Finally, the "permanent" chemical alterations can be slowly repaired, resulting in cell alterations, or even be amplified by biological processes. In the following, we discuss the mathematical description of the physical stage for a biological or physical material in which the atomic constituents can be considered randomly ordered and the incident radiation is a fast ion. The discussion is intended to make clear the collision parameters needed for modeling this problem.

To describe a radiation cascade, we imagine dividing a uniformly ir-

Table 1.1. Events Initiated by Radiation

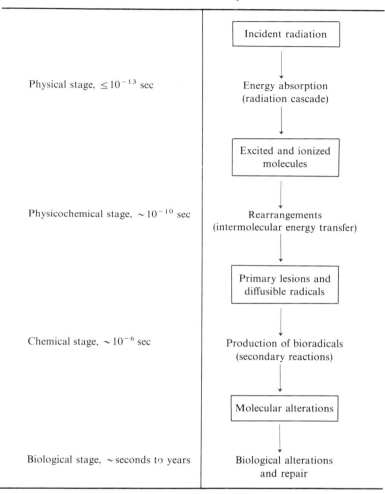

	Incident radiation
Physical stage, $\leq 10^{-13}$ sec	Energy absorption (radiation cascade)
	Excited and ionized molecules
Physicochemical stage, $\sim 10^{-10}$ sec	Rearrangements (intermolecular energy transfer)
	Primary lesions and diffusible radicals
Chemical stage, $\sim 10^{-6}$ sec	Production of bioradicals (secondary reactions)
	Molecular alterations
Biological stage, $\sim$ seconds to years	Biological alterations and repair

radiated material into thin slabs. At each depth fast primaries and second-aries enter and leave the slab, as shown in Figure 1.3, at rates to be determined. These rates are obtained by applying a rather obvious conservation principle. That is, the difference between the radiation intensity entering and leaving the slab must be due to sources or sinks of radiation *within* the slab. For instance, an ion traversing the slab may collide with an atom, losing some of its momentum and setting the target atom in motion, which is now added to the flux of particles.

We define $I_i(\mathbf{p}, z)d^3p$ to be the intensity of particle radiation of type i,

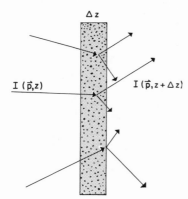

Figure 1.3. Collision cascade. Events leading to a change in the flux of particles of various momenta in a slab of thickness Δz.

with momenta between $\mathbf{p}$ and $\mathbf{p} + d\mathbf{p}$ at the depth z [i.e., $I_i(\mathbf{p}, z)d^3p$ is the number of particles per unit area per unit time having momenta between $\mathbf{p}$ and $\mathbf{p} + d\mathbf{p}$ crossing a unit surface perpendicular to $\mathbf{p}$]. The change in intensity in a slab involves a decrease due to interaction of the incident radiation with the target particles and an increase due to any newly created particles of the same type and momentum. We can define a quantity $\omega_i(\mathbf{p}, \mathbf{p}')$ to be the probability per unit path length of a particle of type i and momentum $\mathbf{p}$ changing to a momentum between $\mathbf{p}'$ and $\mathbf{p}' + d\mathbf{p}'$ in passing through a thin slab. The change in intensity in crossing the slab is

$$\Delta I(\mathbf{p}) = -\text{Loss } (\mathbf{p} \rightarrow \mathbf{p}') + \text{Gain } (\mathbf{p}'' \rightarrow \mathbf{p}) + \text{Sources} \qquad (1.1)$$

In Appendix E the transport equation for a radiation cascade is constructed using Eq. (1.1). The resulting integro-differential equation for the intensity at depth z of radiation of type i has the form

$$\cos \theta \, \frac{\partial I_i}{\partial z}(\mathbf{p}, z) = -\int [I_i(\mathbf{p}, z)\omega_i(\mathbf{p}, \mathbf{p}') - I_i(\mathbf{p}', z)\omega_i(\mathbf{p}', \mathbf{p})] \, d^3p' + \mathscr{S}_i(\mathbf{p}, z) \quad (1.2)$$

where $\cos \theta \equiv \hat{p} \cdot \hat{z}$ accounts for the difference between the direction of motion and the penetration direction. On the left of Eq. (1.2) is the change in the intensity of the radiation; on the right are the loss and gain due to scattering, where the integral indicates a sum over all possible changes in momenta. The $\mathscr{S}_i$ accounts for all other sources and sinks and obviously contains a considerable amount of the physics of processes occurring in materials. A similar equation can be constructed for each type of radiation in the cascade, and these equations will not be independent.

To solve Eq. (1.2), in addition to specifying boundary conditions, one has to describe $\mathscr{S}_i$ and $\omega_i(\mathbf{p}, \mathbf{p}')$ for each type of particle. This means that one has to calculate or measure the occurrence probability of all collisions or energy-loss events that might occur in the material, a formidable task. What is clear immediately is that $\mathscr{S}_i$ and $\omega_i(\mathbf{p}, \mathbf{p}')$ are proportional to the target number density. That is, more events will occur per unit path length if the targets are closely spaced. If the targets are all the same, and one defines the number of targets per unit volume as n_T, then $\omega(\mathbf{p}, \mathbf{p}') d^3p/n_T$ has the dimensions of area, and is referred to as the collision cross section for scattering with a change in the momentum from $\mathbf{p}$ to $\mathbf{p}'$. This is a quantity we will consider in detail throughout the text. A simple loss mechanism or source term, $\mathscr{S}_i$, occurs if, in addition to scattering, the material provides an overall drag force on the radiation, as is the case for charged-particle radiation. Writing the energy loss per unit path length for an atom of type i and momentum $\mathbf{p}$ moving through the atomic electrons as $(dE/ds|_e)_i$, referred to as the electronic stopping power, we have

$$\mathscr{S}_i(\mathbf{p}, z) = -\left(\left|\frac{dE}{ds}\right|_e\right)_i \frac{d}{dE} I_i(\mathbf{p}, z) \tag{1.3}$$

The intensity, $I_i(\mathbf{p}, z)$, tells the state of the cascade at any z. For example, if $\omega_i^b(\mathbf{p})$ is the probability per unit path length of producing a broken bond by a particle of type i and momentum $\mathbf{p}$, then the distribution of broken bonds per unit path length is calculated using I_i:

$$\sum_i \int I_i(\mathbf{p}, z) \omega_i^b(\mathbf{p}) \, d^3p$$

In Eq. (1.4), we have summed over all particle types in the cascade and all momenta. Source terms like $\mathscr{S}_i$ or ω_i^b, which are also collision related, form part of the subject matter of the text.

In the process of describing the cascade by solving for $I_i(\mathbf{p}, z)$, it is easy to lose track of the eventual goal, which is to predict or understand the possible macroscopic results of the cascade. Although the relationship between the initial set of events and the final result is often not clear, describing the cascade, and the immediate results of the cascade, is an important first step, and already much progress has been made in describing the latter stages. For irradiation-induced changes in physical materials, the situation is somewhat simpler. The physical stage is followed by a chemical stage in which the vacancies produced and atoms implanted migrate thermally until equilibrium is established, resulting in a material with new properties or structural damage.

Gas Dynamics

The description of the thermal behavior of gases was the first significant success of the atomic model of matter. This is most vividly characterized via Boltzmann's description of entropy, a quantity which is a direct result of the statistical nature of a universe composed of large numbers of small, mobile, identical components. Although the science of thermodynamics, without the explicit reference to the atomic nature of matter, is a well-understood art in its own right, the fact that its laws can be founded on atomic dynamics is important. It is important not only because of the statistical basis of the laws, but also because the atoms and molecules themselves have structure which produces deviations in even the simplest thermodynamic behavior. For instance, the van der Waals correction to the ideal gas law depends on the size of the atoms of the gas and the fact that they distort somewhat as they collide.

Whereas in our previous example we considered a directed beam on a material, the more general case of a random thermal flux entering a material is also important. The diffusion of gases through materials, including other gases, has been of continuing interest both as a means for separating gases, like the different isotopes of UF_6, and as a mechanism for the distribution of gases in the upper atmospheres of planets. Diffusion is a result of density and/or temperature gradients acting as thermodynamic driving forces. Conserving particles in any volume element of the gas, we give the continuity equation for diffusion in a stationary medium, shown in Appendix E to be

$$\frac{\partial}{\partial t} n_i + \nabla(n_i \mathbf{w}_i) = P_i - L_i \tag{1.5}$$

where n_i is the density of species i, $\mathbf{w}_i$ the mean transport velocity, and P_i and L_i are the production and loss terms. If over a reasonable period of time the average value of n_i at each position remains static, the time-independent form of Eq. (1.5),

$$\nabla(n_i \mathbf{w}_i) = P_i - L_i \tag{1.6}$$

is seen to be related to the earlier time-independent transport equation for a cascade, Eq. (1.2). That is, $n_i \mathbf{w}_i$ is an average flux of particles, $(P_i - L_i)$ is the source/sink term, and the collision term in Eq. (1.2) averages to zero because of the random nature of the particle motions.

The mean flux of particles can be determined from the conservation of momentum. Defining a frictional drag force on the particles proportional to $\mathbf{w}_i$, we write the change in momentum of the particles per unit volume, when $\nabla \cdot \mathbf{w}_i = 0$, as

$$\frac{d}{dt} (m_i n_i \mathbf{w}_i) = - v_i(m_i n_i \mathbf{w}_i) + \mathbf{f}_i \tag{1.7}$$

where m_i is the mass of the particles; v_i is the collision frequency, which indicates the resistance to flow; f_i is the force per unit volume, if any, acting on the particles; and the time derivative is a total time derivative $d/dt =$

$\partial/\partial t + \mathbf{w}_i \cdot \nabla$. In a steady state, the particle flux is proportional to the applied force,

$$n_i \mathbf{w}_i = \frac{1}{v_i m_i} \mathbf{f}_i \tag{1.8}$$

In a planetary atmosphere with density or temperature gradients, subject to the gravitational force, $\mathbf{f}_i$ has the form

$$\mathbf{f}_i = - m_i n_i \mathbf{g} - \nabla p_i \tag{1.9}$$

where p_i is the local thermodynamic pressure and $\mathbf{g}$ is the local gravitational attraction. Neglecting at first any sources or sinks, we have, in diffusive equilibrium, $\mathbf{f}_i = 0$. If we assume an ideal gas, $p_i = n_i k T_i$, where k is the Boltzmann constant and T_i is the temperature of species i, Eq. (1.9) yields the hydrostatic equation for a one-dimensional (flat) atmosphere:

$$n_i(z) = n_i(0) \frac{T_i(0)}{T_i(z)} \exp \left[- \int_0^z \frac{dz}{H_i} \right] \tag{1.10}$$

In Eq. (1.10) z is the altitude and $H_i = k T_i / m_i g$ is referred to as the scale height. This equilibrium requires no special knowledge, however, of atomic interactions. When $(P_i - L_i)$ is not zero, a diffusive flow is maintained, which is the case for a number of molecular species like NO, which are created and destroyed chemically in the upper atmosphere. Calculating the distribution of such species requires a knowledge of molecular collisions which determine both the collision frequency, v_i, in Eq. (1.7) and the reaction chemistry of the source term $(P_i - L_i)$ in Eq. (1.6).

For simple diffusion, in which there are no outside forces and the temperature is constant, the density gradient in Eq. (1.9) is the only driving force. For an ideal gas the equilibrium diffusive flux in Eq. (1.8) can be written

$$n_i \mathbf{w}_i = - D_i \nabla n_i \tag{1.11}$$

where D_i is the diffusion coefficient, $D_i = k T_i / m_i v_i$. The quantity D_i can be directly determined by experiment for some systems, from which v_i can be extracted, or v_i can be calculated from a knowledge of the intermolecular forces involved. Even a simple quantity like v_i is very system dependent, being sensitive to the diffusing species and the background gas molecules. The diffusion concept is quite general and occurs in many phenomena, as, for instance, the migration of vacant lattice sites produced by radiation damage in materials, mentioned in the previous section (e.g., Table 1.1), and the mobility of ions in a material subject to an electric field, ε. For the latter case, the drift velocity of the ions from Eq. (1.8) is $\mathbf{w}_i = \kappa_i \varepsilon$, where the quantity $\kappa_i = q/m v_i$ is the mobility coefficient with, q, the ion charge.

Another quite interesting application in gas dynamics is the study of thermal beams of gases emitted under pressure from nozzles. The early interest in this had to do with rockets and jet engines. However, nozzle beams of a much smaller scale have become practical tools for studying collisions, for use in gas lasers, and for isotope separation. In a nozzle beam, the collisional motion of the initially confined gas becomes forward directed, and the average speed of the particles is, therefore, related to the initial internal energy of the gas. As the particles in the nozzle exit, they eventually reach a region in which they move essentially collision-free. In this region, one is tempted to say that the internal energy has been totally converted to forward directed motion. However, part of the internal energy is stored as vibrational and rotational energy of the gas molecules which, while collisions occur, can be converted to kinetic energy and vice versa. In the collision-free region, the remaining excited molecules may eventually emit photons, but this is a slow process, and in any case this energy can never be converted into beam motion. It is, therefore, of some interest to design nozzle beams in which a maximum amount of this internal energy is converted into beam energy, or one would like to be able to know, *a priori*, what the distribution of excited molecular species in the beam is. The nozzle beam, therefore, can be thought of as an example of a rapid cooling mechanism in which the gas does not end up in a state of thermodynamic equilibrium. In all such nonequilibrium phenomena, the energy-transfer collisions are important and will be considered in the text.

Gas Lasers

A simplified model of a gas laser which converts an input energy into a coherent beam of light is shown in Figure 1.4. Two processes are involved. The first is a means of using the input energy to pump the laser, that is, to rapidly populate a particular excited state of the atoms or molecules, labeled state 1. The second is a mechanism for depopulating the lower state involved in the transition, labeled state 2. The rates for populating and depleting these states are arranged such that the upper level, 1, is more heavily populated at any time than the lower level, 2, a situation referred to as a population inversion. When such an inversion exists, stimulated emission can dominate over spontaneous emission. In stimulated emission, a photon in the vicinity of an excited atom can induce that atom to radiate, if possible, a photon of the same frequency. The advantage obtained is that the individual photon emission processes are related, hence coherent rather than random light is produced.

A number of methods are available for pumping the gas. Pumping can be induced by photons from another source (not coherent), by irradiating the gas with fast electrons, by a rapid heating or expansion of the gas as in a nozzle, by applying external electric fields, or even by chemical mixing of

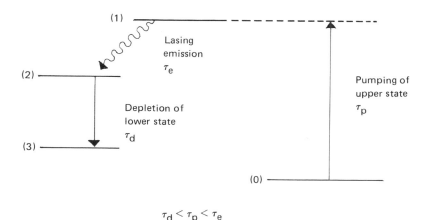

$$\tau_d < \tau_p < \tau_e$$

Figure 1.4. Schematic diagram of energy levels for a laser. The quantities τ are the characteristic lifetimes for each step indicated.

species. It is clear from the outset that the energy input in pumping can never be fully realized as coherent radiation. The laser, like other "engines," is a means of converting energy from one form to another more useful form, coherent radiation, and is, therefore, limited by the usual efficiency problems. That is, one tries to minimize the loss processes to realize the maximum amount of converted energy in the form of coherent radiation. To construct a laser, however, one needs more than a population inversion. Mirrors are often needed on the container walls to obtain amplification. By reflecting the photons through the gas a number of times, a cascade of coherent photons is created from which a beam can be extracted.

To illustrate the pumping, depletion, and loss mechanisms, we consider the well-known CO_2 laser which employs a mixture of N_2 and CO_2. Vibrational excitation of N_2 by collisions with fast electrons is the pumping mechanism,

$$e + N_2(v = 0) \rightarrow e + N_2(v = 1) \tag{1.12a}$$

where $v = 0$ indicates the ground vibrational state and $v = 1$ the first excited level. In a subsequent collision with a CO_2 molecule, the excitation energy is transferred to an excited vibrational state of CO_2 which is of nearly the same energy above the CO_2 ground state

$$CO_2(000) + N_2(v = 1) \rightarrow CO_2(001) + N_2(v = 0) \tag{1.12b}$$

where the three labels indicate the three types of vibrations a linear triatomic molecule can undergo. The reaction rate for vibrational-to-vibrational energy transfer between N_2 and CO_2 depends on the densities of these two species and the frequency with which the colliding molecules exchange energy. The state of CO_2 thus populated lies above three other vibrationally excited states as well as the ground state, shown schematically in Figure 1.5. The laser emission involves two lines or transitions to two vibrational levels. These levels are rapidly depopulated by collisions with ground-state CO_2 and N_2 molecules in which their vibrational energy is transformed to translational energy. Because the population inversion is reduced by collisional depopulation of the upper vibrational state, the rate for this process should be slow compared to those processes creating the inversion. As pumping is controlled by the input energy, and collisions involving small energy transfers are favored, such a situation can be arranged. However, to determine what conditions are required for optimum efficiency depends on our ability to describe the energy transfer and loss collisions.

Although a variety of lasers have been proposed and created based on various pumping and depletion mechanisms, we finish this discussion with a concept which exploits atomic interaction potentials. Whereas many atoms in their ground states combine readily to form molecules, rare gas atoms, for instance, do not. The force between two ground-state argon atoms is predominantly repulsive as the interatomic potential decreases with increasing separation (Figure 1.6). (At large R a weak attractive force does occur

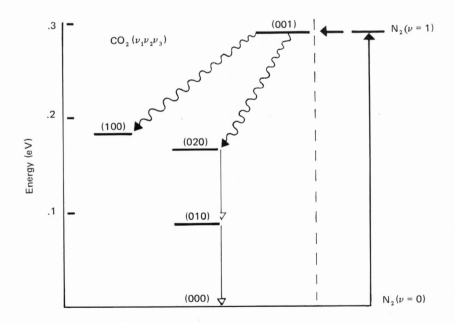

Figure 1.5. Diagram of levels for the CO_2 laser (see also Figure 3.11).

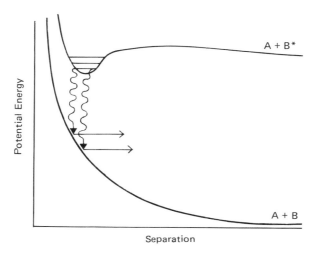

Figure 1.6. Hypothetical potential-energy diagram with a repulsive ground state. Upper bound state is pumped to obtain population inversion.

(Chapter 4), which we ignore here.) However, an excited argon atom with a loosely bound electron is capable of forming a chemical bond with a ground-state argon atom, resulting in an excited Ar_2 molecule indicated by a potential with a minimum. When the excited molecule radiates to the "ground state," the predominant repulsive force associated with this state causes the Ar atoms to separate rapidly. The depletion rate of the lower level is, therefore, related to the separation rate of the atoms as determined by their interatomic force. The population rate is controlled by the excitation of the argon atoms by hot electrons and the collisional binding rate between the excited and ground-state argon atoms. The latter process requires a three-body collision to carry off the extra momentum and must dominate the collisional depopulation of the excited argon atoms and molecules. Because of the vibrational motion in the excited state of Ar, the separation and hence the wavelength at emission varies, as determined by the energy differences in Figure 1.6 between the upper and lower states. Lasing is most efficient if the frequency spread is very small; however, it is often of practical interest to have a laser with a spread in frequency. To determine if such a laser is feasible requires a fairly accurate description of the states involved as well as the collisional population and loss mechanisms. At present, it appears that this process is realizable with systems like Ar + F, which also have predominantly repulsive ground-state potentials. The description of the interaction potentials between atoms is a topic treated in Chapter 4.

Plasmas

Plasmas, often referred to as the fourth state of matter, present another situation in which atomic collision phenomena play an important role. If an outside energy source like ultraviolet light or hot electrons impinges on a gas, the gas will become a composite of ions, electrons, and neutral particles referred to as a plasma. Such plasmas occur, for instance, in electric discharges, in the ionosphere of a planetary atmosphere, and in stars (i.e., naturally occurring thermonuclear reactors) and are employed in the hope of creating man-made thermonuclear reactors as controlled energy sources. The long-range Coulomb interactions among charged particles (in macroscopic numbers) cause the many special properties of this state of matter. Much of plasma physics involves describing the behavior of particles in electric and magnetic fields and this will not be dealt with here. However, when one tries to understand molecular transport processes, equilibrium ratios of particles in a plasma created in a molecular gas, energy dissipation in the plasma, the effect of impurities, etc., one immediately runs into the problem of having to know or to estimate atomic collision frequencies or reaction rates. Much of this we have already described in the previous sections. Transport processes like ion mobilities, diffusion coefficients, and heat conduction are similar in form to those discussed earlier, if one bears in mind that there are three types of particles, ions, electrons, and neutrals, to be considered. Below, we discuss additional topics related to plasmas.

As a first example, we consider the weakly ionized component in the atmosphere produced by solar radiation. In the earth's ionosphere the ions produced initially would be predominantly O^+, N_2^+, and O_2^+, whereas in the atmospheres of Mars and Venus, CO_2^+ would be predominant. These ions are free to interact with the neutral gas particles to form other chemical species. Some important processes are listed below and will be employed in Chapter 6:

$$
\begin{aligned}
CO_2^+ + O &\rightarrow O_2^+ + CO \quad &(1)\\
&\rightarrow O^+ + CO_2 \quad &(2)\\
O^+ + O_2 &\rightarrow O_2^+ + O \quad &(3)\\
O^+ + N_2 &\rightarrow NO^+ + N \quad &(4)\\
N_2^+ + O &\rightarrow NO^+ + N \quad &(5)\\
&\rightarrow O^+ + N_2 \quad &(6)\\
N_2^+ + O_2 &\rightarrow O_2^+ + N_2 \quad &(7)\\
O^+ + CO_2 &\rightarrow O_2^+ + CO \quad &(8)
\end{aligned}
$$

$$(1.13)$$

Because of the rapidity of the above reactions, temporary equilibrium concentrations occur that can be quite different from the initially ionized parent gas atoms and molecules. These concentrations depend on the relative rates for the above reactions as compared to the electron–ion recombination and transport rates. Recombination of O^+ with an electron, resulting in an oxygen atom and a photon, is very slow compared to $O_2^+ + e$ or $NO^+ + e$, producing two neutral atoms, and to reactions like (3) and (4) above. Consequently O^+ is a very short-lived and minor species at lower altitudes but is a relatively long-lived and dominant species at very high altitudes where there are few molecules. In the ionosphere of Mars, because reactions (1)–(3) above are fast compared to the recombination rates, O_2^+ rather than CO_2^+ is the dominant equilibrium molecular-ion species, and similarly, a significant fraction of NO^+ is found in the earth's ionosphere. The rate equation for determining the concentration of any of the species (e.g., O^+) is simply the diffusion equation, Eq. (1.5),

$$\frac{\partial}{\partial t} [O^+] + \nabla([O^+]\mathbf{w}_{O^+}) = k_2[CO_2^+][O] + k_6[N_2^+][O]$$

$$- k_3[O^+][O_2] - k_4[O^+][N_2] - k_8[O^+][CO_2] + \mathscr{S}_{O^+}[O] \qquad (1.14)$$

where the brackets indicate the species number density (throughout the text we will use n_{O^+} and $[O^+]$ interchangeably for this quantity) and $\mathbf{w}_{O^+}$ accounts for horizontal or vertical diffusion. The quantity $\mathscr{S}_{O^+}$ accounts for the production of O^+ and the rate constants k_i indicate the occurrence frequency for the reactions described in Eq. (1.13). These are determined from experiment or a knowledge of the molecular interactions of the species, and will be addressed in the text.

A further example of the role of collisions in plasma-related phenomena is the energy loss produced by impurities in plasmas for thermonuclear reactors. In a weakly ionized plasma, like that discussed above, charge-exchange collisions and chemical reactions occur between the ions and neutrals. However, the plasmas in thermonuclear reactors are fully ionized and confined by magnetic fields. A few escaping high-energy ions colliding with the walls of the container can sputter a number of heavy atoms off the surface. These neutral particles may then enter the confinement region of the plasma where they are partially ionized via charge-exchange collisions with the plasma ions or by the fast electrons. The losses produced are twofold: The partially ionized impurities can radiate repeatedly, and the light ions neutralized by charge exchange can escape the magnetic confinement region. In this book we will be concerned with electron recombination, charge-exchanging collisions of the type described between protons or partially stripped heavy particles and neutrals, as well as energy-transfer collisions on the walls of the containers that cause sputtering. For a fusion reactor, the mean temperature is such that these generally occur at keV energies (average energy $= \frac{3}{2} kT$, $k = 8.6 \times 10^{-5}$ eV/°K), although many of the protons, deuterons, and tritium ions will be moving at much higher energies in order that the fusion reactions occur.

Studies of Interaction Potentials and Surfaces

Forming the very basis of atomic-collision studies is the desire to know and understand the forces between atoms and molecules. By now, the fact that such forces can be determined from quantum mechanics is well known even to the nonscientist. However, the quantum mechanics of interacting atoms involving more than one or two electrons is difficult, and therefore experimental determination of forces, reaction rates, and collision frequencies is still extremely important and will remain so in the forseeable future.

A knowledge of the interaction forces or potentials is required to determine how atoms (or molecules) can bind to form molecules. We have already noted the existence of repulsive nonbinding potentials between atoms as well as binding potentials. A description of these potentials as a function of atomic separation can be obtained from both molecular spectroscopy and atomic collisions (collision spectroscopy), which tend to complement each other. That is, information on the behavior of the potentials at large and small separations is often obtained via collisions, whereas information in the binding region, intermediate separations, is often obtained spectroscopically. In some instances, for predominantly repulsive states or, equivalently, short-lived species, collisions provide a valuable probe at all separations. Recently, experiments involving both atomic beams and laser spectroscopy have been successful for determining potentials for what were difficult to measure molecular interactions. In this text we will consider in some detail the relationship between the intermolecular potentials and the experimentally obtainable quantities, cross sections, collision frequencies, and rate constants.

The use of atoms or ions as probes has also been extended to the studies of aggregates of atoms, like large biological molecules, or atoms on surfaces of solids. The fact that the incident particle not only sputters atoms from the target as mentioned earlier, but occasionally is itself backscattered, has provided a useful probe in surface science. A simple example is the measurement of the thickness of an absorbed layer of molecules on a surface referred to as the substrate. The backscattered ions will have a kinetic energy determined by the mass of the target atom and the thickness of the adsorbed layer. By monitoring the energy loss of backscattered ions and knowing the energy loss rate in the condensed gas, the electronic stopping power of Eq. (1.3), the thickness of the layer can be determined. Such a technique has been used to determine the sputtering rate of layers of frozen gases, a quantity of interest in astrophysics. Similarly, ion beams are used to locate the position of adsorbed atoms on a crystalline lattice or to obtain the depth profile of implanted particles. Lastly, ions and molecular species colliding with a surface may charge-exchange or react with the surface atoms in much the same way they do with gaseous atoms and molecules. Scattering from surfaces therefore provides a means of measuring incident particle–surface interaction potentials and electron distributions in much the same way that intermolecular potentials are determined from molecular collisions.

Summary

Although the list of phenomena we could have considered is extensive, it is hoped that this discussion has been comprehensive enough to provide a general rationale for studying atomic and molecular collisions and has given the reader a sense of what quantities are required to describe macroscopic observations. The art of using collision quantities to model macroscopic observations involves being able to estimate reaction rates or collision frequencies in order to decide in advance which collision processes need to be included in the model. Each of the phenomena discussed can be made as complex as one desires by considering as many interactions between differing species and as many states of these species as possible. This procedure seems to be popular particularly so because of the overall enchantment with the use of big computers. Although such a procedure is valid, and has a certain completeness, it is often more useful to reduce any problem to a consideration of the most important processes, at least as a first step. The first procedure suffers in that the dominant processes or effects are often obscured, and these are the ones which require accurate input information. The latter, simple, procedure has its own pitfalls. Often, estimates predict that a certain phenomenon can occur, but in fact the accumulative effect of the slower collision processes plays a long-term deleterious role. However, it is generally useful to know, as a first approximation, that a particular outcome may occur, as long as one realizes that it is only a first approximation. In the final chapter, I will treat a few of these phenomena described here considering only the dominant atomic and molecular processes. In the following chapter, the notion of a cross section will be used to develop means of estimating rate constants and collisions frequencies, and to extract interaction potentials from experiment.

Suggested Reading

Radiation Effects

H. DERTINGER and H. JUNG, *Molecular Radiation Biology*, Springer-Verlag, New York (1970).

P. D. TOWNSEND, J. C. KELLY, and N. E. W. HARTLEY, *Ion Implantation, Sputtering and Their Applications*, Academic Press, New York (1976).

W. R. NELSON and T. M. JENKINS, *Computer Techniques in Radiation Transport and Dosimentry*, Plenum Press, New York (1979).

Atomic Beam Interactions with Materials and Surface Effects

P. SIGMUND, "Collision Theory of Displacement Damage, Ion Ranges and Sputtering," *Rev. Roum. Phys.* **17**(7), 823 (1972).*

J. JACKSON, J. ROBINSON, and D. THOMPSON, eds., *Atomic Collisions in Solids*, North-Holland, Amsterdam (1980).

G. WOLKEN, JR., in *Dynamics of Molecular Collisions*, Vol. 1, ed. W. H. Miller, Plenum Press, New York (1976), Chapter 5.

Gas Dynamics and Diffusion

J. O. HIRSCHFELDER, F. CURTISS, and R. B. BIRD, *Molecular Theory of Gases and Liquids*, Wiley, New York (1964).

E. A. MASON and T. R. MARRERO, in *Advances in Atomic and Molecular Physics*, Vol. 6, ed. D. R. Bates, Academic Press, New York (1970), Chapter 4.

Gas Lasers

B. A. LENGYEL, *Lasers*, 2nd edn. Wiley, New York (1971), Chapter 9.

CH. K. RHODES, Excimen Lasers in Topics in *Applied Physics, Vol. 30*, Springer-Verlag, Berlin and New York (1979).

Plasmas

F. S. CHEN, *Introduction to Plasma Physics*, second printing. Plenum Press, New York (1977), (first printing, 1974).

Atmospheric and Astrophysical Processes

J. W. CHAMBERLAIN, *Theory of Planetary Atmospheres*, Academic Press, New York (1978).

A. E. S. GREEN and P. J. WYATT, *Atomic and Space Physics*, Addison-Wesley, New York (1965).

C. W. ALLEN, *Astrophysical Quantities*, 2nd edn., Athlone Press of the University of London (Oxford University Press, New York)(1963).

P. G. BURKE and B. L. MOISEIWITSCH, *Atomic Processes and Applications*, North Holland, Amsterdam, Chapters 1–5 (1976).

A. DALGARNO, "Atomic Physics from Atmospheric and Astrophysical Studies," in *Advances in Atomic and Molecular Physics*, Vol. 15, eds. D. R. Bates and B. Bederson, Academic Press, New York, p. 37 (1979).

Interaction Potential Studies

J. O. HIRSCHFELDER, ed., *Advances in Chemical Physics*, Vol. 12, Wiley, New York (1967).

R. B. BERNSTEIN, ed. *Atom-Molecule Collision Theory*, Plenum Press, New York (1979).

* This set of lectures, although difficult to obtain, is extremely useful.

2

Cross Sections and Rate Constants

Introduction

In order to describe in a quantitative way the phenomena discussed in the previous chapter, we need to be able to calculate or measure the required cross sections or rate constants. To begin the discussion, we divide collisions into two classes: elastic collisions (scattering) during which the particles interact (collide) with each other but only their directions of motion and speeds change, and inelastic collisions in which both the motion and the internal energies of the particles are changed. In Table 2.1 are given examples of inelastic collisions which we will consider in this text. Although inelastic collisions are clearly more interesting, we start by discussing experiments which only determine whether or not a particle was deflected.

Experiments that only reproduce the phenomena discussed in Chapter 1 often do not give much information about which individual atomic or molecular-level events are occurring or which of these play a controlling role. Therefore, experiments are concocted which isolate particular events, in this case collisions between atoms. In lieu of observing a single collision between atoms, beams of presumably identical atoms are made to impinge on a target containing a large number of identical atoms generally, though not necessarily, different from those in the beam. By observing a large number of similar events one can deduce information about individual events, at least in an average or statistical sense. In the following section we obtain an operational definition of cross section based on this type of experiment.

Total Cross Sections

Cross section, as the name implies, is an effective area, a physical size associated with the colliding particles. To measure effective areas associated

Table 2.1. Examples of Important Collision Processes[a]

Collision (reaction)			Notation	
(1) $A + B \rightarrow A + B$	$\sigma_{AB \rightarrow AB}$,	σ,	Elastic collision
(2) $A + B \rightarrow$ all possible results	σ_{AB},	σ_T,	Total cross section
(3) $A + B \rightarrow A + B^*$	$\sigma_{AB \rightarrow AB*}$,	σ_Q,	Excitation
(4) $A^* + B \rightarrow A + B$	$\sigma_{A*B \rightarrow AB}$		De-excitation
(5) $A + B \rightarrow A + B^+ + e$	$\sigma_{AB \rightarrow AB^+}$,	σ_I,	Ionization
(6) $A^+ + B \rightarrow A + B^+$	$\sigma_{A*B \rightarrow AB^+}$	σ_{ct},	Charge transfer
(7) $A + B^* \rightarrow A^* + B$	$\sigma_{AB* \rightarrow A*B}$		Excitation transfer
(8) $A + BC \rightarrow AB + C$	$\sigma_{A, BC \rightarrow AB, C}$,	σ_R,	Rearrangement
(9) $A + B + C \rightarrow AB + C$	$\sigma_{A, B, C \rightarrow AB, C}$,	σ_R,	Reaction

[a] The species indicated by A, B, and C can imply a simple atom or ion, or a complex molecular species. The asterisk implies the species is *not* in the lowest state. It may be in a higher electronic state or, for molecules, a higher rotational or vibrational state. The superscript plus implies the species is ionized.

with a simple collision between two atoms or molecules, or between an electron and a molecule, the experimental setup in Figure 2.1 can be employed. A beam of particles, all assumed to have approximately the same speed and direction of motion, is incident on a target made up of atoms or molecules. The target particles may be in a liquid, gaseous, or solid state, and the speeds of the target particles are assumed negligible compared to the incident particle speeds; that is, the temperature of the target is kept low. In traversing the target we allow, for the present discussion, two possible results: (a) particles pass through the target with their directions and speeds essentially unchanged; (b) particles are deflected and slowed (scat-

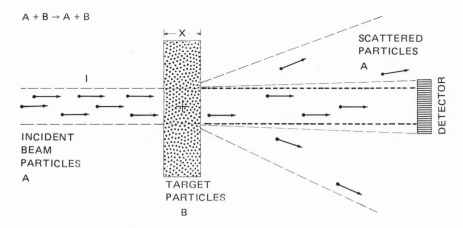

Figure 2.1. Beam experiment to obtain total scattering cross section.

tered) or stopped in the target. Internal energy changes, if they occur, are ignored. Clearly, if the target is very "thick," i.e., dense or long or both, the likelihood of case (b) occurring for any incident particle increases. It may appear remarkable at first that even for the thinnest targets a beam particle *can* pass through an apparently solid material of atoms with negligible deflection. However, before Rutherford's experiments, in which he irradiated a gold target with a beam of α-particles (He^{++}), it was thought that few, if any, particles would be deflected significantly, and the effect of the material would be to slow the particles.

Employing the experimental setup in Figure 2.1, we obtain the attenuation of the beam of unscattered particles by measuring the transmitted beam intensity, I, at the detector as a function of target thickness, X.* The intensity or flow is the rate at which particles strike the detector divided by the detector area. If the target thickness is increased from X to $X + \Delta X$, the change in beam intensity at the detector, ΔI, is proportional to the change in thickness, ΔX, the intensity, $I(X)$, of particles entering ΔX, and the density of atomic targets in the target material, n_B. That is,

$$\Delta I = -\sigma_{AB} n_B I(X)\Delta X \qquad (2.1)$$

where the proportionality constant, σ_{AB}, inserted into Eq. (2.1) has the dimensions of area and is referred to as the total collision cross section for A and B.† In the limit of small ΔX, the above equation has the solution

$$I(X) = I(0)\exp\left[-\sigma_{AB} n_B X\right] \qquad (2.2)$$

Using Eq. (2.2), the cross section can be determined by plotting the measured intensity as a function of $(n_B X)$ on semilog paper and noting the slope at low values of $(n_B X)$.† The quantity σ_{AB} is a measure of the ability of the target particles, B, to deflect the beam particles, A. That is, if σ_{AB} is large, the particles are deflected easily and appear large to each other; the opposite is true if σ_{AB} is small. As the quantity $(n_B X)^{-1}$ is the average cross-sectional area of the target material occupied by each target atom looking in the direction of the beam, then $(\sigma_{AB} n_B X)$ in Eq. (2.2) is the probability of a deflection by any atom.

As atoms and molecules are *not* hard spheres with well-defined boundaries, one of the first results revealed by measurements like those described above is that σ_{AB} depends on the relative speeds of approach of the incident and target particles, as is seen in Figure 2.2. This also indicates the size of the cross sections that will be considered ($\sim 10^{-16} - 10^{-15} cm^2$). Because

---

* For gaseous targets the gas density instead of X may be varied.

† In the experiment described care must be taken to discriminate against particles which have been scattered a number of times and end up moving in the incident-beam direction by keeping the target thin or by also measuring final velocities.

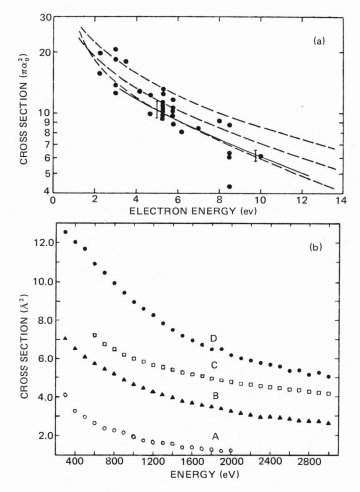

Figure 2.2. Cross section, σ_{AB}, vs incident particle energy: (a) e + H: circles and solid lines—experiment; dashed lines—calculations [from R. H. Neynaber, L. L. Marino, E. W. Rothe, and S. M. Trujillo *Phys Rev.* **124**, 135 (1961)]; (b) He + He: detector apertures: A, 0.57°; B, 0.26°; C, 0.11°; D, 0.056° [from W. J. Savola, Jr., F. J. Eriksen, and E. Pollack, *Phys. Rev A* **7**, 932 (1973)]. For units see Appendix J.

the detector has finite width, as does the beam, it cannot distinguish between particles which are unscattered or particles which are deflected only slightly. In fact, since atoms have diffuse boundaries, the distinction may be meaningless if the angle in question is very small, a point we will return to in Chapter 5. Therefore, the measured cross sections σ_{AB} are often labeled by the effective angular aperture (opening) of the detector, shown in Figure 2.1, particularly in heavy particle collisions, viz., Figure 2.2.

Instead of cross section one often discusses, especially in thermodynamics, the mean free path, $\bar{\lambda}$, of the atoms traversing the material. This is the quantity

$$\bar{\lambda}^{-1} \equiv n_B \sigma_{AB} \qquad (2.3)$$

occurring in the exponent of Eq. (2.2). The mean free path $\bar{\lambda}$ indicates the average distance atoms A travel before colliding with a target atom; or $\bar{\lambda}^{-1}$ is the probability of a collision per unit path length traveled. A target is considered thin, as stated above, if its thickness, X, is much less than $\bar{\lambda}$. A closely related quantity is the total collision frequency

$$\nu \equiv v\bar{\lambda}^{-1} \qquad (2.4)$$

where v, the relative collision velocity, equals the velocity of incident atoms A (v_A in these experiments).

In defining the cross section, via the experiment in Figure 2.1, we have also shown that the intensity of a beam of radiant energy incident on a material (atmosphere, reactor wall, etc.) decreases exponentially with depth and/or number density. If cross sections have been measured in the laboratory, then Eq. (2.2) can be used to estimate the amount of radiant energy transmitted through a material of given thickness. Equation (2.1) is simply an extension of Eq. (1.2) to the case of perpendicular incidence and in which only loss of incident particles is considered.

Inelastic Cross Sections

Attenuation measurements of the undeflected *and* unchanged particles, like those described above, will lead to a determination of σ_{AB}, where σ_{AB} accounts for all scatterings, with and without internal changes. To obtain information about the likelihood of a particular inelastic event, the changed species have to be distinguished one from the other directly, by collection, or indirectly, via electrons (ionization) or photons (excitations) produced. The experiments to be described, therefore, require detection of the *scattered* incident or target particles, or their collision products, often involving a distinct detection scheme for each type of inelastic effect.

As we are interested in what takes place, statistically, in a single collision, the target thickness must be kept small enough that most of the beam particles are not scattered.* This ensures that, if a particle *is* scattered out of the beam, it is, with high probability, due to a single collision between the beam particle and a single target particle. Multiple scattering experiments are more difficult to interpret but are sometimes unavoidable.

---

* In all beam experiments, the beam density should be small enough that the effect of the beam particles on one another is negligible.

We also will require that the number of targets affected at any time be small compared to the number of targets available, so the number density of targets n_B is nearly constant. For this type of experiment sensitive detectors must be employed as the experimenter is generally counting a small number of events. Considering, as an example, the charge transfer reaction, $A^+ + B \rightarrow A + B^+$, measurements may be made of the rate of production of target ions B^+, $d\mathcal{N}_{B+}/dt$, for a given target thickness. This may be accomplished by applying a very small electric field across the target to sweep out and count the number of ions produced.

The rate of production of target ions B^+ is proportional to the target area, $\mathcal{A}$, irradiated by the beam, the thickness ΔX, the number density of targets, and the beam intensity I_{A+} (the number of ions crossing a unit area per unit time), which we assume here is practically constant over the target length. This can be written quantitatively as

$$\frac{d\mathcal{N}_{B+}}{dt} = \sigma_{A^+ + B \rightarrow AB^+} \, n_B \, I_{A+} \Delta X \mathcal{A} \tag{2.5}$$

where the proportionally constant, written $\sigma_{A^+ + B \rightarrow AB^+}$, again has the dimension of area. The quantity $n_B \Delta X \mathcal{A}$ is the number of atoms in the target material, and therefore the cross section is defined, per target atom, as

$$\sigma_{A^+ + B \rightarrow AB^+} \equiv \frac{\begin{array}{c}\text{Number of collisions per unit time}\\ \text{resulting in an electron capture}\end{array}}{\text{Incident intensity}} \tag{2.6}$$

Writing the number of ions produced, $\mathcal{N}_{B+}$, in terms of the density of species B^+ produced, $\mathcal{N}_{B+} = n_{B+} \Delta X \mathcal{A}$, we see that the rate of increase of the product species density, n_{B+},

$$\frac{dn_{B+}}{dt} = \sigma_{A^+ + B \rightarrow AB^+} \, n_B \, I_{A+} \tag{2.7}$$

is independent of the size of the target volume irradiated by the beam.

The experiment as described cannot distinguish between ionizations, $A^+ + B \rightarrow A^+ + B^+ + e$, and charge exchanges, $A^+ + B \rightarrow A + B^+$. This distinction can be made by also counting the electrons produced by ionization or by detecting the scattered beam particle in coincidence with the production of B^+. In some cases the experimenter may be interested only in the number of ionized targets produced by either charge exchange or ionization, in which case the measured cross section is a sum of $\sigma_{A^+ + B \rightarrow AB^+}$ + $\sigma_{A^+ + B \rightarrow A + B^+}$.

The meaning of the quantity $\sigma_{A^+ + B \rightarrow AB^+}$ is not as clear as the cross section defined earlier. Recalling that if *all* the inelastic events are monitored for A^+ incident on B, as well as the elastic scatterings, $\sigma_{A^+ + B \rightarrow A^+ B}$, then of course all the scattering events are accounted for. Now the total rate of scattering of particles $d\mathcal{N}_s/dt$ is the sum of the rates for the individual events, each of which has the form of Eq. (2.5), yielding

$$\frac{d\mathcal{N}_s}{dt} = \left(\sum_j \sigma_{A+B\to j}\right) n_B I_{A+} \Delta X \mathscr{A} \tag{2.8}$$

where j labels the possible results: elastic collision, ionization, change transfer, etc. The total scattering rate is related to the decrease in beam intensity for a thin target, $d\mathcal{N}_s/dt = -\Delta I \mathscr{A}$. Using Eqs. (2.1) and (2.8), we find that the total scattering cross section from the attenuation experiment is the sum of the individual cross sections:

$$\sigma_{A+B} = \sum_j \sigma_{A+B\to j} \tag{2.9}$$

Now the probability, $P_{A+B\to AB+}$, of a charge-transfer collision occurring when A^+ collides with B is the ratio of the scattering rate leading to charge transfer, Eq. (2.5), to the total scattering rate, Eq. (2.8), or

$$P_{A+B\to AB+} = \frac{\sigma_{A+B\to AB+}}{\sigma_{A+B}} \tag{2.10}$$

Therefore, the inelastic cross section indicates not only the effective sizes of the colliding particles, but also the likelihood, or probability, of a given inelastic event occurring. It is often useful to think of an inelastic cross section as proportional to the particle size, indicated by the total cross section, and the probability of the particular inelastic event,

$$\sigma_{A+B\to AB+} = P_{A+B\to AB+}\, \sigma_{A+B} \tag{2.11}$$

Both σ_{A+B} and $P_{A+B\to AB+}$ generally depend on the relative velocity of approach of the beam and target particles.

Lastly, if an inelastic process is thought of as "damaging" the target molecule B, then, integrating Eq. (2.7), we obtain the fraction of the target molecules which remain unaffected after a time t:

$$f = \exp\left[-\sigma_D I t\right] \tag{2.12}$$

where σ_D is the inelastic cross section (e.g., ionization, dissociation, etc.). This expression will be employed in Chapter 6 when discussing radiation effects.

Rate Constants

In Eq. (2.7), the intensity of the beam, I_{A+} (ions/cm^2/sec), can also be written as the *beam* density, n_{A+} (ions/cm^3), times the particle velocity v_{A+}. For our example, v_{A+} is also equal to the relative velocity between A^+ and B and, therefore, we have used instead the symbol v. Now, the production of ions can be explicitly written as being proportional to both the target and incident atom densities,

$$\frac{dn_{B^+}}{dt} = \sigma_{A^+ \to AB^+} v n_{A^+} n_B \tag{2.13a}$$

and the formal distinction between target and incident particle disappears. That is, the experiment with B moving and A^+ standing still will yield the same result as long as the relative speed of approach is the same! This is a general property of collisions occurring in a region where there are no outside effects, like strong electric fields. This type of equation can be applied also to an experiment in which two sets of beams are crossed, as long as v, n_{A^+}, and n_B are calculated carefully. Changing the angle between the crossed beams changes v, allowing one to study the velocity dependence of the cross section. One can further relax the constraint that either set of atoms colliding is in the form of a monoenergetic beam. That is, thermal beams can be used where there is a distribution of velocities, for which case one writes

$$\frac{dn_{B^+}}{dt} = \overline{(\sigma_{A^+ + B \to AB^+} v)} n_{A^+} n_B \tag{2.13b}$$

where the bar means the product v and the cross section are averaged over the velocities occurring in the thermal beam. (Note: This is not usually equal to $\bar{\sigma} \cdot \bar{v}$.) If such an experiment is used to measure the cross section at a single velocity, then a procedure has to be developed for unfolding the average, which usually requires some knowledge of the general behavior of σ with velocity.

A very important class of experiments is carried out for studying chemical reactions in which the gases are simply mixed. For a particular reaction, for example reaction (8) in Table 2.1, which is known to be a result of a binary collision (a bimolecular reaction), one combines known densities of the two species A and BC together at a given temperature and monitors as a function of time the production of C or AB. The production of C clearly depends on the density (availability) of the two initial species and, therefore, we write

$$\frac{dn_C}{dt} = k_{A, BC \to AB, C}(T) n_A n_{BC} \tag{2.14}$$

which is the continuity equation of Chapter 1 without flow, e.g., Eqs. (1.5) and (1.14). In Eq. (2.14), the proportionality constant $k_{A, BC \to AB, C}(T)$ is referred to as the rate constant of the reaction and is dependent on the temperature of the mixture. If the above experiment is performed for a period of time which is very short, so that the changes in the densities n_A and n_{BC} are negligible but the amount of C produced is detectable, then extracting $k_{A, BC \to AB, C}(T)$ is simple and direct. The number density of C as a function of time is

$$n_C(t) \simeq k_{A, BC \to AB, C}(T)\, n_A\, n_{BC}\, t \tag{2.15}$$

where t is the time at which the measurement is taken, and the densities n_A and n_{BC} are assumed to be the initial densities.

Comparing Eq. (2.14) to Eq. (2.13b), it is at once evident that in a binary reaction the rate constant is related to the inelastic (here reactive) cross section via

$$k_{A, BC \to AB, C}(T) = \overline{(\sigma_{A, BC \to AB, C}\, v)_T} \tag{2.16}$$

where, again, the bar implies an average, which, for the example considered here, is over the thermal distribution of collision velocities in the gas, as indicated by the subscript T. It is clear that rate constants contain less detailed information than cross sections obtained from a beam experiment because of the averaging over relative velocities. However, there are many instances for which beam experiments are not required and/or not practical. It is often difficult to produce monoenergetic beams with low velocities, though recently, crossed beam techniques have been used to create low relative velocities of approach as well as nozzle beams.

The above method is particularly well suited to reactions involving more than two bodies for which beam experiments would be extremely difficult. For the three-body reactions $A + B + C \to AB + C$, we can generalize Eq. (2.14):

$$\frac{dn_{AB}}{dt} = k_{A, B, C \to AB, C}(T) n_A\, n_B\, n_C \tag{2.17}$$

In this particular reaction, particle C acts as a catalyst and, as its density does not change, the time dependence of the production of n_{AB} is similar to that for the two-body reaction. This three-body reaction requires that species A and B be *both* close enough to C for the reaction to proceed. Therefore, the effective cross-sectional area of the catalyst C will determine the size of the three-body rate constant. Many three-body chemical reactions can be thought of for which particle C does not play a passive role. That is, its structure is important in determining the size of the reaction rate, and the state of C may change during the reaction.

Another quantity of interest in collisions is the average energy-loss rate of the incident particle, which was invoked in our discussion of random collision cascades. The energy-loss rate associated with a particular inelastic process, $AB \to j$, can be written

$$-\frac{dE}{dt}\bigg|_{AB \to j} = v_A\, \overline{\Delta E_j}\, \sigma_{AB \to j}\, n_B$$

by the arguments used to obtain Eq. (2.13a). The quantity $\overline{\Delta E_j}$ is the average energy loss of incident particle A in a collision during which the process $AB \to j$ occurs. Summing over all possible processes, as in Eq. (2.8), one obtains the total energy-loss rate or the more frequently used quantity, the energy loss per unit path length,

$$-\frac{dE}{ds} = n_B \sum_j \overline{\Delta E_j} \, \sigma_{AB \to j} \equiv n_B S_{AB} \qquad (2.18)$$

where we used $dE/dt = v_A(dE/ds)$. The coefficient of n_B in Eq. (2.18) is often referred to as the stopping cross section, S_{AB}, and dE/ds the stopping power of the material, a quantity directly measurable with an experimental arrangement like that in Figure 2.1. Instead of counting the number of particles deflected out of the incident beam in that experiment, one monitors the loss of kinetic energy of all the exiting particles as a function of material thickness and then averages. The net energy loss for any particle is conceptually broken into two categories, that involved in the deflection of A by B, due primarily to the nuclei, and that transferred to the electrons of the material. These are referred to separately as the elastic-nuclear energy, loss and the inelastic energy loss, and can be separated because of the large mass differences between electrons and nuclei. These ideas were implicit in our construction of Eqs. (1.2) and (1.3) as we separated collisions which changed the momentum of the particle from the electronic drag on the incident particle.

The energy-loss rate is of interest in other than beam experiments. If a small amount of a "hot" gas, A, is mixed into a dense background gas B, then the rate at which A equilibrates in B is determined by dE/dt averaged over the velocity distribution of the mixture. Atoms A can be "heated" selectively using a laser of the appropriate frequency, where we assume B is dense enough that collision processes are faster than radiative processes. Another example is a weakly ionized plasma in which one heats the electrons and ions, as for example solar ultraviolet (UV) radiation heating of the ionospheric plasma discussed in Chapter 1. The subsequent equilibration of the electrons and ions is determined by dE/dt.

Cross Section Calculations

The concept of a cross section, as discussed earlier, involves an effective size and an average probability of a particular event. We imagine a moving particle A approaching an initially stationary particle B (Figure 2.3), the situation occurring in the experiment discussed earlier. Clearly, if a large number of atoms are incident on a large number of targets as in Figure 2.1, some will approach rather closely, even head-on, and others will pass quite

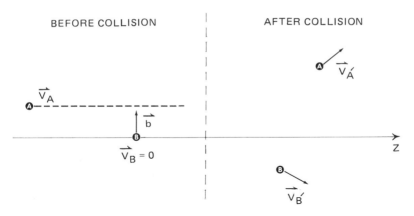

Figure 2.3. Collision parameters: b is the impact parameter; primed quantities are the final velocities.

far away. As the likelihood of a collision will clearly depend upon the closeness of approach, it is customary to introduce the concept of an impact parameter, b. For the beam experiment (Figure 2.3), b is the perpendicular distance from the center of the target to the path the incident particle would take it if were *not* deflected—the dashed line. The impact parameter, along with the velocity of approach, is sufficient to characterize the initial conditions for objects which are spherically symmetric. For nonspherical objects, as in collisions involving most molecular species, the relative orientations of the colliding particles need to be specified as well as b and v. Initially, we will discuss spherically symmetric systems.

On each trajectory, labeled b, there is a probability, $P_{AB}(b)$, that a deflection will occur resulting in either an elastic or inelastic scattering of the two particles. This depends, obviously, on how closely the objects approach as in the collision between two spheres represented in Figure 2.4. As we have already stated, the concept of cross section is statistical. Therefore,

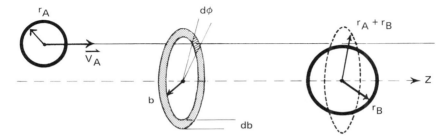

Figure 2.4 Collision of spheres. All incident particles A passing through the ring experience a similar impulse. Dotted area of radius $r_A + r_B$ indicates cross section for the collision.

we assume it is equally likely, when large numbers of beam and target particles are involved, that A approaches B with any impact parameter, b. As the objects are spherically symmetric, it does not matter on which side A approaches B as long as the value of b is the same. Imagining a thin ring of width db, circumference $2\pi b$, and area $2\pi b\, db$ in the plane perpendicular to v_A, through which the center of A passes; then, in the limit that db is small, trajectories for all particles passing through the ring are similar. For the collision in Figure 2.4, if $b \le r_A + r_B$, the two objects will collide and be scattered, whereas for $b > r_A + r_B$ they miss. Therefore, any particle passing through a circle of radius $r_A + r_B$ in Figure 2.4 will be scattered, implying that the effective cross-sectional area the two objects present to each other is $\sigma_{AB} = \pi(r_A + r_B)^2$.

This area also can be thought of as being made up of a sum of the areas of those rings, discussed above, for which a collision will occur. That is,

$$\sigma_{AB} = \int_0^{r_A + r_B} 2\pi b\, db \tag{2.19}$$

or, more generally, noting that the collision probabilities are

$$P_{AB}(b) = 1, \qquad b \le (r_A + r_B)$$
$$= 0, \qquad b > (r_A + r_B)$$

we write

$$\sigma_{AB} = 2\pi \int_0^\infty P_{AB}(b)\, db \tag{2.20}$$

In this expression the nature or size of the particles is contained in the quantity $P_{AB}(b)$. For particles which behave deterministically, such as classical particles, P_{AB} is either zero or one, as above. Therefore, for forces of infinite range, this calculation yields an infinite cross section! For particles that interact non-deterministically, such as quantum-mechanical particles, P_{AB} depends on b and σ_{AB} may be finite even for such forces (viz. Fig. 2.2). In any particular calculation, for comparison with experiment, the integral cannot be carried out to infinity as there are always neighboring target atoms which may be closer to A than the target of interest. Therefore, an upper limit, d, related to the average distance between the target atoms (e.g., $\{n_B \Delta X\}^{-1} = \pi d^2$) should be imposed on the integral in Eq. (2.20). At this distance the collision ceases to be binary. For the thin-target experiments described earlier, an alternative upper limit is b_m, the impact parameter beyond which all deflections are smaller than the angular resolution of the detector. Computationally, the ideal situation is that the upper limit is large enough for P_{AB} to be zero for all b greater than either d or b_m. Now one can integrate out to b_m, d, or infinity and obtain the same result.

The concept of an impact parameter is obviously a very classical concept, in which it is assumed that the path of the incident particle can be determined. A closely related quantity, the angular momentum, L, is often used as this is a quantity which can be discussed in both classical and quantum-mechanical descriptions of collisions. For the collision in Figure 2.3, the angular momentum of A about the initial center of B is

$$\mathbf{L_A} = M_A \cdot \mathbf{v_A} \times \mathbf{R} \tag{2.21}$$

where M_A is the mass of the moving object A and $\mathbf{R}$ is the vector distance between the particles. It is easily seen that the impact parameter is now equal to

$$b = L_A / M_A v_A \tag{2.22}$$

Therefore, in all the expressions above, L_A can be used to replace b.

For a large number of incident and target particles, the probability of scattering per unit time per scattering center is the probability per unit time that a particle A will pass within an impact parameter $b \leq (r_A + r_B)$ of a target B. This can be written as $I\pi(r_A + r_B)^2$ or $I\sigma_{AB}$. Now the density of scattered particles per unit time, for a target number density n_B, is

$$\frac{dn_s}{dt} = \sigma_{AB} I n_B \tag{2.23}$$

hence σ_{AB} calculated above is consistent with the total cross section in Eq. (2.8). To determine cross sections for separate inelastic or elastic events, we note that, for particles entering at any impact parameter (or with any angular momentum), there is a probability of a given inelastic event occurring. Therefore, a particle passing through the element of area $2\pi b\, db$ about a target will cause, for example, an ionization with a probability $P_{AB \to AB^+}(b)$, implying that

$$\sigma_{AB \to AB^+} = 2\pi \int_0^\infty P_{AB \to AB^+}\, b\, db \tag{2.24}$$

The sum of probabilities of all possible scattering events, elastic and inelastic, caused by particles passing at an impact parameter b equals the total probability of a scattering in Eq. (2.20),

$$P_{AB}(b) = \sum_j P_{AB \to j}(b) \tag{2.25}$$

maintaining the relationship established earlier between the total cross section and the separate inelastic cross sections.

In many chemical reactions due to binary collisions, for instance reaction (8) in Table 2.1, it is known that the reaction will take place with a given probability $\bar{P}_{A, BC \to AB, C}$ *if* the particles pass within a certain critical distance of each other, i.e., approach each other within a given impact parameter b_R. When this is the case the reaction probability is written as

$$P_{A, BC \to AB, C} = \begin{cases} \bar{P}_R, & b \leq b_R \\ 0, & b \geq b_R. \end{cases}$$

Now the reaction cross section becomes, using Eq. (2.24),

$$\sigma_R = \bar{P}_R \pi b_R^2 \tag{2.26}$$

This has the general form described in Eq. (2.11), as πb_R^2 is often the geometric total cross section.

The usefulness of Eq. (2.26) is that physical arguments may be used frequently to obtain either $\bar{P}_R$ or b_R, allowing one to avoid a detailed calculation of σ_R. We will refer to this form frequently when we discuss applications of cross-section calculations later.

In calculating cross sections in the manner described above, the problem reduces to a determination of impact-parameter probabilities for inelastic and elastic scattering. We can generalize the above results to consider nonspherically symmetric particles by introducing a set of relative orientation angles $\Omega_M = (\theta_M, \phi_M)$. If the outside fields are imposed to select a single orientation, then $P_{AB}(b)$ is replaced by $P_{AB}(b_j, \Omega_M)$ and a cross section is obtained for each orientation. Generally all orientations are equally likely, in which case the quantity $P_{AB}(b)$ which we have been discussing is an average over all orientations. Similar considerations apply to inelastic cross sections.

Before leaving this section we estimate some characteristic sizes of the quantities we have been describing. For atoms and molecules, quantities like $(r_A + r_B)$ and b_R discussed above are of the order of 10^{-8} cm. For example, the mean radius of a ground-state hydrogen atom is 0.53×10^{-8} cm and the mean separation of the nuclei in a hydrogen molecule is 0.74×10^{-8} cm. Such characteristic sizes imply total cross sections of the order of 10^{-15} to 10^{-16} cm^2 (viz. Figure 2.2). On the other hand, the probability of a reaction or an inelastic effect, $\bar{P}_R$, will depend on the ratio (τ_R/τ_c), where τ_c is the length of time the particles are close together (the collision time), and τ_R is a characteristic reaction time. Characteristic reaction times are related to natural periods of motion of the composite particles in the targets. For ionization collisions, the orbital period of the outer-shell electrons, $\tau \sim 10^{-16}$ sec, determines the reaction times. For chemical reactions, the vibrational period, $\tau \sim 10^{-13}$ sec, or rotational period, $\tau \sim 10^{-11}$ sec, is the important characteristic times. When the natural periods match the collision time, $\tau_R/\tau_c \sim 1$, one expects that the reaction probability would be optimum. Based on this notion, ionizing collisions would occur efficiently for atoms and molecules with keV (10^3 eV, 1 electron volt = 1.602×10^{-12} ergs) to MeV (10^6 eV) energies depending on the masses of the particles involved and ionization energies; i.e., inner-shell ionization requires higher energies. For incident electrons the range would be of the order of 5 to 50 eV. Chemical reactions, on the other hand, are generally efficient at meV (10^{-3} eV) to eV energies. Extending this, we note that the mean kinetic energy of a molecule at room temperature is about $\frac{1}{40}$ eV, suggesting a bimolecular reaction rate, $k(T) \sim \overline{(\sigma_R v)_T}$ of the order of 10^{-11} to 10^{-12} cm^3/sec, depending on the masses of the reactants, if $\bar{P}_R$ is near unity. For an inefficient bimolecular reaction $k(T)$ may be considerably smaller. On the other hand, in reactions involving electrons (e.g., $e + O_2^+$), $k(T)$ would be larger as the electron speeds are larger. A three-body reaction [e.g., Eq. (2.17)] can be thought of as two separate encounters (reactions) of A and B with C during the collision time τ_c; $k_{A, B, C} \sim k_{AC} k_{BC} \tau_c \sim 10^{-33}$ cm^6/sec. Such estimates are very useful in designing experiments or making preliminary models of collision phenomena and will be used at various points in the text.

Angular Differential Cross Sections

In our previous discussion we only needed to know *if* an incident particle was deflected. For any net force between the two particles along a given trajectory the incident particle will be scattered. However, to describe the probability of being scattered into a particular angle, it is not sufficient to know there is a force, but the details of interaction have to be known. Conversely, knowing the angular distribution of scattered particles, one should be able to learn something about the details of the interaction and not just the extent of the particles.

The cross sections defined previously, which do not differentiate between particles scattered into different angles, are referred to as integrated cross sections, implying that all scattered particles are counted. Sometimes the word total is used, though here we use total when referring to the sum of all integrated cross sections. Employing an experimental setup like the schematic diagram in Figure 2.5, one can measure the number of particles scattered into a particular angular region. For this arrangement, the distance between the target and the detector, R, is assumed to be large compared to the dimensions of the target volume, and the angular position of the detector is labeled by θ and φ shown. The detector spans an angular region (solid angle) about θ and φ indicated by the angles $\Delta\theta$ and $\Delta\varphi$ in Figure 2.5. The element of solid angle, $\Delta\Omega = \Delta a/R^2$, Δa being the detector area and R the distance to the detector, is $\Delta\Omega = \sin\theta \, \Delta\theta\Delta\varphi$. By moving the

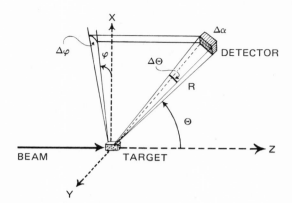

Figure 2.5. Experiment for measuring angular differential cross section. The detector area is $\Delta a = (R\Delta\Theta)(R \sin \Theta \, \Delta\varphi)$.

detector to all angular positions for a fixed R, all the scattered particles can be counted, and the detector will have covered an area $4\pi R^2$ or a total solid angle of 4π.

Returning to Eq. (2.8), we measure the number of particles scattered per unit time. Now, however, we consider only those scattered into a given solid angle, $\Delta\Omega_A$, where the subscript indicates that we are collecting particles of type A. Based on the previous discussion, the angular differential cross section, $d\sigma/d\Omega|_{AB}$ for the scattering of A on B is defined by

$$\frac{d\sigma}{d\Omega}\bigg|_{AB} \Delta\Omega_A = \frac{\text{Number of particles per unit time}}{\text{Incident intensity}} \qquad (2.27)$$

By summing over all angular regions, the integrated cross section is recovered:

$$\sigma_{AB} = \int \frac{d\sigma}{d\Omega}\bigg|_{AB} d\Omega_A = \int_0^{2\pi} d\varphi_A \int_0^\pi \sin \theta_A \, d\theta_A \frac{d\sigma}{d\Omega}\bigg|_{AB} \qquad (2.28)$$

When there are no outside fields the orientation of the x, y axis is arbitrary, and therefore $d\sigma/d\Omega|_{AB}$ is independent of φ_A. Consequently, one often writes, using $d \cos \theta_A = -\sin \theta_A \, d\theta_A$,

$$\sigma_{AB} = 2\pi \int_{-1}^{1} d \cos \theta_A \, \sigma_{AB}(\theta_A) \qquad (2.29)$$

where $\sigma_{AB}(\theta_A)$ and $(d\sigma/d\Omega)|_{AB}$ are identical for this case. We use these symbols interchangeably in the following, unless there is an explicit dependence on the azimuthal angle. It should be emphasized that the absence of the azimuthal angle for this quantity occurs when there is no *outside* field or structure and has nothing to do with the shape of the target particles. If there are no forces or lattices orienting the target molecules, then we assume their orientations are random and, on the average, all azimuthal angles are equivalent.

If, in addition to collecting scattered atoms, the detecting apparatus can differentiate between atoms which have experienced or caused an internal change, then an angular differential cross section for each inelastic process can be defined. The sum of these cross sections will, of course, equal $d\sigma/d\Omega|_{AB}$. Extending the definition in Eq. (2.6), we define an angular differential cross section for charge transfer (per target atom):

$$\frac{d\sigma}{d\Omega}\bigg|_{A+B \to AB^+} \Delta\Omega_A \equiv \frac{\text{Number of particles per unit time scattered into } \Delta\Omega_A \text{ having captured an electron}}{\text{Incident intensity}} \quad (2.30)$$

If the internal change of interest is a change in the state of the struck target particle, the experimenter may choose to measure the distribution of scattered targets or note a property of A which can be associated with the event of interest in B, as for example an additional loss of kinetic energy beyond that associated with a simple deflection.

Calculation of Angular Differential Cross Sections

To calculate an angular differential cross section, we imagine that a particle which entered the collision region as part of a beam, and was detected as being scattered into some angle, followed a particular trajectory in response to the forces between it and a target particle (Figure 2.6). This

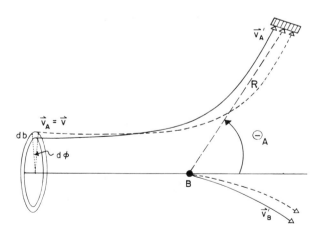

Figure 2.6. Collision of particles A and B. All particles A passing through the ring have similar trajectories and experience very nearly the same deflection Θ_A. The angle Θ_B, not indicated, is the angle $\mathbf{v}_B'$ makes with the axis.

could be a force, either repulsive or attractive, which acted gradually over the full trajectory, as in Figure 2.6, or an abrupt collision, as in Figure 2.4. For simplicity we again consider collisions between spherically symmetric particles.

For each impact parameter b with azimuthal angle ϕ, there is a probability of being scattered into an angular region $d\Omega$ about θ and φ which we write as $p_{AB}(b, \theta) d\Omega$. For spherical particles, the azimuthal angle φ equals the incident azimuthal angle ϕ as there are no torques, and the labels on θ are dropped, as we can consider either the scattered incident or target particles. The probability of being scattered into any angle, $P_{AB}(b)$, which we discussed earlier, is a sum of these probabilities:

$$P_{AB}(b) = \int\int p_{AB}(b, \theta) \, d\Omega$$

$$= 2\pi \int_{-1}^{1} p_{AB}(b, \theta) \, d\cos\theta \qquad (2.31)$$

From the probability per unit solid angle of a scattering, $p_{AB}(b, \theta)$, we can write an expression for the angular differential cross section. Using Eqs. (2.31) in Eq. (2.20) and comparing it to Eq. (2.29), we find

$$\sigma_{AB}(\theta) = 2\pi \int_{0}^{\infty} p_{AB}(b, \theta) \, b \, db \qquad (2.32)$$

This expression merely states that the scattering into an angular region about θ is a sum of contributions from all impact parameters for which a scattering into the angle θ is likely, as indicated by p_{AB}. Again we have only postponed the basic problem of obtaining the probability density p_{AB} from the details of the interaction. Carrying the above one step further, if we are able to distinguish in our calculation between inelastic and elastic events, we can write, for the ionization cross section,

$$\sigma_{AB \to AB^+}(\theta) = 2\pi \int_{0}^{\infty} p_{AB \to AB^+}(b, \theta) b \, db \qquad (2.33)$$

Equation (2.32) simplifies further when the interaction is a purely repulsive or attractive force which at any instant during the collision depends only on the distance between the two particles. Now, each impact will result in a scattering into a particular angle θ. A relationship between the scattering angle and impact parameter, referred to as the deflection function, $\Theta(b)$, can be calculated from the forces of interaction using Newton's laws. In such a deterministic model $\Theta(b)$ has a single value for each impact parameter, b, and it is clear that p_{AB} is zero for this example except when $\cos\theta = \cos\Theta(b)$. This type of density function is called a delta function, and, in Appendix A, we discuss the behavior of such functions. Before completing this approach, we return to the defining equation for the differential cross section.

We note that the number of beam particles per unit time entering a ring of area $2\pi b\, db$ about the center of a given target atom (see Figure 2.4 or 2.6) is $I2\pi b\, db$. For our example, this is also the *rate* of scattering into the solid angular region associated with $\cos\theta = \cos\Theta(b)$. Using Eq. (2.27), we obtain the calculated differential cross section from the expression

$$\sigma_{AB}(\Theta)2\pi\sin\Theta\, d\Theta = \frac{(2\pi b\, db\, I)}{I}$$

Simplifying and rewriting we have

$$\sigma_{AB}(\Theta) = \left|\frac{b}{\sin\Theta}\frac{db}{d\Theta}\right| \tag{2.34}$$

where the absolute value signs are included to assure that the cross section is positive. We stated that $\Theta(b)$ was single valued. If, *in addition*, at each measuring angle, $\cos\theta$, there is only one b for which $\cos\theta = \cos\Theta(b)$, then the differential cross section $\sigma_{AB}(\theta)$ will be the calculated classical cross section $\sigma_{AB}(\Theta)$ in Eq. (2.34). However, not infrequently the calculated deflection function, $\Theta(b)$, will have both positive and negative values and there may be more than one value of b for which $\cos\theta = \cos\Theta(b)$. That is, experimentally we cannot distinguish between negative and positive deflections as we cannot see on which side of the target atom a particular incident particle approaches a particular target particle. The measured cross section should, therefore, be estimated as a sum of contributions from all b yielding the same deflection cosine

$$\sigma_{AB}(\theta) = \sum_i \{\sigma_{AB}[\Theta(b)]\}_{\cos\theta = \cos\Theta(b)} \tag{2.35}$$

where $\sigma_{AB}(\Theta)$ is calculated as in Eq. (2.34) when the deflection function $\Theta(b)$ is known. In the following sections, we review classical kinematics and obtain cross sections for some simple examples using Eqs. (2.34) and (2.35).

Before ending this discussion, we return to our earlier approach, that of a deterministic collision for which p_{AB} is a delta function,

$$p_{AB}(b,\theta) = \frac{1}{2\pi}\delta[\cos\theta - \cos\Theta(b)]. \tag{2.36}$$

Substituting this expression into Eq. (2.31) and using the properties of a delta function discussed in Appendix A, we find that the impact parameter probability of a collision is unity, $P_{AB}(b) = 1$, as discussed earlier. Substituting Eq. (2.36) into the expression for the differential cross section, Eq. (2.32), we obtain the result in Eq. (2.35). This is left as a problem for the reader. Based on this approach to the problem, statistical models can be immediately envisaged. As a simple example, we note that when an atom collides with a molecule, the molecular orientations are random. For any particular orientation, the deflection function $\Theta(b, \phi, \Omega_M)$ and corresponding azimuthal deflection function $\varphi(b, \phi, \Omega_M)$ are determined from classical mechanics, where Ω_M indicates the molecular orientation. After integrating over all azimuthal scattering angles φ, the classical density function for the molecular collision becomes

$$p_{AB}(b, \Theta) = \frac{1}{2\pi} \langle \delta[\cos \theta - \cos \Theta(b, \phi, \Omega_M)] \rangle_{\Omega_{M, \phi}} \qquad (2.37)$$

where the brackets indicate averaging over the random orientations of the molecule and the azimuthal impact angle ϕ. One can also imagine models for which the density function will not be a delta function but will have a width associated with the fact that the impact parameter cannot be precisely defined. Extending this notion, then, at larger impact parameters, p_{AB} will not generally integrate to unity because the boundaries of the atom are diffuse. Such a model, though nondeterministic, would still be classical in form and is an attempt to incorporate quantum-mechanical effects. In Chapter 3 we will discuss the quantum mechanical solution of the problem.

Collision Kinematics: Elastic Collisions

Here we apply two conservation laws, the conservation of energy and momentum, derived from Newton's equations of motion, to the collision of two bodies A and B, shown in Figure 2.6, with masses M_A and M_B. The velocities before the collision are v_A and v_B, with $v_B = 0$, and, after they have separated by a large distance, they are v'_A and v'_B. The conservation laws are independent of the details of the forces and apply to all trajectories which produce the same deflection. We define the deflection angles of A and B, Θ_A and Θ_B, as the angles the final velocity vector make with the incident velocity vector. With these definitions, conserving energy, $\mathscr{E}$, yields

$$\mathscr{E} = \tfrac{1}{2}M_A v_A^2 = \tfrac{1}{2}M_A v_A'^2 + \tfrac{1}{2}M_B v_B'^2 \qquad (2.38)$$

The conservation of momentum tells us that the vectors v_A, v'_A, and v'_B all lie in the same plane, and that in that plane,

$$M_A v_A = M_A v'_A \cos \Theta_A + M_B v'_B \cos \Theta_B$$
$$0 = M_A v'_A \sin \Theta_A - M_B v'_B \sin \Theta_B \qquad (2.39)$$

If we take the M_A, M_B, and v_A as given initial quantities, the three equations above relate the four quantities v'_A, v'_B, Θ_A, and Θ_B. Therefore, conservation laws alone cannot describe the collision; we require a force equation involving the interaction. However, some simple relationships can be obtained first.

Table 2.2. Angular Range of Θ_A

	$0 < \lvert\Theta_B\rvert < \pi/2,$	$E_A = \frac{1}{2M} M_A v_A^2, \mu = M_B/M_A$	
Heavy target	$M_B \gg M_A$	$\mu \gg 1$	$0 \leq \lvert\Theta_A\rvert < \pi$	$\dfrac{T}{E_A} \simeq \dfrac{2}{\pi}(1 - \cos\Theta_A)$
Equal masses	$M_B = M_A$	$\mu = 1$	$0 \leq \lvert\Theta_A\rvert < \pi/2$	$\dfrac{T}{E_A} = \sin^2\Theta_A$
Light target	$M_B \ll M_A$	$\mu \ll 1$	$0 \leq \lvert\Theta_A\rvert < \mu$	$\dfrac{T}{E_A} \simeq \dfrac{1}{\mu}\Theta_A^2$

Using Eqs. (2.38) and (2.39), we see that the final scattering angles are related by the expression

$$\tan\Theta_A = \frac{\mu \sin 2\Theta_B}{1 - \mu \cos 2\Theta_B} \tag{2.40}$$

where $\mu = M_B/M_A$ is the mass ratio. It is clear from the collision of spheres, shown in Figure 2.4, that in collisions for which orbiting of the particles does not occur, $0 < \lvert\Theta_B\rvert < \pi/2$. Orbiting will be treated later. Equation (2.40) now places restrictions on the angular range of Θ_A, expressed in Table 2.2. For angles outside the kinematically allowed range, $\sigma(\Theta_A) = 0$. It is seen from Table 2.2 that forward scattering is generally preferred and backscattering can occur only in those collisions for which $\mu \to \infty$. Measuring the angular range of elastically scattered particles A is, therefore, a means of estimating the masses of the target particles B.

Postponing the discussion of realistic forces between atoms we note that in the example of hard-sphere collisions (Figure 2.4) the impulse is along the axis connecting their centers at contact. This implies that the

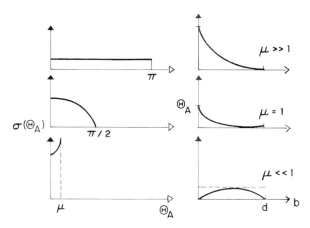

Figure 2.7. Cross sections, $\sigma(\Theta_A)$, and deflection functions, Θ_A, for the collision of hard spheres of radius r_A and r_B, where $\mu = M_B/M_A$ and $r_A + r_B = d$.

scattered target moves off at an angle of Θ_B such that $\sin\Theta_B = b/(r_A + r_B)$. Using Eq. (2.40), we can determine the deflection function for the incident particle, $\Theta_A(b)$. As Θ_A depends on the relative masses of the incident and target particles, so will the elastic cross calculated using Eq. (2.34). Writing $d = (r_A + r_B)$, the collision diameter, the differential cross sections for the collision of spheres are plotted in Figure 2.7 for three limiting cases. The angular dependence of the cross section is shown to be strongly dependent on the mass ratio μ. For light particles incident on heavy target particles the differential cross section is isotropic, a rather special result for an impulsive collision, which is employed, too often, in making macroscopic calculations. It should also be noted that $\sigma(\Theta)$ is nonzero at $\Theta = 0$. This may at first seem surprising on recalling the experimental measurements of the cross section in which we excluded transmitted beam particles. However, $\sigma(\Theta)$ is the cross section per unit solid angle and, therefore, must account for those small-angle collisions in an angular region about $\Theta = 0$. Using Eq. (2.27), we see that the number of particles scattered per unit time into a given angular region is proportional to $\sigma(\theta)\sin\theta\,\Delta\theta$, where $\Delta\theta$ is the aperture. As θ goes to zero the number of scattered particles decreases, in this example, even though, for the case $\mu \ll 1$, the differential cross section may be quite large. It is common to plot the quantity $\rho(\theta) = \theta\sin\theta\,\sigma(\theta)$ to account roughly for these angular factors.

From the conservation laws, a relationship between the elastic energy loss, $T = \frac{1}{2}M_B v_B'^2$, and the scattering angles can be obtained. Using Eq. (2.38) and (2.39), we find

$$\cos\Theta_B = (T/\gamma E_A)^{1/2}$$
$$\cos\Theta_A = \frac{\{1 - [(1+\mu)/2]T/E_A\}}{[(1-T)/E_A]^{1/2}}$$

(2.41)

where $\gamma = 4M_A M_B/(M_A + M_B)^2$, $E_A = \frac{1}{2}M_A v_A^2$, and the maximum energy transfer is $T_M = \gamma E_A$. For the equal-mass case, $\gamma = 1$, all the energy may be transferred, whereas, for a large mismatch in particle masses, either $\mu \gg 1$ or $\mu \ll 1$; then $\gamma \ll 1$, and only a fraction of the energy may be transferred in an elastic collision. Table 2.2 also contains estimates of T for the cases being considered. Because the energy transfer is of considerable interest and is simply related to Θ_A, it is often useful to discuss a cross-section differential in elastic energy transfer. Defining $d\sigma/dT$ via

$$\frac{d\sigma}{dT}\Delta T \equiv \frac{\begin{array}{c}\text{Number of particles scattered per unit time with}\\ \text{elastic energy transfer between } T \text{ and } T + \Delta T\end{array}}{\text{Incident intensity}}$$

we can relate $d\sigma/dT$ and $\sigma(\Theta_A)$:

$$\frac{d\sigma}{dT} = \sigma(\Theta_A)\left|\frac{d\cos\Theta_A}{dT}\right|$$

(2.42)

This yields, for the elastic collisions of spheres, the simple result $d\sigma/dT = (1/\gamma E_A)\pi d^2$, which is independent of T. In the following section we will reconsider the kinematics in a coordinate system for which the mass factors play a less important role as in the energy transfer cross section.

Center of Mass System

The kinematics of the collision were shown to play an important role in determining the angular differential cross section. Thus far we have concentrated on the collision between a moving atom, A, and a stopped particle, B. However, it was pointed out that the integrated cross sections and the rate constants depended only on the relative speed between the particles if there were no outside fields, and therefore the ratio of the masses M_A and M_B was not important. It would be useful, therefore, to discuss the collision in a reference system for which the kinematics could be described using only the relative velocity.

The center of mass of the colliding particles, a distance $\mathbf{R}_c$ from some fixed origin, is located on a line connecting A and B, as shown in Figure 2.8. The quantities $\mathbf{r}_A$ and $\mathbf{r}_B$ in Figure 2.8 are the positions of A and B as measured from the center of mass, C. The usefulness of using C as an alternate origin is immediately apparent as the velocities of A and B, $\mathbf{r}_A$ and $\mathbf{r}_B$, are both proportional to $\mathbf{v}$, the relative velocity. The coordinate system centered on C is referred to as the center-of-mass (CM) coordinate system, and that with the fixed origin, the laboratory coordinate system. Relationships between CM and laboratory variables are summarized in Table 2.3 and the derivations are left as a problem for the reader. From Table 2.3 it is seen that in the CM system quantities of physical interest are given in terms of $\mathbf{v}$ and a quantity $m = M_A M_B/(M_A + M_B)$, the reduced mass. In the CM system the momenta of A and B are equal and opposite, or the total momentum is zero, hence it is also referred to as the zero-momentum system. For the case we are considering, there are no outside fields and the

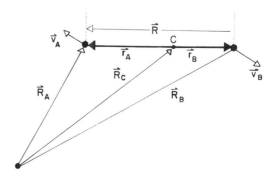

Figure 2.8. Positions of A and B in the laboratory ($\mathbf{R}_A$, $\mathbf{R}_B$) and CM ($\mathbf{r}_A$, $\mathbf{r}_B$) coordinate systems. The point C is the location of the center of mass.

Table 2.3. Relationships between Laboratory and CM Variables

Positions	Velocities
$\mathbf{R}_c = (M_A \mathbf{R}_A + M_B \mathbf{R}_B)/(M_A + M_B)$	$\mathbf{V}_c \equiv \dot{\mathbf{R}}_c = (M_A \mathbf{v}_A + M_B \mathbf{v}_B)/(M_A + M_B)$
$\mathbf{R} = \mathbf{R}_A - \mathbf{R}_B = \mathbf{r}_A - \mathbf{r}_B$	$\mathbf{v} \equiv \dot{\mathbf{R}} = \mathbf{v}_A - \mathbf{v}_B = \dot{\mathbf{r}}_A - \dot{\mathbf{r}}_B$
$\mathbf{r}_A = \mathbf{R}_A - \mathbf{R}_c , \; \mathbf{r}_B = \mathbf{R}_B - \mathbf{R}_c$	$M_A \dot{\mathbf{r}}_A = -M_B \dot{\mathbf{r}}_B = M\mathbf{v}$

Laboratory quantities	CM quantities
$M = M_A + M_B$	$m = M_A M_B/(M_A + M_B)$
$\mathscr{E} = \frac{1}{2}M_A v_A^2 + M_B v_B^2 = \frac{1}{2}MV_c^2 + E$	$E = \frac{1}{2}M_A \dot{r}_A^2 + \frac{1}{2}M_B \dot{r}_B^2 = \frac{1}{2}mv^2$
$\mathscr{P} = M_A \mathbf{v}_A + M_B \mathbf{v}_B = M\mathbf{V}_c$	$P = M_A \dot{\mathbf{r}}_A + M_B \dot{\mathbf{r}}_B = 0$
$\mathscr{L} = M_A \mathbf{R}_A \times \mathbf{v}_A + M_B \mathbf{R}_B \times \mathbf{v}_B = M\mathbf{R}_c \times \mathbf{V}_c + \mathbf{L}$	$L = M_A \mathbf{r}_A \times \dot{\mathbf{r}}_A + M_B \mathbf{r}_B \times \dot{\mathbf{r}}_B = m\mathbf{R} \times \mathbf{v}$

forces between the two bodies must be equal and opposite, or $\dot{\mathbf{R}}_c$ is a constant. Since the laboraotry and the CM coordinates move relative to each other at constant velocity, as we know from classical mechanics, examining the motion in the CM system does not affect the forces.

The kinematics of the collision in the CM system are indicated in Figure 2.9 using the momentum vectors of the particles. Since the net momentum is zero before the collision, it is also zero after the collision. If, in addition, the collision is elastic, the magnitudes of the momenta remain the same. For the case $\mathbf{v}_B = 0$, $\mathbf{v}$ and $\mathbf{V}_c$ are in the same direction, which simplifies the transformation between the laboratory and CM coordinates after the collision. For this case the final and initial velocities all lie in the same plane, the collision plane, as shown in Figure 2.9. When $\mathbf{v}_B \neq 0$, the transformation back to the final laboratory velocities is also straightforward. In general, we will deal with the $\mathbf{v}_B = 0$ case.

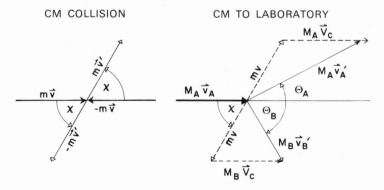

Figure 2.9. Kinematics of collision indicated in the CM system. To obtain laboratory momentum and angles the velocity of the center of mass, $\mathbf{V}_c$, is added to the CM motion of each particle. Here we assume $v_A = v$ and $v_B = 0$ initially.

When the colliding systems are spherically symmetric, only the CM deflection angle, χ, is needed to describe the collision, and it is clear from Figure 2.9 that χ takes on all values between 0 and π. The relationships between the laboratory quantities and the CM angle, from Fig. 2.9, are

$$\Theta_B = \tfrac{1}{2}(\pi - \chi)$$

$$\tan\Theta_A = \frac{\mu\sin\chi}{1 + \mu\cos\chi} \tag{2.43}$$

$$T = \gamma E_A \sin^2\chi/2, \qquad E_A = \tfrac{1}{2}M_A v_A^2, \qquad v_B = 0$$

Equations (2.43) are equivalent to Eqs. (2.40) and (2.41). From Figure 2.9 it is apparent that particles A and B follow equivalent trajectories in the CM system. Further, from Eq. (2.43) it is seen that if $\mu \gg 1$, $\Theta_A = \chi$. Therefore, in the CM system the collision can be treated as if a particle of mass m, the reduced mass of the two particles, and velocity, $\mathbf{v}$, the relative velocity, collides with an infinitely heavy, hence stationary, target. This will be exploited in subsequent discussions.

The forces and the equations of motion are used to obtain a CM deflection function, $\chi(b)$, for a reduced-mass particle and, hence, an angular differential cross section, in the CM system:

$$\sigma(\chi) = \left| \frac{b}{\sin\chi} \frac{db}{d\chi} \right| \tag{2.44}$$

The calculated CM cross section can be changed to the laboratory system, Eq. (2.34), by the transformation

$$\sigma(\Theta_A) = \sigma(\chi)\left| \frac{d\cos\chi}{d\cos\Theta_A} \right| = \sigma(\chi)\frac{(\mu^2 + 2\mu\cos\chi + 1)^{3/2}}{\mu^2|\mu + \cos\chi|} \tag{2.45}$$

if the collision is between spherically symmetric particles. By using the relationships in Eq. (2.43) the elastic-energy-transfer cross section and the CM cross section are simply related,

$$\frac{d\sigma}{dT} = 2\pi\sigma(\chi)\left| \frac{d\cos\chi}{dT} \right| = \frac{4\pi}{\gamma E_A}\sigma(\chi) \tag{2.46}$$

For the impulsive collision of spheres considered earlier, $\sigma(\chi) = d^2/4$. That is, the CM cross section is isotropic for this example regardless of the particle masses. The nonisotropic laboratory cross sections shown in Figure 2.7, therefore, are a result *only* of the particle kinematics. This is the essence of the CM calculation—to separate the calculation of the kinematical factors from the details of the interaction, the later determining the CM cross section.

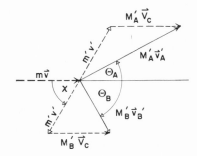

Figure 2.10. Kinematics of inelastic collision in the CM system and transformation to laboratory frame indicated by dashed vectors. Solid vectors are final laboratory velocity. Here $v_A = v$ and $v_B = 0$ initially.

Kinematics of Inelastic Collisions

The kinematics of an inelastic collision are shown in Figure 2.10 using the CM system, and the results are summarized in Table 2.4. During the collision we allow the internal energies of each particle, ε_A and ε_B, to change as well as their masses. For nonrelativistic collisions the energy and mass are conserved separately. Defining Q as the net change in internal energy (see Table 2.4), we specify the type of inelastic collision as follows: $Q > 0$, an endothermic collision;* $Q = 0$, an elastic collision if no internal changes have occurred, or resonant if the net internal changes in A and B yield $Q = 0$; $Q < 0$, an exothermic collision; and $M_A \neq M'_A$ and $M_B \neq M'_B$, a rearrangement collision. The change in internal energy Q is essentially a CM quantity, as can be seen in Table 2.4. That is, it effects a corresponding change in the center-of-mass kinetic energy. Conversely, in order for an endothermic reaction to occur, the kinetic energy in the CM system must exceed the threshold energy for a given reaction. Below this energy the inelastic cross section is zero.

The relationship between Θ_A and χ, the CM scattering angle, has the same form as that found for elastic collisions, except the internal energy change modifies the mass ratio. For instance, a large increase in the internal energy to a higher or more loosely bound state (Q positive, endothermic) restricts the laboratory scattering to the forward direction. A large decrease in the internal energy to a more tightly bound state, that is, internal energy converted to kinetic energy (Q negative, exothermic), may permit significant backscattering. The latter is like an explosion occurring when the particles collide. For the collision of two soft spheres, that is, spheres which dissipate an amount of energy Q, the center-of-mass cross section is $\sigma_Q(\chi) = \frac{1}{4}d^2$, as in the elastic case. Again the CM cross section only indicates the nature of the

---

* The convention for the sign on Q is often the opposite to that used here. We prefer Q, like T, to be positive when the initial kinetic energy is lost or transferred. Minus Q, as used here, is often referred to as the energy defect of the collision.

Table 2.4. Inelastic Collision Kinematics

Laboratory system

$$\mathcal{E} = \tfrac{1}{2} M_A v_A^2 + \tfrac{1}{2} M_B v_B^2 + \varepsilon_A + \varepsilon_B = \tfrac{1}{2} M_A' v_A'^2 + \tfrac{1}{2} M_B' v_B'^2 + \varepsilon_A' + \varepsilon_B'$$

$$= \tfrac{1}{2} M V_c^2 + E + \varepsilon_A + \varepsilon_B$$

$$M = M_A + M_B = M_A' + M_B'$$

$$\mathcal{P} = M_A \mathbf{v}_A + M_B \mathbf{v}_B = M_A' \mathbf{v}_A' + M_B' \mathbf{v}_B' = M\mathbf{V}_c$$

CM system

$$E = \tfrac{1}{2} m v^2 = \tfrac{1}{2} m v'^2 + Q \qquad [Q = (\varepsilon_A' + \varepsilon_B') - (\varepsilon_A + \varepsilon_B)]$$

$$m = M_A M_B / M, \qquad m' = M_A' M_B' / M$$

$$\mathbf{P} = 0$$

Transformations for $v_B = 0$

$$\tan \Theta_A = \frac{(\mu \mu' g)^{1/2} \sin \chi}{1 + (\mu \mu' g)^{\frac{1}{2}} \cos \chi}; \qquad g = 1 - Q/E$$

$$T = \frac{\gamma E_A}{4} \frac{M_B'}{M_B} \left[1 + \frac{\mu}{\mu'} g - 2 \left(\frac{\mu}{\mu'} g \right)^{1/2} \cos \chi \right]; \qquad E_A = \tfrac{1}{2} M_A v_A^2$$

$$\cos \Theta_A = \frac{(M_A'/M_A)^{1/2} \{ \tfrac{1}{2}(1 + M_A/M_A') - [(1 + \mu')/2](T/E_A) - \tfrac{1}{2} Q/E_A \}}{[1 - (T + Q)/E_A]^{1/2}}$$

interaction, and the kinematics are determined by Q and the mass ratio using the relationships in Table 2.4.

It is seen in Table 2.4 that $\cos \Theta_A$ depends on both T and Q, the amount of elastic energy transfer to particle B and the net internal energy change. This suggests a simple scheme for measuring the inelastic cross sections. At each scattering angle the net change in kinetic energy of the beam particle, $\Delta E = T + Q$, can be measured, as we discussed earlier when considering the energy loss rates. Such a procedure is straightforward using the apparatus in Figure 2.11 if the scattered particle is an ion. In atomic and molecular systems, a large number of Q values are possible for any particular collision, and, therefore, one measures, at each angle, an energy spectrum of scattered particles. For very fast collision, for which $Q/E \ll 1$, T and $\cos \Theta_A$ have the same relationship we found for elastic collisions. Therefore, at each scattering angle, T can be calculated independent of Q and the spectrum ΔE can be converted directly to a spectrum in Q. For low-energy collisions, the extraction of Q at each angle and E is only slightly more complicated. The measured energy spectrum, therefore, yields

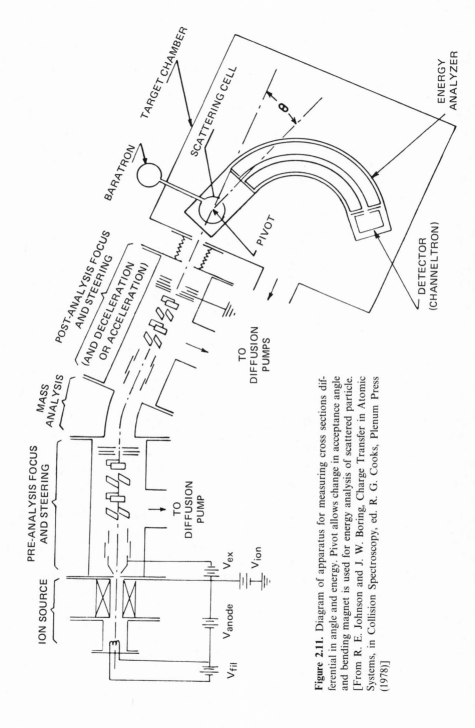

Figure 2.11. Diagram of apparatus for measuring cross sections differential in angle and energy. Pivot allows change in acceptance angle and bending magnet is used for energy analysis of scattered particle. [From R. E. Johnson and J. W. Boring, Charge Transfer in Atomic Systems, in Collision Spectroscopy, ed. R. G. Cooks, Plenum Press (1978)]

a double-differential cross section, one differential in angle *and* inelastic energy loss, defined, per target atom B, as

$$\frac{d^2\sigma_{AB}}{d\Omega\, dQ}\, \Delta\Omega\Delta Q \equiv \frac{\begin{array}{c}\text{Number of particles scattered per unit}\\ \text{time between } \Omega \text{ and } \Omega + \Delta\Omega, \text{ and with energy}\\ \text{loss between } Q \text{ and } Q + \Delta Q\end{array}}{\text{Incident intensity}} \quad (2.47)$$

Clearly, summing over all angles and energy losses, one recovers the total cross section σ_{AB}. In atomic systems the internal energy changes are often discrete, except for the ionization process. However, as the instrument has finite apertures, both in angle and energy, the spectrum will be broadened and can generally be treated as a continuum, as suggested by the definition in Eq. (2.47). In Figure 2.12 is shown, as an example, results of an energy-analyzed scattering equipment for the reaction $He^{++} + Ne \rightarrow He^+ + Ne^+$, where the scattered He^+ is monitored. The scale for Q has been identified, and the plot is the scattering count rate, which is proportional to the double-differential cross section. There are seen to be groups of closely spaced discrete states which are populated with varying probabilities as a function of changing scattering angle and, hence, impact parameter. For the harder collisions shown in Figure 2.12, which correspond to large deflections, not only will more elastic energy, T, be transferred, but it becomes increasingly likely that higher states are excited. That is, larger amounts of internal energy tend to be absorbed in hard collisions—a general characteristic of inelastic collisions. For the collision in Figure 2.12, at small scattering angles exothermic collisions are favored. This is *not* a general feature of inelastic collisions; rather, it is determined by the detailed nature of the atomic interaction, i.e., $p_{i \rightarrow j}(b, \theta)$ of Eq. (2.33). These topics will be reconsidered later. In the following we use the laws of classical mechanics to calculate the deflection function $\chi(b)$.

The Classical Deflection Function

To obtain the relationship between χ, the CM scattering angle, and b, the impact parameter, we will consider a collision between a particle of mass m, the reduced mass of the system, and a fixed target. The initial relative velocity, v, changes as a function of time in response to the force between the particles. This force depends on the separation of the particles, their orientations and relative velocities, and is an appropriate CM quantity. We will treat the case of a velocity-independent force which can be expressed in terms of an interaction potential, $V(R)$, such that $\mathbf{F} = -\nabla V$. As with the earlier discussion, we will initially consider spherically sym-

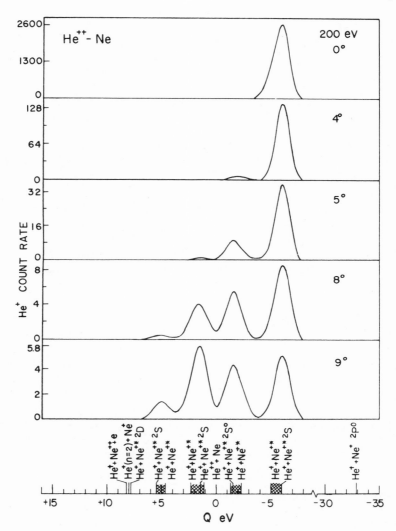

Figure 2.12. Energy spectra of scattered He$^+$ ions from He^{++} + Ne → He$^+$ + Ne$^+$ collisions. Laboratory scattering angle indicated on right. [From R. E. Johnson and J. W. Boring, Charge Transfer in Atomic Systems, in *Collision Spectroscopy*, ed. R. G. Cooks, Plenum Press (1978), p. 126.]

metric incident and target particles, so that V is a function of the separation, R, only.

For this case, the collision takes place in a plane, and it is useful to express the relative velocity at any time in terms of radial and angular components,

$$\mathbf{v} \equiv \dot{\mathbf{R}}(t) = \dot{R}\hat{R} + R\dot{\alpha}\hat{\alpha} \tag{2.48}$$

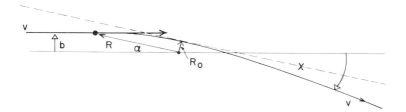

Figure 2.13. Trajectory for reduced-mass particle with velocity v and impact parameter b subject to a central attractive force. The quantity χ is the CM scattering angle, R the distance to the fixed center, α the angular position of the particle along the trajectory, and R_0 the distance of the closest approach. Dashed line, tangent at R_0, indicates symmetry of the first and second halves of the collision.

where α is the angle between $\hat{R}$ and the collision axis in Figure 2.13. The CM angular momentum, from Table 2.3, is written as

$$L = mR^2\dot{\alpha} = mvb \tag{2.49}$$

and the energy as

$$E = \tfrac{1}{2}m\dot{R}^2 + \tfrac{1}{2}mR^2\dot{\alpha}^2 + V(R) = \tfrac{1}{2}mv^2 \tag{2.50}$$

where mvb and $\tfrac{1}{2}mv^2$ are the initial values for these conserved quantities. This collision can be characterized by an initial energy or velocity (E or v) and by an initial angular momentum or impact parameter (L or b). The potential energy in Eq. (2.50) is assumed to be zero at large separations. We will also have occasion in future sections to write the energy in terms of a radial momentum $p(R) = m\dot{R}$ and the angular momentum L,

$$E = \frac{1}{2}\frac{p^2(R)}{m} + \frac{1}{2}\frac{L^2}{mR^2} + V(R) \tag{2.51}$$

As L and E are conserved quantities, Eq. (2.51) explicitly shows that the radial momentum or radial velocity is only a function of the separation R during the collision. That is, rearranging Eq. (2.51), one obtains

$$p(R) = m\dot{R} = \pm p_0 \left[1 - \left(\frac{b}{R}\right)^2 - \frac{V(R)}{E} \right]^{1/2} \tag{2.52}$$

where Eq. (2.49) was used to replace L by the impact parameter and $p_0 = mv$ is the initial momentum of the reduced-mass particle.

Equation (2.52) can be integrated to obtain the radial position as a function of time during the collision. In the experiments described, however, this information is not obtainable. The quantity of interest is the net deflection, the CM scattering angle, $\chi(b)$, which is calculated by integrating α in Eq. (2.49) to obtain the angular excursion of R. Based on the relationship between α and χ indicated in Figure 2.13, the CM deflection function is:

$$\chi = \pi - \int_{-\infty}^{\infty} \dot{\alpha}\, dt$$

$$= \pi - \int_{-\infty}^{\infty} \frac{vb}{R^2}\, dt \qquad (2.53)$$

where R is a function of time and v and b are constants. From Eq. (2.52), it is seen that the radial velocity (or momentum) decreases as the particles approach, becoming zero at the distance of closest approach, and then increases as the particles recede. The distance of closest approach, R_0, for each impact parameter occurs when $\dot{R} = 0$, implying

$$1 - (b/R_0)^2 - V(R_0)/E = 0 \qquad (2.54)$$

Changing variables in Eq. (2.53) and using Eq. (2.52), we calculate the CM deflection angle as

$$\chi(b) = \pi - 2b \int_{R_0}^{\infty} \frac{dR}{R^2} \frac{1}{[1 - (b/R)^2 - V/E]^{1/2}} \qquad (2.55)$$

where the contributions to χ before and after $R = R_0$ are equal, as indicated in Figure 2.13. Specifying the potential energy and performing the integration in Eq. (2.55), we see that $\chi(b)$, the CM deflection function, depends on the initial relative velocity v (or E) and b (or L), as stated earlier. The deflection function can now be used to obtain the CM cross section $\sigma(\chi)$ using Eq. (2.44).

For a number of potential functions (cf. Problem 2.10), the above expression for $\chi(b)$ can be integrated analytically. This is the case for the simple coulomb interaction between charged particles. However, for most realistic potentials a numerical scheme, like that described in Appendix B, must be employed. Examination of Eq. (2.55) shows that the relationship between $\chi(b)$ and the potential V is not very transparent. However, it is clear that the major contributions to the integral come from values of R near R_0. Here the denominator in the integral approaches zero, implying that the behavior of $V(R)$ near R_0 determines the size of the deflection. This fact will prove useful in estimating the deflection functions and hence cross sections.

To further examine the relationship between V and χ, it is instructive to consider collisions in which V/E is small throughout the collision, that is, large b collisions which lead to small-angle scatterings. This can be handled directly by expanding the denominator in Eq. (2.55), which we leave as a problem for the reader. Alternatively, a small deflection can be thought of as being due to a small net impulse, Δmv, roughly perpendicular to the direction of motion. Now χ has the form

$$\chi \sim \frac{\Delta mv}{mv} = \frac{\int_{-\infty}^{\infty} F_{\perp}\, dt}{mv} \qquad (2.56)$$

where $F_\perp$ is the component of force perpendicular to the direction of motion. For large impact parameters the path can be closely approximated by a straight line, allowing us to write $R^2 = R_0^2 + Z^2$, where Z is along the trajectory. Further, the speed can be assumed to be nearly constant, so $Z \simeq vt$. On substitution into Eq. (2.56), $\chi(b)$ can be written as

$$\chi(b) \sim -\frac{d}{dR_0} \frac{1}{2E} \int_{-\infty}^{\infty} V \, dZ \tag{2.57a}$$

Changing the integration variable yields

$$\chi(b) \sim -\frac{1}{E} \int_0^\infty \left(\frac{dV}{dR}\right) \frac{R_0 \, dR}{R[1 - (R_0/R)^2]^{1/2}} \tag{2.57b}$$

This result is often referred to as the classical impulse approximation or the momentum approximation to the deflection function. For small-angle scattering the penetration is small and one generally replaces R_0 by b in the above expressions.

In Appendix B, the expression in Eq. (2.57a) is evaluated for a number of potential functions. For the inverse power law potentials, $V(R) = C_n/R^n$, using Eq. (2.57a), we obtain the deflection function as

$$\chi(b) \simeq a_n[V(R_0)/E] \tag{2.58}$$

with

$$a_n = n \int_0^{\pi/2} \sin^n \alpha \, d\alpha = (\pi)^{1/2} \frac{\Gamma[(n + 1)/2]}{\Gamma(n/2)}$$

where $\Gamma(x)$ is the gamma function (see Appendix B). For $n = 1$, $a_n = 1$; for $n = 2$, $a_n = \pi/2$; etc., with $a_n \xrightarrow{n \to \infty} (\pi n/2)^{1/2}$. This class of potentials is particularly useful since realistic atomic and molecular interaction potentials, which we will describe in Chapter 4, can be written in terms of inverse power laws at large R. For the simplest case, the interaction between two charged particles (the coulomb potential), $n = 1$ for all R and $C_1 = q_A q_B$, where q_A and q_B are the charges of the colliding particles. Even when a potential does not have the power-law form, it often can be fitted locally to a power law near the distance of closest approach. Employing this notion, Lindhard suggested a generally useful expression for estimating the classical deflection function in the impulse approximation. The so-called "magic" formula for $\chi(b)$ for an arbitrary, monotonically increasing or decreasing potential is

$$\chi^2(b) \simeq -\frac{3}{4E^2} b^{1/3} \frac{d}{db} [V^2(b)b^{2/3}] \tag{2.59}$$

which for power laws implies $a_n \sim [(3n - 1)/2]^{1/2}$ in Eq. (2.58).

The velocity dependence of $\chi(b)$ in Eq. (2.57) is quite simple and explicit. The quantity $\tau \equiv \chi(b)E$ calculated in the impulse approximation can be seen to depend only on the impact parameter and is independent of the coordinate system. That is, using Eq. (2.43), τ also can be written in terms of laboratory quantities if the scattering angles are small, $\tau \sim E_A \theta_A$. The reduced cross section, defined as $\rho = \chi \sin \chi \, \sigma(\chi)$, is written, using Eq. (2.44), as

$$\rho \sim \frac{1}{2} \frac{db^2}{d \ln \tau} \tag{2.60}$$

for small χ. The quantity ρ now depends only on τ, and hence b, and is also independent of the coordinate system when χ is small, being equal to the reduced cross section $\rho \simeq \Theta_A \sin \Theta_A \, \sigma_{AB}(\Theta_A)$ defined earlier. Therefore a ρ vs τ plot of experimental data, like that shown in Figure 2.14, should be very nearly energy independent and directly related to the form of the interaction potential. For the power-law potentials, using Eq. (2.58) in Eq. (2.60) and setting $R_0 = b$, we obtain ρ as

$$\rho \sim \frac{1}{n}\left(\frac{a_n C_n}{\tau}\right)^{2/n} \tag{2.61a}$$

For the data in Fig. 2.14, n is seen to increase with decreasing τ (increasing b).

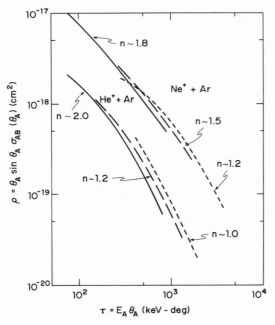

Figure 2.14 Plot of ρ vs τ from experimental data for three values of E_A: ———, 25 keV; ——— 50 keV; ------, 100 keV. For ρ, θ_A is in radians. Approximate power laws are indicated by n. The reader should replot this using CM angles and energies. [From E. N. Fuls, P. R. Jones, F. P. Ziemba, and E. Everhart, *Phys. Rev.* **107**, 704 (1957).]

Using Eq. (2.46), the equivalent energy transfer cross section is

$$\frac{d\sigma}{dT} \sim \frac{\mathscr{C}_n}{E^{1/n}} \frac{1}{T^{1+1/n}} \qquad (2.61b)$$

$$\mathscr{C}_n = \frac{\pi}{n} \left(\frac{M_A}{M_A + M_B} C_n^2 a_n^2 \right)^{1/n}$$

If a potential is thought to be monotonically increasing or decreasing, Eq. (2.61) provides a scheme for extracting the behavior of $V(R)$ directly from experimental data! Plotting ρ on a log-log scale, like the data in Figure 2.14, and using Eq. (2.61), we can determine the power law dependence, n, and C_n for $V(R)$ at the point of closest approach, b in this example. The value of b is then obtained from Eq. (2.58) or (2.59). More sophisticated analytic inversion methods are available but have very limited usefulness. The general approach for χ not small or V not well approximated by power laws is to parametrize a potential and adjust the parameters via some numerical scheme, like a least-squares fit to the cross section or ρ vs τ data. Picking the most physically, appropriate potential form, the one with the fewest parameters and best fit, is part of the art of atomic physics. When nonclassical effects are involved, like interference, which we will discuss in the next chapter, other potential extraction techniques become available.

In Figure 2.15 two simple interaction potentials are characterized, allowing us to examine explicitly the relationship between potential, deflection function, and cross section. For a potential yielding a purely repulsive force between atoms, $\chi(b)$ is always positive and has a maximum value π

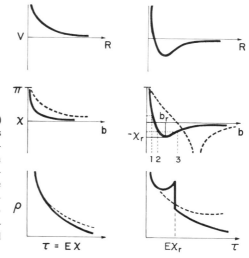

Figure 2.15. Deflection function (χ vs b) and angular differential cross section (as $\rho = \chi \sin \chi \, \sigma(\chi)$ vs $\tau = E\chi$) for two interaction potentials: repulsive potential on left and long-range attractive plus short-range repulsive on right. Rainbow angle is indicated is χ_r. Three impact parameters, b_1, b_2, and b_3, contributing to same $|\chi|$ for $|\chi| < |\chi_r|$. Results for high-energy (solid line) and low energy (dashed line) are indicated.

corresponding to backscattering in the CM system. At each value of $\cos \chi$ one impact parameter contributes to the cross-section calculation. The differential cross section, as is evident in Eq. (2.61), is singular at $\chi = 0$, which is a characteristic of classical cross sections derived from potentials with infinite range. (The hard-sphere example discussed earlier had a finite range and no singularity at $\chi = 0$.) It was of course clear from the discussion of total cross section that if $P_{AB}(b) = 1$ out to infinity, the angular differential cross section *must* be singular at $\chi = 0$.

For the case of a purely attractive potential (not shown) there is also a singularity at $\chi = 0$. However, there is no restriction on the maximum scattering angle as the particles may orbit each other before receding. For the long-range attractive and short-range repulsive potential, shown in Figure 2.15, which is a typical intermolecular potential, the shape of the deflection function, χ, at large b is that predicted by Eq. (2.58) or (2.59). At smaller b, because of the change in slope of the potential, the deflection function passes through a minimum value, χ_r, at the impact parameter labeled b_r. At this angle the classical cross section, defined in Eq. (2.44), is also singular as $d\chi/db = 0$. The quantity χ_r is called the rainbow angle, as this enhancement in the cross section is similar to the effect that produces rainbows in light scattering. When $\chi_r < \pi$, then, for angles larger than χ_r, only one impact parameter contributes to the classical deflection function, as in the previous example. For angles smaller than χ_r three impact parameters are associated with the same value of $\cos \chi$, as indicated in Figure 2.15. Trajectories for this potential are shown in Figure 2.16. For high energies, when the impulse approximation applies, the rainbow angle can be obtained from Eq. (2.58). Using a potential of the Lennard-Jones form,

$$V = \frac{C_{2n}}{R^{2n}} - \frac{C_n}{R^n} \tag{2.62}$$

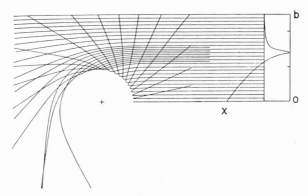

Figure 2.16. Classical trajectories corresponding to a collision of a reduced-mass particle with a fixed force center. Interaction is a long-range attractive and short-range repulsive potential. Deflection function χ vs b is shown on the right. [From H. Pauly, in *Atom–Molecule Collision Theory*, ed. R. B. Bernstein, Plenum Press, New York (1979), p. 127.]

one finds that $b_r > R_m$, the position of the minimum in V, and

$$\tau_r = E\chi_r \simeq \frac{1}{4} \frac{(a_n C_n)^2}{a_{2n} C_{2n}} \tag{2.63}$$

which depends on the relative strengths of the repulsive and attractive interactions. As E decreases significantly, although Eq. (2.63) is not valid, it is clear that χ_r eventually becomes greater than π and additional impact parameters will contribute to the calculation of cross section at all angles. At still lower energies, values of χ less than -2π occur, implying the particles orbit each other during the collision. Therefore, backscattering in the laboratory system becomes a very likely process, and many impact parameters contribute to all scattering angles.

If the attractive part of the potential decreases faster than $1/R^2$, then at very low energies there will be an impact parameter, b_0, at which the particle can be classically trapped. That is, the radial acceleration is zero: $\ddot{R} = 0$, at the distance of closest approach, i.e., $\dot{R} = 0$. The effective potential is plotted in Figure 2.17 for three impact parameters at a low energy: $b > b_0$, $b = b_0$, and $b < b_0$. When the attractive part of the potential dominates, b_0 is easily estimated for power laws:

$$b_0 \simeq \left(\frac{n}{2}\frac{C_n}{E}\right)^{1/n} \bigg/ \left(\frac{n-2}{n}\right)^{(n-2)/2n} \tag{2.64}$$

e.g., for $n = 4$,

$$b_0 \simeq \left(4\frac{C_4}{E}\right)^{1/4}$$

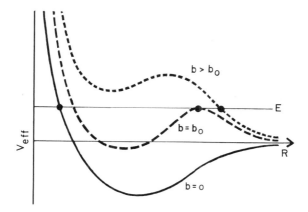

Figure 2.17. Effective potential, $V_{\text{eff}} = V(R) + L^2/2mR^2 = V(R) + Eb^2/R^2$, for three impact parameters (or angular momenta), with CM energy E: $b = 0$, $V_{\text{eff}} = V(R)$; $b = b_0$, orbiting occurs at distance of closest approach; $b > b_0$. Distances of closest approach indicated by dots. The quantity R is the internuclear separation.

This expression is quite useful for estimating ion–molecule reaction cross sections for which the long-range interaction goes as R^{-4}. As can be seen from Figure 2.17, if $b < b_0$, the particles will not only orbit at low energies, but will have a small value of R_0. Since the two molecules spend a relatively long time at close proximity, a reaction is likely for $b < b_0$, whereas for $b > b_0$ they never approach closely and the collision time is relatively short. The reaction cross section from Eq. (2.26) is therefore $\sigma_R = \pi \bar{P}_R b_0^2$, where $\bar{P}_R$ is the reaction probability, often called the steric factor. Employing Eq. (2.64), we obtain the energy dependence of the cross section, $\sigma_R \alpha E^{-2/n}$, the oft-used Langevin result for determining molecular reaction rates which we will consider further in Chapters 5 and 6. From Eq. (2.16) and Problem 2.3 the temperature dependence of the molecular reaction rate is found to be $k_R \alpha T^{(-2/n + 1/2)}$, which is temperature independent for $n = 4$.

The impulse approximation discussed above is easily extended to molecular targets, i.e., multiple impulses. For the collision A + BC, an atom colliding with a diatomic molecule, atom A receives impulses from each of the target atoms during its passage. If the effect of molecular binding between B and C can be ignored during the collision, the magnitude of the net laboratory deflection is obtained from a binary encounter between A and each target atom, as in Eq. (2.56):

$$\Theta_A^2(b, \Omega_M) \sim \left[\frac{(\Delta M_A\, v_A)_{AB} + (\Delta M_A\, v_A)_{AC}}{M_A\, v_A} \right]^2 \qquad (2.65)$$

The deflection therefore depends not only on b and v but also on the molecular orientation, Ω_M. It is tempting now to average Θ_A over the random orientations of the molecule. However, the quantity that *should* be averaged is the cross section, or the scattering probability into each angular region, which was described in Eq. (2.37). For many collisions it is useful to look at two limiting cases. For b large compared to the separation of the molecular atoms the impulses are very nearly in the same direction and the molecular target becomes like an atomic target of mass $M_A + M_B$. Now the earlier results apply directly. On the other hand, at very high energies, measurable deflections only occur if A makes a close encounter with either B or C. Therefore Θ_A is essentially due to a single impulse from one of the target atoms, again allowing the use of the previous results. In this limit the molecular gas is equivalent to a gas mixture of unbounded atoms A and B. That is, their separations can be assumed to be large and random. The two limiting cases, referred to respectively as the united atom and binary-encounter approximations, have been extensively used to avoid the difficulties inherent in treating a true molecular target, and they apply over a broad range of energies and impact parameters.

The classical deflection function for inelastic collisions requires a knowledge of the energy loss mechanism, which of course is quantum mechanical. If it is assumed that the inelastic energy transfer or transition occurs at a single point or a series of points along the classical trajectory, then the equations of motion can be integrated between these points with the CM energy changing in steps at each transition point. For instance, an assumption consistent with the impulse approximation is to describe the transition as a single change in energy occurring at the distance of closest approach, where $\dot{R} = 0$, i.e., the relative motion is the slowest. The net CM deflection is now a sum of two deflections, one on the approach and another as the particles recede. However, for fast heavy-particle collisions, the only transitions occurring with significant probability are those for which $E \gg Q$. In this case, as the change in energy due to a transition is negligible, the transition probability $P_{AB \rightarrow j}(b)$ and the cross section $\sigma_{AB}(\theta)$ can be determined separately, the latter generally via classical mechanics, as described above, and the probabilities via quantum mechanics (Chapter 4).

Before leaving this section, we note that there exists an alternative starting point for deriving the equations of motion and the classical deflection function. Defining a quantity called the classical action, S_c, as

$$S_c = \int_{\text{path}} \left[\frac{1}{2m} p^2 - V(R) \right] dt$$

where the integral is carried out over the path of the particle, we obtain the equations of motion by assuming that the classical trajectory is that path for which S_c is a minimum. This approach is referred to as the principle of least action and is discussed in standard texts on classical mechanics. The action, therefore, is a property of a moving classical particle and, in fact, is unusual in that it is an accumulative property for which the history of the path plays a role. This property is important in wave mechanics, and we only note here that S_c can also be written in the form

$$S_c = \int_{\text{path}} [\mathbf{p} \cdot \mathbf{dR} - E \, dt] \tag{2.66}$$

a form we will encounter in the following chapter.

Classical Inelastic Cross Sections

Before considering the wave-mechanical description of collisions, we can employ the classical kinematics developed above to approximate the inelastic collision cross sections. In an inelastic collision, the component particles of the target molecule acquire or lose energy and momentum in addition to that required for them to follow the overall motion of the target. The appropriate classical calculation involves solving the equations of motion during the collision for the interaction between the incident particle and all the particles, (electrons and nuclei) that make up the target atom or molecule. This formidable, many-body problem has been considered by a number of people, but more useful information often comes from analyzing two limiting cases. These are the same limits considered above when analyzing the deflections by molecules. One such limit applies to very large impact parameters for which the target electrons and nuclei appear and behave as an aggregate during the collision. Here one immediately runs into difficulties trying to describe the aggregate effects classically. An approach developed in early quantum mechanics was to consider the individual target particles as oscillating classically, about the CM of the target. The collision between an incident ion and such a target is described in Appendix C.

In the other limit, appropriate when the impact parameter is small, one considers the interaction between each component of the target and the incident particle separately. During the collision, the interaction times are assumed short enough that the effect of the particles of the target on one another is small, that is, all aggregate effects in the target are ignored. This case is referred to as the classical binary encounter approximation (BEA), as the many-body problem is reduced to two-particle collisions. This is merely an extension of our definition of a binary cross section in the beginning of the chapter where we ignored the effect of neighboring particles when describing collisions in gaseous or solid targets. In a gas one has the option of varying the mean distance between the target particles. However, for the present problem, as for solid targets, these distances cannot be controlled, implying the BEA will be valid only when the distance of closest approach between the target component in question and the incident particle is small compared to the mean separation of target particles during the collision. In this method the net effect of the collision is a sum of the effects for each component of the target, as in Eq. (2.65); here, however, we are concerned with energy transfer *to* the target.

A binary cross section can be constructed as in Eq. (2.34) with the exception that the target particles, like the electrons in an atom, are not stationary but will have a distribution of velocities. Rather than develop the general case, we assume the incident particle is moving at a speed large compared to the speed of the target particle of interest. That is, the target particles are assumed to be at rest and our previous results apply directly. The energy-transfer cross section between the incident particle A and particle i in target B is, from Eq. (2.46),

$$\frac{d\sigma}{dQ} = \frac{4\pi}{\gamma_i E_A} \, \sigma(\chi_i) \tag{2.67}$$

where $Q = \gamma_i E_A \sin^2 (\chi_i/2)$, with χ_i the CM angle for this collision, and $\gamma_i = 4 M_A M_i/(M_A + M_i)^2$. We use the symbol Q to remind the reader that the energy transferred to the electrons results in inelastic energy loss.

Considering any electron which changes energy by an amount greater than the ionization limit, I_B, as ionized, we write the inelastic cross section for ionization approximately as

$$\sigma_{AB \, , AB} \approx N_B \int_{I_B}^{Q_{max}} dQ \, \frac{d\sigma}{dQ} \tag{2.68}$$

Here $Q_{max} = 2m_e v_A^2$, where m_e, the electron mass, is small compared to M_A, and N_B is the number of electrons on B. If the incident particle is a bare ion, A^+, then the force between the ion and electron is obtained from the coulomb potential, $V(R) = -Z_A e^2/R$, where the charge on A is $q_A = Z_A e$, with e the magnitude of the electronic charge. The CM cross section for this potential is $\sigma(\chi) = (Z_A e^2/2m_e v^2)^2(1/\sin^4 \chi/2)$ [cf. Problem 2.10 or Eq. (2.61b)], which is equivalent to Eq. (2.61a) if χ is small. Assume, further, that atom B has electrons in a number of shells, where N_{Bi} and I_{Bi} are the number of electrons and the ionization energy for shell i. Using the above in Eqs. (2.67) and (2.68), we obtain the simple Thomson cross section for ionization of B by ion A^+

$$\sigma_{A^+ B \to A^+ B^+} \sim 2\pi \frac{(Z_A e^2)^2}{m_e v_A^2} \sum_{i, \text{shells}} N_{Bi} \left(\frac{1}{I_{Bi}} - \frac{1}{2m_e v_A^2}\right) \tag{2.69}$$

This result has the basic behavior of an inelastic cross section involving electronic processes. It decreases at high energies and goes through a maximum when the velocity of the incident particle is comparable to some mean speed of the electrons in the atom. That is,

$$v_A^{-2} = \frac{2}{N_B} \sum N_{Bi} \left(\frac{2I_{Bi}}{m_e}\right)^{-1}$$

at the maximum, where the quantity in brackets is roughly the mean-square speed of the electrons in the ith shell. Further, the cross section becomes small well above the threshold energy (the threshold energy for the ith shell is $E = I_{Bi}$), which is a result of the large difference in mass between A and the electrons.

Of course at low and intermediate energies the above result is suspect as we cannot assume the electrons are stationary, unless, of course, the energy transfer required for an ionization is "large". Large energy transfers require a small distance of closest approach, favoring a binary encounter approach. The above method applies quite well for a screened interaction. For instance, when A is a neutral atom, the long-range force is weak and significant energy transfer occurs only in close collisions. It is also applicable to the treatment of molecular dissociation as the heavy particles generally behave in a classical manner. For the reaction

A + BC → A + B + C the BEA cross section would be written

$$\sigma_{A, BC \to A, B, C} \sim \int_{[(M_B + M_C)/M_C]D_{BC}}^{\gamma_{AB}E_A} \left(\frac{d\sigma}{dT}\right)_{AB} dT$$

$$+ \int_{[(M_B + M_C)/M_B]D_{BC}}^{\gamma_{AC}E_A} \left(\frac{d\sigma}{dT}\right)_{AC} dT \qquad (2.70)$$

where, for instance, the power law expressions for $d\sigma/dT$ of Eq. (2.61b) can be used to evaluate the integrals. In Eq. (2.70) D_{BC} is the dissociation energy of BC, and the mass factors indicate that the energy in the CM system of the molecule must be greater than D_{BC} for dissociation to occur. For comparison with our earlier result, note that if C is an electron, B a nucleus, and D_{BC} the ionization potential, it is, of course, very difficult to ionize an atom via energy transfer to the nucleus.

Another useful quantity defined in Eq. (2.18), the stopping power, is also easily evaluated in the BEA approximation. In this approximation the inelastic energy loss to the electron, Q, as well as the energy transferred elastically to the atom, T, are continuous and, therefore, the sum in Eq. (2.18) is replaced by an integral. Treating these energy transfers separately, one obtains a stopping power for target material B of the form

$$-\frac{dE}{dx} \simeq n_B \left(N_B \int_{Q_{min}}^{Q_{max}} Q \frac{d\sigma}{dQ} dQ + \int_{T_{min}}^{T_{max}} T \frac{d\sigma}{dT} dT\right) \equiv n_B S_{AB} \qquad (2.71)$$

where Q_{min} is the minimum excitation energy of the target electrons and T_{min} is generally zero. The two contributions in Eq. (2.71) are referred to as the electronic and nuclear stopping powers, and the terms in brackets the stopping cross sections. The stopping cross sections for the target constituents, calculated from the coulomb potential as in Eq. (2.69), have the form

$$S_{coul} = 4\pi \frac{(Z_A e^2)^2}{\Delta E_{max}} \ln \frac{\Delta E_{max}}{\Delta E_{min}} \qquad (2.72)$$

where ΔE is either Q or T. Because of the large differences in masses, the electronic stopping power dominates at high energies. This may seem strange at first as $T_{max} \gg Q_{max}$. However, the maximum energy transfer occurs only in close collisions. The result in Eq. (2.72) implies that the larger impact parameter collisions, in which the electrons receive energy more efficiently, dominate.

Before leaving this example, we note that a certain ambiguity exists in the classical calculation which can be exploited. In the ionization calculation, Q_{max} and Q_{min} correspond to energy transfer for the smallest and largest distances of closest approach leading to an ionization. If the smallest distance of closest approach $(R_o)_{min} = 2(Z_A e^2/m_e V_A^2)$ for the coulomb collision becomes less than the uncertainty in the separation between A and B, as determined by the Heisenberg uncertainty principle, then the collision in this region is nonclassical. The BEA approximation, therefore, should be cut off at a Q smaller than Q_{max}. Also, if the largest distance of closest approach exceeds the screening length of the neighboring particles in the target, then the BEA does not apply and Q should be cut off at a value larger than $Q_{min} = I_B$. Two different screening lengths can be of importance in determining this cutoff. The most obvious, a static screening, involves the mean distance to the neighboring particles. The other, a dynamic screening, is obtained by requiring that the collision time be short compared to the characteristic periods of the particle in the target for the BEA to apply. That is, if this condition is not met, the target particles should be treated as an aggregate. If ω_c is a characteristic angular frequency, defining the collision time roughly as R_0/v_A, this screening length for slow or distant collisions is v_A/ω_c, often called the adiabatic screening length. When considering an ionization of a small atom, ω_c is the binding frequency of the electrons, as the nucleus screens the interaction between the target electron and A for slow collisions. In the dissociation of a molecule, ω_c is the molecular binding frequency, as the screening is due to the other atoms bound in the target. For ionization of heavy atoms or molecules involving many electrons the most important screening is due to the neighboring electrons. This has been shown by Lindhard and others to be characterized by the so-called plasma frequency, which depends on the electron density. The origin of these effects will become clear from the discussion in the following sections in which we consider the wave-mechanical treatment of collisions. However, it should be evident that carefully choosing the upper and lower limits on the integral in Eq. (2.68) allows one to extend the usefulness of the classical BEA method.

Exercises

2.1. What is the order of magnitude of the geometric cross section for an atom incident on each of the following targets: atom, benzene ring, RNase molecule, DNA molecule, cell.

2.2. The rate equation, Eq (2.14), for a species n_C depends on the changing densities n_A and n_{BC} in the reaction $A + BC \rightarrow AB + C$. Solve the equation in general and obtain the limiting case in Eq. (2.15).

2.3. For a reaction cross section of the form $\sigma_R = \bar{P}_R \pi b_R^2$, where the cutoff impact parameter, b_R, is $b_R = C/E^x$, show that the temperature dependence for the reaction cross section of Eq. (2.16) is $k_R \propto T^{(-2x+1/2)}$. Assume both the reacting particles have Maxwell–Boltzmann velocity distributions:

$$f(\mathbf{v}_i) = (m_i/2\pi kT)^{3/2} \exp(-m_i v_i^2/2kT)$$

2.4. Assume the reaction $A + BC \rightarrow AB + C$ proceeds with a 50% probability if the centers of the particles approach each other with a radius less than 0.5×10^{-8} cm. If the masses are $M_A = 10$, $M_B = 5$, and $M_C = 20$, in atomic mass units, calculate and plot the temperature dependence of the reaction rate for $Q = 0.01$ eV, $Q = 0.1$ eV, and $Q = 1$ eV using Eqs. (2.16) and (2.26).

2.5. For a deterministic collision with p_{AB} a delta function, as in Eq. (2.36), derive the expression for the angular differential cross section in Eq. (2.34).

2.6. Obtain an expression for the laboratory differential cross section for the collision of hard spheres using the CM cross section $\sigma(\chi) = \frac{1}{4}d^2$ for the three limiting cases $\mu \gg 1$, $\mu = 1$, $\mu \ll 1$ shown in Figure 2.7 i.e., Eq. (2.45).

2.7. Derive the relationships between the laboratory qunatities $\mathscr{E}$, $\mathscr{P}$, $\mathscr{L}$ and CM quantities E, P, L in Table 2.3.

2.8. Derive the relationship between the laboratory angle Θ_A and the CM angle χ for the general inelastic collision in Table 2.4. Express Θ_B in terms of χ also.

2.9. For an ionization of a stationary argon atom, which requires about 15.8 eV, estimate the laboratory threshold energy for an incident proton, argon ion, and a uranium ion. Estimate the "threshold" energy for the Thomson cross section, Eq. (2.69).

2.10. Derive the exact classical deflection function and the CM cross section for the coulomb interaction potential $V(E) = q_A q_B/R$, i.e., $\sigma(\chi) = (q_A q_B/4E)^2/\sin^4 (\chi/2)$ using Eq. (2.55).

2.11 Using the Gauss–Mehler quadrature for $\chi(b)$ given in Appendix B, for 5, 10, and 20 integration points, calculate $\chi(b)$ and compare it to the exact and impulse results for power-law potentials $n = 1, 2, 4$.

2.12. Use the data in Figure 2.14 along with Eqs. (2.58) and (2.61a) to extract approximate potentials for the $He^+ + Ar$ and $Ne^+ + Ar$ systems: plot result.

2.13. For a Lennard-Jones form of the potential, Eq. (2.62), relate the measured rainbow angle to the position and depth of the potential minimum.

Suggested Reading

Much of the content of this chapter is well presented in many texts on atomic and molecular collisions or classical mechanics, a few of which are listed below. Additional specific references are also given on some topics.

General

H. W. MASSEY, *Atomic and Molecular Collisions*, Halsted Press, New York (1979), Chapter 1.

M. R. C. MCDOWELL and J. P. COLEMAN, *Introduction to the Theory of Ion–Atom Collisions*, North-Holland, Amsterdam (1970), Chapter 1.

E. W. MCDANIEL, *Collision Phenomena in Ionized Gases*, Wiley, New York (1964), Chapters 1 and 4.

J. B. HASTED, *Physics of Atomic Collisions*, 2nd edn., American Elsevier, New York (1972), Chapters 2–4.

J. T. YARDLEY, *Introduction to Molecular Energy Transfer*, Academic Press, New York (1980).

H. GOLDSTEIN, *Classical Mechanics*, Addison-Wesley, Cambridge, Massachusetts (1959), Chapter 3.

M. S. CHILD, *Molecular Collision Theory*, Academic Press, New York (1974), Chapters 1 and 2.

Effects of Neighboring Atoms in Molecules or Solids

P. SIGMUND, *Phys. Rev. A*, **14**, 996 (1976).

P. SIGMUND, *K. Dan. Vidensk. Selsk. Mat. Fys. Medd.*, **39** (11) 1 (1977).

F. BESENBACHER, J. HEINEMEIER, P. HVELPLUND, and H. KNUDSEN, *Phys. Rev. A*, **18**, 2470 (1978).

Classical Deflection Function Expressions

J. LINDHARD, V. NIELSEN, and M. SCHARFF, *K. Dan. Vidensk. Selsk. Mat. Fys. Medd.*, **36**(10) (1968).

F. T. SMITH, R. P. MARCHI, and K. G. DEDRICK, *Phys. Rev.* **150**, 79 (1966).

Classical Stopping Power Calculation

J. D. JACKSON, *Classical Electrodynamics*, 2nd edn. Wiley, New York (1975), Chapter 13.

N. BOHR, *K. Dan. Vidensk. Mat. Fys. Medd.*, **18**(18) (1948).

3

Waves and Trajectories

Introduction

The angular differential-cross section calculations in the previous section were based on a classical model of the collision. This was a classical model in that a well-defined trajectory was used to describe the locations of the colliding particles as a function of time. Such a model could be modified and made statistical by associating a number of trajectories with every impact parameter and by weighting each trajectory, as in Eq. (2.32); nevertheless the model remains classical. In the first half of this century it became clear that beams of atoms and molecules do exhibit statistical behavior. However, the statistical properties exhibited were those associated with wave-like phenomenon, e.g., interference and diffraction. It is well established now that the wave-mechanical description of atomic interactions is the more fundamental description. The particle or trajectory (i.e., classical) model of atomic interactions bears roughly the same relationship to the wave-mechanical description of atomic collisions that geometric optics (ray description of light) bears to the wave description of the interaction of light with matter. The criteria for the validity of trajectory and ray approximations to their respective wave-mechanical phenomena are analogous. If the wavelength of light is small compared to the distances over which changes in the material occur, then the passage of light through the material can be described quite accurately using rays. On the other hand, when the changes in the material are abrupt, connection formulas, like Snell's law in geometric optics, can be developed between rays in adjacent regions. The ray description is also useful if the source of light is very incoherent and/or the detectors (e.g., the eye) are insensitive to small spatial or time differences, that is, significant averaging occurs.

The marriage of the ray and wave methods is most easily envisaged starting with Huygens' principle, a special case of the principle of superposition in wave mechanics. This principle states that any moving wave front can be thought of as a sum of an infinite number of point sources of waves (wavelets) which are emitted in all directions. The wave can be reconstructed from these wavelets at a later time and at another position (e.g., after it interacts with the material) by recombining the wavelets. Of course each wavelet in the sum will have changed both in phase and amplitude at the new point. The sum of wavelets so constructed consists of contributions from wavelets traveling from every point on the initial wave front via every possible path through the material to the point of interest, which is a rather cumbersome sum. This sum can be simplified considerably as only groups of wavelets traveling along certain paths contribute significantly at a given point and the contributions from the remainder of the wavelets essentially cancel. Those paths from which significant contributions are obtained are the paths, by which a ray would travel between the source and the point of interest as calculated from geometric optics. Associated with each group of contributing waves, that is with each ray, is a phase and amplitude determined by the path of the ray through the material. Hence calculating the paths of the rays can be useful for describing wave phenomena.

A scheme similar to the above can be constructed to describe the interaction of a beam of atoms with a material made up of atomic targets. The rays for which significant contributions occur are the classical trajectories calculated in the previous chapter. Now, however, the interaction potential replaces the index of refraction in determining the effect of the target medium on the incident waves. In the following we develop such a method for describing atomic collisions without being completely rigorous at all points. The reader is referred to a number of excellent discussions of the method in the literature. The procedure is referred to by a variety of names (the JWKB approximation, the eikonal approximation, the semiclassical method) which imply slightly different starting points and applications. However, they all have the same physical basis in that they are short-wavelength approximations. We will also consider a first-order wave-mechanical approximation, which is patently not classical, to complete the description of atomic interactions. The discussion begins with a consideration of the wave equation and some solutions to this equation. This is followed by a discussion of the wave-mechanical model of the angular scattering experiments of Chapter 2. Readers who are familiar with quantum mechanics may find the material in the following two sections somewhat elementary; if so, they are encouraged to bypass them.

The Wave Equation

The description of oscillatory motion (wave motion) is very nearly the same* whether the phenomena of interest are acoustic oscillations, light waves, mechanical waves in material, etc. The wave motion is obtained as a solution to a set of differential equations. These equations are expressions of the "laws" for wave motion in much the same way that a set of differential equations can express Newton's laws. The solutions to the equations of motion for objects obeying Newton's laws ("classical" particles) are the positions and momentums of the particles at any time. For a group of objects or a medium whose motion is determined by wave equations, the solutions are oscillatory and are expressed in terms of the amplitudes and phases of the oscillations at any time. These solutions often are written as imaginary exponentials with amplitudes and phases, and the intensity is the absolute value squared of the amplitude of the exponential.

A plane wave with intensity I_0 moving in a direction indicated by the wave number **K** has the functional form

$$\psi(\mathbf{R}, t) = I_0^{1/2} \exp\left[i(\mathbf{K} \cdot \mathbf{R} - \omega t)\right] \tag{3.1}$$

For a monochromatic (single frequency) beam of light, this wave function expresses the behavior of the electric and magnetic fields which comprise the light beam, where $K = 2\pi/\lambda$ and $\omega = 2\pi\nu$ are related by $c = \omega/K$; c is the speed of light, λ the wavelength, and ν the frequency. When such waves are used to describe atomic particles, the wave function is a probability amplitude. That is, the absolute value squared of the amplitude is proportional to the *probability* of finding a particle at a particular position and time.† In a uniform beam of atoms, by definition, there is an equal chance of finding an atom at any position in the beam, and the number moving across a surface per unit time determines the intensity, I_0. A plane wave, therefore, also can be taken to describe a monoenergetic beam of particles where the quantities K and ω have the following physical interpretation:

$$\omega = E/\hbar, \qquad K = p/\hbar \tag{3.2}$$

with $\hbar = h/2\pi$, h being Planck's constant, the unit of action in quantum mechanics. This plane wave is a solution to the differential equation

$$-\frac{\hbar^2}{2m}\nabla^2\psi = i\hbar\frac{\partial}{\partial t}\psi, \qquad \nabla^2 = \frac{\partial^2}{\partial x^2} + \frac{\partial^2}{\partial y^2} + \frac{\partial^2}{\partial z^2} \tag{3.3}$$

which is the Schrödinger equation for particles moving in empty space. On substitution of the plane wave in Eq. (3.1) into Eq. (3.3), one obtains the standard relationship between p and E for a free particle, i.e., $E = p^2/2m$.

---

* We confine our treatment here to scalar, or single component waves. Light waves in fact have two polarizations but can be treated as single component waves in many instances.

† In fact, for light, the functions can also be thought of as probability functions for photons.

Upon interacting with a material, the momentum, hence the wave number, changes in response to the potential of interaction, which plays the role of the refracting medium in particle collisions, modifying the form of the wave to produce scattering. The potential energy of interaction is included linearly in the full wave equation,

$$\left[-\frac{\hbar^2}{2m} \nabla^2 + V(\mathbf{R}) \right] \psi(\mathbf{R}, t) = i\hbar \frac{\partial \psi}{\partial t} (\mathbf{R}, t) \tag{3.4}$$

and the solution $\psi(\mathbf{R}, t)$ is no longer a simple plane wave. For a time-independent potential the total energy of the particles in the beam is conserved, in which case the time-dependent phase factor of the wave function can be separated,

$$\psi(\mathbf{R}, t) = \psi(\mathbf{R}) \exp[-i\, Et/\hbar] \tag{3.5}$$

Now, after substitution into Eq. (3.4), we see that $\psi(\mathbf{R})$ is a solution of

$$\left[-\frac{\hbar^2}{2m} \nabla^2 + V(\mathbf{R}) \right] \psi(\mathbf{R}) = E\psi(\mathbf{R}) \tag{3.6}$$

where the term in brackets is referred to as the energy or Hamiltonian operator. It is often instructive to rearrange the time-independent equation in the form

$$\left[\nabla^2 + \frac{p^2(\mathbf{R})}{\hbar^2} \right] \psi(\mathbf{R}) = 0 \tag{3.7}$$

where $p^2(\mathbf{R}) = 2m[E - V(\mathbf{R})]$ is the classical momentum. Before leaving this section we note that when the potential is constant, or even slowly varying, over any region, then an approximate solution to Eq. (3.7) is

$$\psi(\mathbf{R}) \simeq \exp\left[\frac{i}{\hbar} \int_{\text{path}} \mathbf{p}(\mathbf{R}) \cdot d\mathbf{R} \right] \tag{3.8a}$$

where the integration is over a path through the region. Now the wave function, including the time dependence, has the form (or can be written as a sum of terms with the form)

$$\psi(\mathbf{R}, t) \simeq \exp[iS_c/\hbar] \tag{3.8b}$$

where the phase is $S_c/\hbar$ and

$$S_c = \int_{\text{path}} [\mathbf{p}(\mathbf{R}) \cdot d\mathbf{R} - E\, dt] \tag{3.9}$$

is just the classical action of Eq. (2.66). The phase, therefore, is the ratio of the classical action to $\hbar$. The classical form of the phase above will be exploited in obtaining approximate solutions to the wave equation for describing particle scattering.

Solutions to the Wave Equation

Before considering the scattering problem, we look briefly at some solutions to the wave equations that are used in the following sections. Solutions to Newton's equations of motion are well understood even by the nonscientist: the particles follow trajectories in response to the forces of interaction. Forms for the solutions to the wave equation also are familiar to the reader although the mathematical expressions may not be. We discuss in the following the time-independent wave equation and leave much of the verification of the results as problems at the end of the chapter.

Simple Wave Equation

The plane-wave solution $\psi(\mathbf{R}) = \exp(i\mathbf{K} \cdot \mathbf{R})$, which we have been discussing, is most easily obtained from the wave equation by treating the Laplacian, ∇^2, of Eq. (3.3) in rectangular coordinates. The Laplacian can also be written in terms of the radial and angular coordinates, and spherical polar coordinates, i.e.,

$$\nabla^2 = \frac{1}{R} \frac{\partial^2}{\partial R^2} R + \frac{1}{R^2} \nabla^2_{\theta, \phi}$$

where

$$\nabla^2_{\theta, \phi} = \frac{1}{\sin \theta} \frac{\partial}{\partial \theta} \sin \theta \frac{\partial}{\partial \theta} + \frac{1}{\sin^2 \theta} \frac{\partial^2}{\partial \phi^2}$$

Now the solutions may be written in terms of radial and angular functions that are not as familiar: $j_l(KR) \cdot Y_{lm}(\theta, \phi)$ and $n_l(KR) \cdot Y_{\ell m}(\theta, \phi)$, where the j_l and n_l are the spherical Bessel functions, and $Y_{lm}(\theta, \phi)$ the spherical harmonics. These are separately solutions of

$$\nabla^2_{\theta, \phi} Y_{lm} = l(l + 1) Y_{lm} \tag{3.10}$$

and

$$\left[\frac{1}{R} \frac{d^2}{dR^2} R + \left(K^2 - \frac{l(l + 1)}{R^2} \right) \right] \cdot y_l(KR) = 0 \tag{3.11}$$

where $y_l = j_l$ or n_l, and the reader should verify that the product satisfies the full time-independent wave equation when there is no potential. For particles the quantity in brackets,

$$\left(K^2 - \frac{l(l+1)}{R^2}\right) = \frac{p_0^2(R)}{\hbar^2}$$

is simply related to the radial momentum of Eq. (2.52), $p(R)$, in the absence of a potential if $l(l+1)\hbar^2$ is identified with the angular momentum,

$$l(l+1)\hbar^2 = L^2 = (p_0 b)^2 \tag{3.12}$$

Therefore, in wave mechanics the parameter l, from the angular part of the solution, replaces the impact parameter of classical mechanics. At very large R the radial functions j_l and n_l have the form

$$j_l(KR) \xrightarrow{R \to \infty} \frac{1}{KR} \sin(KR - l\pi/2)$$

$$\tag{3.13}$$

$$n_l(KR) \xrightarrow{R \to \infty} -\frac{1}{KR} \cos(KR - l\pi/2)$$

These describe spherical waves the intensities of which decrease as $1/R^2$. For instance, writing

$$(j_l \pm in_l) \xrightarrow{R \to \infty} \frac{1}{KR} \exp[\pm i(KR - (l+1)\pi/2)] \tag{3.14}$$

we can construct outgoing and incoming radial waves when the time-dependent phase factor is included (Prob. 3.1). The angular functions, Y_{lm}, change the shape of these outgoing or incoming waves.

As we are often interested in spherically symmetric scattering problems, the wave functions of interest may be independent of ϕ. For this situation, the Legendre polynomials, $P_l(\cos\theta)$, are used to describe the angular dependence of the wave function. The Legendre polynomials are proportional to Y_{l0} and these also satisfy Eq. (3.10). The first few Legendre polynomials have the form

$$P_0 = 1 \qquad P_1 = \cos\theta \qquad P_2 = (3\cos^2\theta - 1)/2 \tag{3.15}$$

Since the plane wave and spherical wave functions both solve the same wave equation, it is evident that one set can be written in terms of the other. For instance, for a plane wave incident on a spherical potential it is useful to express the plane wave in terms of the spherical functions as

$$\exp[i\mathbf{K}\cdot\mathbf{R}] = \sum_{l=0}^{\infty} i^l(2l+1)P_l(\cos\theta)j_l(KR) \tag{3.16}$$

when solving a scattering problem.

Point Source

If a point source (sink) located at $\mathbf{R}'$ is emitting (absorbing) waves, the solutions to the wave equation are outgoing (incoming) spherical waves centered on $\mathbf{R}'$ (Figure 3.1). The amplitude of such waves must decrease as the inverse distance from the source, $|\mathbf{R} - \mathbf{R}'|$, and must be proportional to the square root of the source strength, $\mathscr{S}$. Therefore, a point-source wave function can be written as

$$\psi(\mathbf{R}, t) = G(\mathbf{R}, \mathbf{R}') \exp[-iEt/\hbar]$$

with

$$G_{\pm}(\mathbf{R}, \mathbf{R}') = \frac{\mathscr{S}}{(4\pi)^{1/2}} \frac{\exp[\pm iK|\mathbf{R} - \mathbf{R}'|]}{|\mathbf{R} - \mathbf{R}'|} \tag{3.17}$$

where the free particle relationship holds, $E = \hbar^2 K^2/2m$. The form of Eq. (3.17) is constructed so that the number of particles per unit time crossing an arbitrary surface enclosing the source is a constant, i.e.,

$$\int_{\text{surf}} ds\,|\psi(\mathbf{R}, t)|^2 = \mathscr{S}^2.$$

Conversely, the flux of particles at a distance $|\mathbf{R} - \mathbf{R}'|$ from the source is $\mathscr{S}^2/4\pi|\mathbf{R} - \mathbf{R}'|^2$, the source strength divided by the area of the sphere over which the particles are spread. The reader should verify that the plus and minus signs in Eq. (3.17) correspond to outgoing and incoming waves respectively. Representing the point source by the ubiquitous delta function, we write the wave equation for a point source as:

$$(\nabla^2 + K^2)G(\mathbf{R}, \mathbf{R}') = \mathscr{S}\delta(\mathbf{R} - \mathbf{R}') \tag{3.18}$$

Because this equation differs from the simple wave equation at only one point, $\mathbf{R}'$, the solutions to the two equations have the same form at large R, as can be seen by comparing Eqs. (3.14) and (3.17).

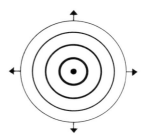

Figure 3.1. Outgoing spherical waves from a point source decrease in intensity as they recede from the source.

Extended Sources

Extended sources can be thought of as a sum of distributed point sources. The function $\mathscr{S}(\mathbf{R})$ in the wave equation

$$[\nabla^2 + K^2]\psi(\mathbf{R}) = \mathscr{S}(\mathbf{R}) \tag{3.19}$$

is a source (sink) of outgoing (incoming) waves with a form determined by the shape of the source. Using Eq. (3.18), we can regard $\psi(\mathbf{R})$ in Eq. (3.19) as a superposition of spherical waves from each point of the source, yielding

$$\psi(\mathbf{R}) = \psi_0(\mathbf{R}) + \frac{1}{4\pi} \int \frac{\exp[\pm iK|\mathbf{R} - \mathbf{R}'|]}{|\mathbf{R} - \mathbf{R}'|} \mathscr{S}(\mathbf{R}') \, d^3R' \tag{3.20}$$

where $\psi_0(\mathbf{R})$ is an arbitrary solution of the homogeneous equation, the simple wave equation we discussed earlier. If $\mathscr{S}(\mathbf{R})$ is a point source, $\mathscr{S}(\mathbf{R}) = \mathscr{S}\delta(\mathbf{R} - \mathbf{R}')$, the second term in Eq. (3.20) reduces to the function in Eq. (3.17). It is often useful to also decompose the source of waves into characteristic shapes of antennae or multipole moments. Using the Legendre polynomials or spherical harmonics, for example writing $\mathscr{S}(\mathbf{R}) = \sum C(R)P_l(\cos\theta)$, we see that the first term corresponds to a spherical source, the second to a dipole source, and the third to a quadrupole source, etc.

In the wave equation given in Eq. (3.6) the interaction potential is the source of scattered waves. The strength of this source depends not only on $V(R)$ but also on the number of particles incident on the target, that is, it depends on the wave function itself. Comparing Eq. (3.19) to Eq. (3.6), we obtain the source function for scattering:

$$\mathscr{S}(\mathbf{R}) = \frac{2m}{\hbar^2} V(\mathbf{R})\psi(\mathbf{R}) \tag{3.21}$$

On substitution of $\mathscr{S}(\mathbf{R})$ into Eq. (3.20), the solution to $\psi(\mathbf{R})$ is seen to depend on $\psi(\mathbf{R})$, and it appears we have not gotten very far! However, such an expression is quite useful when the incident wave is not strongly perturbed by the scattering.

One-Dimensional Equations with Potentials

If the motion is along a path, e.g., a plane wave, the Laplacian can be reduced to a one-dimensional second derivative $\nabla^2 = d^2/ds^2$, where s is measured along the path. The wave equation is

$$\left(\frac{d^2}{ds^2} + \frac{p^2(s)}{\hbar^2} \right)\psi(s) = 0 \tag{3.22}$$

with $p^2(s) = 2m[E - V(s)]$. If $V(s)$ is a constant, then $\psi(s)$ has the plane wave form, $\psi(s) = \exp\{\pm i\,[p(s)/\hbar]\}$. In Figure 3.2 a crude model for inter-

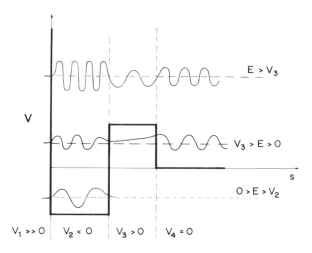

Figure 3.2. Potential (bold line) with wave solutions for three different particle energies, E: $0 > E > V_2$, a bound solution; $V_3 > E > 0$, free wave penetrates barrier; and $E > V_3$, wave slowed in region 3 and reflected at $s = 0$. Wavelength is proportional to momentum, amplitudes indicate probability of finding a particle in each region.

atomic potentials in one dimension is shown, in which $V(s)$ has a different constant value for four regions of s. In regions where $V > E$, p becomes imaginary and the solutions have the form $\psi(s) = \exp[\pm |p(s)/\hbar| s]$. These are increasing and decreasing exponentials, that is, they are no longer oscillatory in s. Solutions to Eq. (3.22) for all s are obtained, as is usual, for second-order differential equations, by matching the amplitudes and slopes of the functions on the boundaries of each region. For the largest energy shown, the wave oscillates everywhere with a wavelength determined by the classical momentum p. When the particle slows, as in region 3, the wavelength increases.

The wave amplitudes in Figure 3.2 depend on how the solutions match at the boundaries of each region and, therefore, the amplitudes are very dependent on the particle energy, E. These amplitudes give the probability of finding particles in each region, *or* they indicate, on the average, how long a single particle spends in a given region. Two hypothetical examples are shown for particles incident from the right. For lower values of E, $V_3 > E > 0$, the particles may pass through a region where $V > E$. This, of course, cannot occur classically. Classical particles of this energy incident from the right would never enter regions 1, 2, or 3, (viz. Figure 2.17), whereas the wave-mechanical particles not only may enter region 2 but may spend considerable time there *if* the amplitude is large. An important class of phenomena in the scattering of waves, referred to as resonance phenomena, occur when the incident particle spends a long time in region 2, in which case there is a considerable delay in the return scattering of the incident wave. When this occurs in three dimensions, the character of scattering is quite different from simple, repulsive, potential scattering as the particles lose all information about the incident beam direction. For energies $E < 0$, solutions to Eq. (3.22) represent bound particles. Because the waves must exponentially decay in *both* regions where $V > E$, unlike the above cases, there are solutions to Eq. (3.22) for only a few values of E, which is a familiar quantum-mechanical result, i.e., bound systems have discrete energy levels.

The WKB Method

An approximate solution to the one-dimensional potential is given by the Wentzel–Kramers–Brillouin (WKB) method. When the momentum $p(s)$ is slowly varying in s, the solutions must be similar to those of the preceding section [see also Eq. (3.8a)]. The functions

$$\psi(s) \simeq \frac{1}{[p(s)]^{1/2}} \exp\left[\pm \frac{i}{\hbar} \int p(s) \, ds \right] \tag{3.23a}$$

are approximate solutions to the *one*-dimensional equation which reduce to the previous results when the effective potential is a constant. As with the previous case, solutions are obtained by matching the amplitudes and derivatives of the solutions at the boundaries of the damped regions, that is, at the turning points. However at a turning point, $p(s) = 0$, the approximate solution above is not defined. The procedure developed for solving this problem is to expand the potential in the vicinity of the turning point, $V(s) \simeq V(s_0) + dV/ds|_{s_0}(s - s_0)$, where s_0 is the position of a boundary at which $p(s_0) = 0$. Solutions in the vicinity of the turning point are used to match the oscillatory to the damped solutions on either side of s_0. In the oscillatory region the WKB solutions have the form

$$\psi_{\text{WKB}}(s) = \frac{C}{[p(s)]^{1/2}} \sin\left[\int_{s_0}^{s} \frac{p(s') \, ds'}{\hbar} + \frac{\pi}{4} \right] \tag{3.23b}$$

For free particles, that is, scattering, solutions exist for every energy. For bound solutions, when there are two turning points, matching occurs at both boundaries and solutions can be shown to exist only for

$$\int_{s_0}^{s_0'} [p(s)/\hbar] \, ds' - \pi/2 = n\pi$$

This is just the quantization condition for determining bound levels that was one of the early expressions of quantum mechanics.

The solutions outlined above will be used in the following sections to describe the scattering of a beam of atoms by atomic targets. As a first step we determine the form of the scattering solution at large R, which establishes the boundary conditions.

Scattering of a Plane Wave

In Figure 3.3, the wave description of a scattering experiment is represented. Incident on the target medium is a plane wave of intensity I_0. As the medium is "thin" a large fraction of the beam intensity will be transmitted and a small fraction will be scattered. The transmitted beam is indicated by a plane wave with intensity I_t in Figure 3.3. At distances large compared to its size, the target in Figure 3.3 appears to be very nearly a point source of scattered waves located, for our purposes, at the origin of the coordinate system. The scattered-wave intensity will decrease as $1/R^2$, or the amplitude of the scattered wave will decrease like $1/R$, as in Eq. (3.14), where R is the distance from the target to the detector. The scattered wave has an oscillatory part which depends on the momentum of the outgoing particles. Lastly, this amplitude also depends on the direction of observation (θ_A, φ_A) and is proportional to the incident intensity I_0. Therefore, the wave function for particles A after the beam passes through the target material and at large distances from the target is a sum of scattered and transmitted components, with the intensities assumed to be normalized to the number of target atoms:

$$\psi_f(\mathbf{R}) \rightarrow I_t^{1/2} \exp\left[i\mathbf{p}_A \cdot \mathbf{R}/\hbar\right] + I_0^{1/2} f_{AB}(\theta_A, \varphi_A) \frac{\exp(ip'_A R/\hbar)}{R} \quad (3.24)$$

The factor $f_{AB}(\theta_A, \varphi_A)$, called the scattering amplitude, accounts for the angular dependence of the scattering, and p'_A is the final momentum of the scattered particles.

In Eq. (3.24) the exponential part of the scattered wave contains information only about the wave motion after scattering, indicating waves receding with a velocity proportional to p'_A. The scattering amplitude, f_{AB}, i.e., the angular dependence, contains information about the passage of the beam through the material. Based on the probability interpretation of the

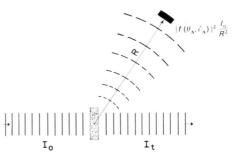

Figure 3.3. Incident plane wave of intensity I_0 on a thin target with transmitted wave of intensity I_t. Scattered wave intensity decreases as $1/R^2$, where R is the distance to the detector and depends on angle of scattering.

wave function, $I_0 |f_{AB}(\theta_A, \varphi_A)|^2 \Delta\Omega_A$ is the probability per unit time that a particle A will be scattered into a detector subtending a solid angle, $\Delta\Omega_A$, about (θ_A, φ_A), where, as before, $\Delta\Omega_A$ is area of the detector divided by R^2. Applying the definition in Eq. (2.27), we relate the angular differential cross section to this scattering amplitude by

$$\frac{d\sigma_{AB}}{d\Omega_A} = |f_{AB}(\theta_A, \varphi_A)|^2 \qquad (3.25)$$

When there are no outside fields or lattice effects, the above cross section would be averaged over φ_A. Further, if the target and incident particles are spherically symmetric, the calculated scattering amplitudes depend only on θ_A, which reduces the wave-mechanical problem to the same level that we treated classical scattering.

There is also, of course, an outgoing wave of scattered, target particles, not indicated in Figure 3.3, with wave amplitude decreasing as $1/R$ and a scattering amplitude $f_{AB}(\theta_B, \varphi_B)$. In the CM system, the motion of the incident and target particles are simply related; therefore, the subsequent discussion will refer to a reduced-mass particle scattered from an infinitely heavy target. At large distances from the target the wave equations reproduce the classical relationships between the energy and momentum, viz. Eqs. (3.2) and (3.3). This implies that the particle kinematics discussed earlier can be used to transform the cross sections between the laboratory and CM systems here also. Further, because the target is assumed thin, i.e., single-scattering limit, $I_0 \sim I_t$ and these factors are dropped from the expression in Eq. (3.24). Based on the above discussions, the asymptotic wave function in the CM system for the scattering problem is

$$\psi(\mathbf{R}) \rightarrow \exp(i\mathbf{p}_0 \cdot \mathbf{R}/\hbar) + f(\chi) \frac{\exp(ip_0 R/\hbar)}{R} \qquad (3.26)$$

which now applies to all directions including the incident-beam direction. The scattering amplitude $f(\chi)$ can be obtained by solving the wave equation, Eq. (3.6), subject to the boundary conditions imposed by Eq. (3.26). That is, the correct or physical solution must have the asymptotic form we have been discussing.

The expected form for the scattering amplitude, based on the discussion in the introduction, is

$$f(\chi) \propto \int d^2b \, a(\mathbf{b}, \chi) \exp[i\alpha(\mathbf{b}, \chi)] \qquad (3.27)$$

That is, the total scattering amplitude at any angle is made up of amplitudes, $a(\mathbf{b}, \chi)$, from all regions on the wave front, each amplitude contributing with a phase $\alpha(\mathbf{b}, \chi)$ determined by the "path," i.e., the point of origin and the scattering angle. Based on the probability interpretation of the wave function, the absolute value squared of the amplitude, $|a(\mathbf{b}, \chi)|^2$, is proportional to a scattering probability like that in Eq. (2.32), $p(\mathbf{b}, \chi)$. It is important to remind the reader that neither $p(\mathbf{b}, \chi)$ nor $a(\mathbf{b}, \chi)$ is a measurable quantity, as an area on the wave front having atomic dimensions cannot be isolated. Both quantities are of theoretical interest only. What is measured is the net contribution at each angle or the differential cross section, and the wave and trajectory models for calculating this quantity differ. In the wave method, the amplitudes are summed over impact parameters as in Eq. (3.27), whereas in the trajectory model the probabilities, which are the amplitudes squared, are summed as in Eq. (2.31). The magnitude of the wave-mechanical sum should fluctuate, therefore, about the trajectory sum as the angle is varied, which is the general result seen if the experimental resolution is sufficient.

The remainder of this chapter will be devoted to determining wave-mechanical expressions for the scattering amplitude, hence the differential cross section, for elastic collisions. Although inelastic collisions will be considered in detail in Chapter 4, here we briefly examine the asymptotic form for the wave function when inelastic processes occur. Labeling, as in Chapter 2, the final CM momentum p_j for an inelastic reaction $AB \rightarrow j$, we obtain the asymptotic wave function as

$$\psi(\mathbf{R}) \rightarrow \exp\left[i\mathbf{p}_0 \cdot \mathbf{R}/\hbar\right] + \sum_j f_{AB \rightarrow j}(\chi) \frac{\exp\left[ip_j R/\hbar\right]}{R} \tag{3.28}$$

That is, the scattered wave is a sum of amplitudes associated with each possible inelastic process. The separate scattering processes are often referred to as "channels" for scattering. A differential, inelastic cross section can now be written in terms of the inelastic scattering amplitudes, using the definition in Eq. (2.30) as a basis. The CM wave function, as given in Eq. (3.28), is a probability amplitude normalized to unit density in the incident beam. Therefore, writing the incident intensity as being proportional to the incident velocity, p_0/m, we see that the scattered intensity at the detector for process j is proportional to $(p_j/m)|f_{AB \rightarrow j}(\chi)|^2 \Delta\Omega$. The inelastic CM cross section, therefore, is

$$\sigma_{AB \rightarrow j}(\chi) = \frac{p_j}{p_0} |f_{AB \rightarrow j}(\chi)|^2 \tag{3.29}$$

The momentum factors in Eq. (3.29) explicitly account for the flux change in the CM system when the particles lose or gain momentum inelastically.

Scattering Amplitude for Potential Scattering

In this section an expression for the scattering amplitude is obtained, based on the partial-wave expansion of the plane wave described in Eq. (3.16). This is used to develop a semiclassical approximation to the scattering amplitude which relates directly to the notion of classical trajectories for the particles. As stated earlier, there are a number of approaches to develop such an expression. The one outlined here was first described by Ford and Wheeler.

We saw in Eq. (3.16) that the plane wave, i.e., the solution to the simple-wave equation, can be decomposed into a sum of radial waves times the Legendre polynomials which contain the angular factors. The presence of a potential modifies this plane-wave form, producing scattering. However, if the potential depends only on R, then only the radial part of the wave function will be affected by its inclusion in the wave equation. The general wave function, in the presence of a potential, can be written in a form similar to the plane-wave form in Eq. (3.16),

$$\psi(\mathbf{R}) = \sum_{l=0}^{\infty} b_l \mathcal{R}_l(R) P_l(\cos \chi) \qquad (3.30)$$

where the radial functions $\mathcal{R}_l$ and the coefficients b_l must be determined. These radial wave functions are solutions of an equation similar to Eq. (3.11),

$$\left[\frac{1}{R} \frac{d^2}{dR^2} R + \frac{p^2(R)}{\hbar^2} \right] \mathcal{R}_l(R) = 0 \qquad (3.31)$$

where $p^2(R) = p_0^2[1 - L^2/p_0^2 R^2 - V(R)/E]$ is the radial momentum with $L^2 \equiv l(l + 1)\hbar^2$ as in Eq. (3.12). The reader should verify that $\psi(\mathbf{R})$ in Eq. (3.30) is a solution of the full wave equation if $\mathcal{R}_l(R)$ satisifes Eq. (3.31) using the fact that the Legendre polynomials are orthogonal functions. That is,

$$\int_{-1}^{1} P_l(x) P_{l'}(x) \, dx = \frac{2}{2l + 1} \delta_{ll'} \qquad (3.32)$$

where

$$\delta_{ll'} = 1, \quad l = l', \quad \text{and} \quad \delta_{ll'} = 0, \quad l \neq l'$$

The radical equation (3.31), which contains the potential, can take the form of the one-dimensional equation discussed earlier, Eq. (3.22), if one makes the replacement $\psi(R) \to R\mathcal{R}_l(R)$ in that equation. In regions where $p(R)$ changes slowly, by analogy with the earlier results, e.g., Eqs. (3.23a) and (3.23b), we can write an approximate solution for the radial functions

$$\mathcal{R}_l(R) \simeq \frac{1}{R} \left[\frac{1}{p^{1/2}(R)} \exp \left(\pm i \int \frac{p(R)}{\hbar} \, dR \right) \right] \qquad (3.33)$$

Solutions to the radial equation, which are shown schematically in Figure 3.4, have roughly the same form as those described in Figure 3.2. For the

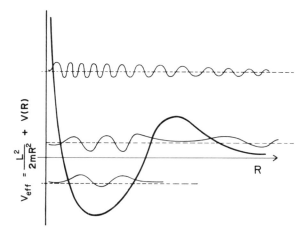

Figure 3.4. Effective potential for the radial equation, $V_{eff} = L^2/2mR^2 + V(R)$, vs radial distance R. Solutions like those in Figure 3.2: bound and scattered waves.

scattering problem we are interested in solutions that are oscillatory at large R.

At large R, where $V(R) \rightarrow 0$, the radial wave functions must solve the simple radial wave equation, Eq. (3.11), and, therefore, the function $\mathcal{R}_l(R)$ will be some combination of the spherical Bessel function, j_l and n_l, described earlier [Eq. (3.13)]. As these functions differ only in phase at large R we can write the asymptotic form of the radial wave functions quite generally as

$$\mathcal{R}_l(R) \xrightarrow{R \rightarrow \infty} \frac{1}{KR} \sin(KR - l\pi/2 + \eta_l) \tag{3.34}$$

which is just a sum of j_l and n_l functions. Such functions differ in phase from the j_l functions, which were used in Eq. (3.16) to describe the plane wave, by an amount η_l. Clearly, if the potential is zero everywhere, this phase shift will be zero and the wave function in Eq. (3.30) will have the form of a plane wave in Eq. (3.16). The phase shift therefore contains the information about the effect of the potential on the wave and depends on K (i.e., energy).

The asymptotic form for the wave function in Eq. (3.34) can now be related to the boundary condition on $\psi(\mathbf{R})$ for the scattering problem

$$\psi(\mathbf{R}) \xrightarrow{R \rightarrow \infty} \exp[i\mathbf{K} \cdot \mathbf{R}] + f(\chi) \frac{\exp[iKR]}{R}$$

from Eq. (3.26). Using the partial wave expansion of $\psi(\mathbf{R})$, Eq. (3.30), and $\exp[i\mathbf{K} \cdot \mathbf{R}]$, Eq. (3.16), and the form of the radial functions, $\mathcal{R}_l(R)$ and

$j_l(KR)$, we can now express the scattering amplitude in terms of the phase shifts, η_l,

$$f(\chi) = \frac{1}{iK} \sum_{l=0}^{\infty} [(l + 1)/2] P_l(\cos \chi)[\exp(2i\eta_l) - 1] \qquad (3.35)$$

which we leave as a problem for the reader. This procedure, used by Faxén and Holtzmark, reduces the calculation of the scattering amplitude, and hence the cross section, to the calculation of phase shifts acquired by each partial wave in the course of scattering.

Although the scattering problem is far from solved, the form of Eq. (3.35) is striking. We first note that when the phase shifts are zero in Eq. (3.35) no scattering occurs. Further, the second term in the sum can be evaluated separately using the completeness of the Legendre polynomials,

$$\sum_{l=0}^{\infty} [(l + 1)/2] P_l(\cos \chi) = \delta(\cos \chi - 1)$$

This expression has a value only when $\chi = 0$, which, in Eq. (3.35), essentially subtracts the unscattered, transmitted wave. There can, of course, be waves "scattered" into $\chi = 0$ which are shifted in phase, indicating that an interaction with the target occurred. In constructing the classical differential cross section, no distinction was made between particles "scattered" into $\chi = 0$ and transmitted particles; here, however, they are distinguished by a phase shift.

Another feature of the scattering amplitude in Eq. (3.35) is that it is expressed as a sum of multipole moments or antenna shapes. It is seen that each moment is stimulated by a different region of impact parameter, as indicated in Figure 3.5, if the usual identification $b \simeq (l + 1/2)\hbar/p_0$ is made. If the colliding atoms, hence the potential, have some physical size, $d = r_A + r_B$, then one would not expect significant contributions to the scattering amplitude for $b > d$ or, equivalently, $l > d/\lambda$. For fast collisions, that is, short wavelengths ($\lambda \ll d$), a large number of l values or multipole moments contribute to the scattering. In this case, a change in l of 1 implies a corresponding change in b of $\Delta b \simeq \hbar/p_0 = \lambda$, which is small. Therefore, for fast collisions the sum in Eq. (3.35) can be accurately replaced by an integral

$$f(\chi) \simeq \frac{K}{i} \int_0^{\infty} b \, db \, P_{l(b)}(\cos \chi)\{\exp[2i\eta(b)] - 1\}, \qquad \lambda \ll d \qquad (3.36)$$

where some approximation to the P_l should be used which interpolates between neighboring l values. This expression is now in a form similar to the one we surmised earlier for the scattering amplitude, Eq. (3.27). In the example described, the phase shifts should go to zero rapidly as b exceeds the range d of the potential.

In the opposite extreme, for very slow collisions or long wavelengths,

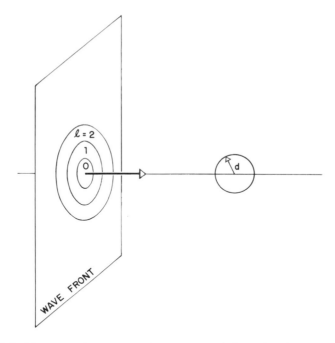

Figure 3.5. Incident wave front approaches a target of dimension d. Regions of impact parameter on the wave front indicated by $l = 0$, 1, 2, where $b \simeq [(l + 1)/2]/K$. Each region produces scattering with different angular characteristics or multipole moments, P_l.

$\lambda \gg d$, only small values of l contribute significantly to the scattering amplitude in Eq. (3.35), and it often can be well represented by a single term

$$f(\chi) \sim \frac{1}{K} \sin \eta_0 \exp{(i\eta_0)} \qquad \lambda \gg d \qquad (3.37)$$

This expression represents the same range of impact parameters as above, but now they are described by a single phase shift. That this should be so is not surprising. If the wavelength is long compared to the size of the interaction region, the change in phase will be insensitive to the particular b. In this limit, the scattering amplitude, hence the CM differential cross section, is independent of the details of the potential. It is also independent of angle, a result which occurred classically only for special forms of the potential, e.g., hard sphere.

Before evaluating the phase shifts, we note that the integrated cross section has the form

$$\sigma = 2\pi \int_{-1}^{1} | f(\chi) |^2 \, d \cos \chi = \frac{8\pi}{K^2} \sum_{l=0}^{\infty} [(l + 1)/2] \sin^2 \eta_l \qquad (3.38)$$

This result is obtained using the orthogonality of the Legendre polynomials expressed in Eq. (3.32). In the long-wavelength limit, a single term again dominates the sum in Eq. (3.38). In the short-wave length limit, the integrated cross section can be written

$$\sigma \simeq 2\pi \int_0^\infty db \, b P(b) \tag{3.39a}$$

as in Eq. (2.20), with

$$P(b) = 4 \sin^2 \eta(b) \tag{3.39b}$$

If the phase shift varies randomly, or very rapidly, for $b < d$, then $\sin^2 \eta(b)$ can be approximated by its average value, implying $P(b) \to 2$; this is twice the classical result. We note another important distinction between the classical and wave-mechanical cross sections. For potentials of infinite range, the classical cross section was infinite. However, because of the subtraction of the transmitted beam in Eq. (3.35), the wave-mechanical scattering cross section may be finite. This can be seen most easily by assuming that the phase varies rapidly for $b < b_0$ and then becomes small for $b > b_0$, where b_0 is a value to be determined. Now the cross section in Eq. (3.39) can be written

$$\sigma \simeq 2\pi b_0^2 + 8\pi \int_{b_0}^\infty \eta^2(b) b \, db \tag{3.40a}$$

If the phase shift goes to zero rapidly for large b, for example,

$$\eta(b) \xrightarrow{b \to \infty} C/b^m, \qquad m > 1$$

then the cross section would be finite with a value

$$\sigma \simeq 2\pi b_0^2 \left[1 + \frac{2}{m-1} \eta^2(b_0) \right] \tag{3.40b}$$

This is referred to as the Massey–Mohr approximation to the integrated cross section and will be used later to evaluate σ. Using identical arguments, the angular differential cross section will also be finite at $\chi = 0$ if σ is finite. Since $P_l(1) = 1$, the reader can verify that the scattering amplitude in Eq. (3.36) can be written

$$f(0) \simeq \frac{iK}{4\pi} \sigma + K \int_0^\infty b \, db \sin 2\eta(b) \tag{3.41}$$

where σ is the cross section from Eq. (3.39). The first term is simply related to the total cross section, and the second term is generally small, as the average value of the sine function is zero. The relationship between σ and the imaginary part of $f(0)$ is often referred to as the optical theorem. From Eq. (3.41), if σ is finite, $f(0)$ and $\sigma(\chi = 0)$ are finite, and vice versa.

The remaining problem in the calculation of σ and $f(\chi)$ is the evaluation of phase shifts, $\eta(b)$, for any given potential. This is dealt with in the following section. To understand the forms for $\chi(b)$ and the differential cross section obtained in the following section, it is not absolutely necessary to follow the derivations used. However, these arguments, which are like those used in optics, will give the reader a better feeling for the relationship between the wave and trajectory methods for calculating cross section.

The Semiclassical Approximation

The short-wavelength limit of the total cross section in Eq. (3.39a) is similar in form to the classical calculation, except at large impact parameters where the phase shifts go to zero. However, the classical and wave-mechanical differential cross sections are determined quite differently even for fast collisions: one is a sum of probabilities and the other a sum over probability amplitudes. These two sums can yield the same results only when a small region of impact parameters dominates the contribution to scattering at any angle. This was assumed always to be the case for the classical trajectory calculation, as a delta function was used to describe the scattering probability $p(b, \chi)$. To examine the wave-mechanical amplitude in Eq. (3.36) we employ the following integral representation of the oscillatory functions P_l:

$$P_l(\cos \chi) \sim \left(\frac{\chi}{\sin \chi}\right)^{1/2} \frac{1}{2\pi} \int_0^{2\pi} \exp\{-i[(l+1)/2]\chi \cos \phi\} \, d\phi \qquad (3.42)$$

which gives a value for all l and is valid for χ not too close to π and l large. Variations of this form can be used to represent scattering near $\chi = \pi$.

Using the above expression, we see that the scattering amplitude in Eq. (3.36) becomes

$$f(\chi) \simeq \frac{K}{2\pi i} \left(\frac{\chi}{\sin \chi}\right)^{1/2} \int d^2b \, \exp[-iKb\chi \cos \phi]$$

$$\times \{\exp[2i\eta(b)] - 1\} \qquad (3.43)$$

where ϕ is assumed to be the azimuthal angle on the wave front. In this form the integrand is a purely oscillatory function. That is, the amplitude, $a(\mathbf{b}, \chi)$, postulated in Eq. (3.27), is a function of the angle χ, and the integral over $\mathbf{b}$ involves only the phase factors in the exponential terms. When the phase changes rapidly as a function of $\mathbf{b}$ the integral oscillates rapidly and the positive and negative contributions to the integral tend to cancel. Conversely, in regions of b for which the phase varies slowly, the integrand will add constructively, yielding a nonnegligible contribution to the integral. If the phase has a well-defined maximum or minimum in b (a stationary point), this will be a region of slowly varying phase in which constructive interference occurs in the integral. Such a situation is indicated schematically in Figure 3.6 where the integral of an oscillatory function is seen to accumulate most of its value in the region where the phase is stationary, here a maximum in b.

The method of stationary phase for evaluating an integral with an oscillating integrand, also referred to as the saddle point method, is described in Appendix F. Briefly, one finds the stationary points, those points for which the derivative of the phase with respect to the variable is zero, and integrates in a region around these points. For the integral in Eq. (3.42) the stationary points in ϕ are 0 (or 2π) and π. Physically this means that the major contributions at any scattering angle χ occur when the azimuthal angle in the wave front is equal to the azimuthal angle of scattering. In our discussion we have implicitly assumed, as there are no external fields and the results depend on χ only, that our azimuthal angle of scattering is 0 or π. Both angles occur because experiments cannot distinguish between negative and positive scattering angles, χ. Applying the saddle point method to the evaluation of P_l in Eq. (3.42), we obtain

$$P_l \sim \frac{1}{2i\{\pi/2[(l+1)/2]\sin\chi\}^{1/2}}$$
$$\times \left(\exp\{i\chi[(l+1)/2] + i\pi/4\} - \exp\{-i\chi[(l+1)/2] - i\pi/4\} \right)$$

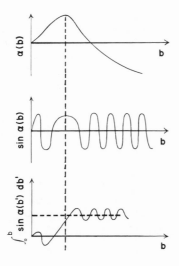

Figure 3.6. Extremum (here a maximum) in the phase $\alpha(b)$ produces a stationary region in the oscillatory function $\sin\alpha(b)$. This leads to an accumulation in the integral, $\int_0^b \sin\alpha(b')\,db'$.

or

$$P_l \sim \frac{1}{\{\pi/2[(l+1)/2]\sin\chi\}^{1/2}} \sin\{\chi[(l+1)/2] + \pi/4\} \tag{3.44}$$

when l is large and χ is not close to *either* 0 or π. The additional constraint at $\chi = 0$ comes because the two separate stationary phase values, $\phi = 0$ and $\phi = \pi$, merge at $\chi = 0$. That is, $\chi = 0$ is a diffraction region. The method is only valid when the points of the stationary phase can be treated as distinct, contributing regions of constructive interference.

That $P_l(\cos\chi)$ in Eq. (3.44) is an approximate solution to the angular equation, Eq. (3.10), and that the normalization conditions in Eq. (3.32) hold, can be verified by substitution. Whereas the functions P_l repeat for $\chi \to \chi - 2\pi s$, s an integer, the exponential expressions in Eq. (3.44) change by $\exp[\pm i\pi s]$ for integer l, which we include to allow the possibility of large angles contributing significantly to the cross section. Using Eq. (3.44) in Eq. (3.36), we can express the scattering amplitude in the form

$$f(\chi) \approx -\left[\frac{1}{2\pi}\frac{K}{\sin\chi}\right]^{1/2} \times \int_0^\infty b^{1/2}\,db\,\{\exp[i\alpha_+(s)] - \exp[i\alpha_-(s)]\}, \quad \chi \neq 0$$

with phases

$$\alpha_\pm(s) = 2\eta(b) \pm (\chi - 2\pi s)Kb \pm \pi[(s+1)/4] \tag{3.45}$$

where, as indicated above, s is an arbitrary integer. The stationary points in b are found by setting the derivative of the phases, $\alpha_\pm(s)$ in Eq. (3.45), to zero:

$$2\frac{d\eta}{db} \pm K(\chi - 2\pi s) = 0 \tag{3.46}$$

For each χ and arbitrary integer s, all values of b for which Eq. (3.46) is satisfied will contribute significantly to the scattering amplitude, and, hence, to the cross section calculated at a given χ. The classical trajectory condition for determining which impact parameters contribute at a given scattering angle is determined by the classical deflection function, $\chi(b)$. By analogy, Eq. (3.46) can be thought of as defining a semiclassical deflection function in the CM system,

$$\chi^{sc}(b) = \frac{2}{K}\frac{d\eta(b)}{db} \tag{3.47}$$

That is, for a given b, significant scattering will occur into those angles χ for which $\cos\chi = \cos\chi^{sc}(b)$ or $\chi = \pm\chi^{sc} + 2\pi s$. Applying the saddle point method to the integration over b, we see that the scattering amplitude becomes

$$f^{sc}(\chi) \simeq \sum_{\text{all stationary points}} \mp \left(\frac{b}{\sin \chi^{sc}} \frac{db}{d\chi^{sc}} \right)^{1/2}_{\pm, s} \exp \left[i\alpha_{\pm}(s) \pm (i\pi/4) \right] \quad (3.48)$$

where each term in the sum is evaluated at the appropriate stationary point indicated by the labels $\pm$, s in Eq. (3.48).

Before discussing this result the reader should be cautioned that we have not as yet stated how to calculate $\eta(b)$, hence $\chi^{sc}(b)$. The form of Eq. (3.48) is, however, propitious looking. The scattering amplitude is represented as a sum of contributions from a few points on the wave front, the key word being *few*. If there are a large number of stationary points for each χ, then there is little advantage in calculating this approximate sum rather than the exact sum in Eq. (3.35). Each contributing term in Eq. (3.48) has an amplitude, proportional to the square root of a classical-like cross section, and the ubiqitous phase factor that is characteristic of wave scattering. The validity of this sum depends on the stationery points being well separated, in both azimuthal angle ϕ and impact parameter b, and the wavelength being small. In the limit that only one impact parameter contributes to the sum, i.e., there is only one point of stationery phase, the semiclassical cross section has the same *form* as the classical cross section,

$$\sigma^{sc}(\chi) = |f^{sc}(\chi)|^2 = \left| \frac{b}{\sin \chi^{sc}} \frac{db}{d\chi^{sc}} \right| \quad (3.49)$$

and the phase factor does not appear. When more than one impact parameter contributes, the semiclassical cross section exhibits interference due to the differences in the phase factors.

A short-wavelength expression for the radial phase shift, $\eta(b)$, can be obtained from the WKB approximation in Eq. (3.23b). The result though is self-evident. That limit, in which one distinct region of impact parameter is associated with each scattering angle, is by definition the classical trajectory limit for scattering. Therefore, one might reasonably expect the classical and semiclassical cross sections to be equivalent. In order that the semiclassical cross section in Eq. (3.49) be equal to the classical cross section in Eq. (2.44), $\chi^{sc}(b)$ must equal the CM classical deflection function $\chi(b)$. As the classical deflection function can be obtained from a stationary condition on the classical action, S_c in Eq. (2.66), and χ^{sc} is obtained from the stationary condition on $\alpha_{\pm}(s)$, then $\alpha_{\pm}(s)$ must be simply related to the classical action. This gives us a way of calculating the radial phase shift.

Requiring the phase shift to be zero when no scattering occurs, we postulate that the phase factors in Eq. (3.48) are proportional to the difference between the classical action, S_c, for a particle deflected into angle χ along a trajectory determined by the interaction potential, and the classical

Figure 3.7. Trajectories for calculating the classical action in Eq. (3.50a): S_c (solid line, with potential) and S_0 (dashed lines, no potential). In wave method, trajectories indicate paths of constructively interfering spherical waves from the wave front.

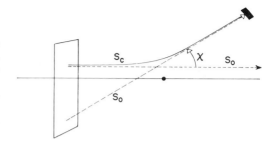

action, S_0, for a particle moving along a straight-line trajectory, as if there were no target, as shown in Figure 3.7. Using Eq. (2.66) and the equations of motion, Eqs. (2.49) and (2.50), we obtain the difference in action, $A(b, \chi)$:

$$A(b, \chi) = S_c - S_0$$

$$= 2 \left[\int_{R_0}^{\infty} p(R)\, dR - \int_{b}^{\infty} p_0(R)\, dR \right] - L\chi(b) \qquad (3.50a)$$

Comparing Eqs. (3.45) and (3.50a), we can identify the radial phase shifts to within a constant as

$$\eta^{sc}(b) = \left[\int_{R_0}^{\infty} p(R)\, dR - \int_{b}^{\infty} p_0(R)\, dR \right] \Big/ \hbar \qquad (3.50b)$$

It is left as a problem for the reader to verify that this is precisely the result found by matching the asymptotic function, $R\mathcal{R}_l$, using Eq. (3.34), with the oscillatory WKB function in Eq. (3.23b) at large R. The reader should also verify that using η^{sc} from Eq. (3.50b) in Eq. (3.47) produces the classical expression for the deflection function, thereby allowing us to drop the distinction between $\chi^{sc}(b)$ and $\chi(b)$.

From Eq. (3.48) the semiclassical cross section can be written in the form

$$\sigma^{sc}(\chi) = \left| \sum_{j} \sigma_j^{1/2}(\chi) \exp\{i[(A_j)/\hbar + \gamma_j]\} \right|^2 \qquad (3.51)$$

where $\sigma_j(\chi) = |b\, db/\sin \chi\, d\chi|_j$ is a classical trajectory cross section, $A_j(\chi, b)$ is the corresponding classical action of Eq. (3.50a), and j labels the stationary condition, i.e., the particular classical trajectory leading to scattering into χ. The extra phase factor γ_j is determined from the square root of the sign of $[(b/\sin \chi)(db/d\chi)]_j$ in Eq. (3.48) and the additional constant factors in Eq. (3.45) and (3.48) that are appropriate to each trajectory.

Equation (3.51) reduces to the classical result if one impact parameter contributes at any scattering angle or if significant experimental averaging occurs. For rainbow scattering, which we discussed earlier (that is, when the deflection function has a mimimum), more than one impact parameter may contribute at a given scattering angle (Figure 2.16). Classical and semi-classical cross sections for this case are compared in Figure 3.8. The semi-classical cross section is seen to be oscillatory for $\chi < \chi_r$ (the rainbow angle). In this region, three impact parameters contribute to scattering at each angle, whereas for $\chi > \chi_r$, the simple classical behavior of the cross section is evident. Actually for angles very close to χ_r, two of the impact parameters merge, as can be seen from Figure 2.15, and the condition for the validity of Eq. (3.51) breaks down. Although this is a diffraction region, a procedure similar to the above can be applied to a narrow region around χ_r. Oscillations of the type exhibited in Figure 3.8 can be exceedingly helpful (Chapter 5) in determining the interaction potential for a collision, as the frequency of the oscillations depends on differences in the classical action for trajectories contributing to the same scattering angle.

Before proceeding, we will review the results obtained in this section. Because of the large masses involved in atomic collisions, short-wavelength methods can be employed down to quite low energies, ~ 0.1 eV, whereas for incident electrons they apply only for "fast" collisions (v greater than the bound electron orbital velocities). In these methods the scattering amplitude is obtained as an integral over contributing amplitudes from all positions on the wave front labeled by an impact parameter. Each source point on the wave front contributes with a different phase. This phase can be approximated by calculating the classical action associated with a path from the point of interest on the wave front to the detector. The integrated cross section is also a sum over all contributing points on the wave front, which, when the potential is spherically symmetric, reduces to an evaluation of the radial component of the classical action. These sums can be simplified using

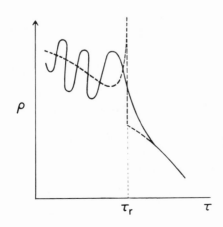

Figure 3.8. Schematic of angular differential cross section, $\rho(\chi) = \chi \sin \chi\, \sigma(\chi)$ vs $\tau = E\chi$, for a potential with a minimum. Heavy dashed line, classical-trajectory calculation (Figure 2.14); solid line, wave-mechanical calculation; light dashed line, rainbow angle, $\tau_r = E\chi_r$. Oscillations occur in region where more than one impact parameter contributes.

the random-phase approximation for regions of impact parameters in which the phase changes rapidly, and the stationery-phase approximation for regions of constructive inference. In the differential cross section, the latter are associated with the classical trajectories determined by the interaction potential.

The Born Approximation

From our earlier discussions, it is clear that scattering at small angles dominates the integrated scattering cross section, particularly at high energies. Small angles are equivalent to large impact parameters for which the slope of the classical deflection function is small. In this region, the separation between those impact parameters contributing significantly to the scattering amplitude is small compared to the wavelength, i.e., $\Delta b \sim |db/d\chi| \Delta \chi \ll \lambda$, even for large energies. In such a region the stationary phase approximation is not applicable and the diffraction effects must be calculated by summing over the impact parameter in Eq. (3.36) or (3.43). In the special case of very low energies, that is, very long wavelengths, the diffraction approximation applies at *all* impact parameters and scattering angles, results in the simple expression in Eq. (3.37).

For scattering at small angles, in fast collisions, the angular momentum is large throughout the collision and the ratio V/E is small. The classical impulse approximation can, therefore, be used to estimate the semiclassical phase shift. Following the procedure developed in Chapter 2, Eqs. (2.51) through (2.57), when $V(R)$ is small, the phase shift becomes from Eq. (3.50b)

$$2\eta^{sc}(b) \simeq -\frac{1}{\hbar} \int_{-\infty}^{\infty} V(R)\, dt \qquad (3.52)$$

where, as in our earlier discussion, $R^2 = R_0^2 + Z^2$ with $Z = (p_0/m)t$. Now Eq. (3.52) used in Eq. (3.47) is seen to be consistent with Eq. (2.57a) for $R_0 = b$. Assuming all impact parameters and azimuthal angles contribute to the scattering at χ, we evaluate the scattering amplitude. For small angles, $\chi \simeq \sin \chi$, and the change in momentum is written $\Delta p \equiv 2p_0 \sin \chi/2 \simeq p_0 \chi$ [e.g., Eq. (2.56)]. With these approximations Eq. (3.43) becomes

$$f(\chi) \simeq \frac{K}{2\pi i} \int d^2b \, \exp\left[(-i\Delta\mathbf{p}\cdot\mathbf{b})/\hbar\right]\{\exp\left[2i\eta(b)\right] - 1\} \qquad (3.53)$$

If, in addition, the phase shift is small, the exponential can be expanded keeping only the first two terms. Using Eq. (3.52) for $\eta^{sc}(b)$, we write (3.53) as

$$f(\chi) \simeq \frac{m}{2\pi\hbar^2} \int d^3R \, \exp\left[(-i\Delta\mathbf{p}\cdot\mathbf{b})/\hbar\right] V(R) \qquad (3.54)$$

where the cylindrical coordinates for the volume element d^3R are $d^2b\,dZ$. This useful expression can be evaluated for a variety of potentials, and some examples are given in Appendix D. Before considering this, we obtain the same expression from another, more general, starting point. The following exercise is not strictly academic, however, as it will provide an insight into the treatment of inelastic, small-angle scattering, to be discussed later, and will allow us to generalize the result above to scatterings in which the potentials are not spherically symmetric.

A solution to the wave equation, when the potential is viewed as a source of scattered waves, has the form

$$\psi(\mathbf{R}) = \exp\left[i\mathbf{K}\cdot\mathbf{R}\right]$$
$$+ \frac{1}{4\pi}\int \frac{\exp\left[ik\,|\,\mathbf{R}-\mathbf{R}'\,|\right]}{|\,\mathbf{R}-\mathbf{R}'\,|}\left(\frac{2m}{\hbar^2}\right)V(\mathbf{R}')\psi(\mathbf{R}')\,d^3R' \qquad (3.55)$$

obtained by combining Eqs. (3.20) and (3.21). The wave function in Eq. (3.55) consists of a transmitted component, that is, a solution of the simple wave equation, and a scattered component. The intensity of the scattered wave depends not only on the interaction potential but also on the intensity of the wave itself. This merely reflects the obvious: There is no source of scattered waves unless there is a beam of particles incident on the target.

The solution of an integral equation such as that in Eq. (3.55) is quite difficult. Born, however, suggested an iterative procedure for obtaining an approximate solution, which involves repeated substitution of the expression for $\psi(\mathbf{R})$ in Eq. (3.55) back into the integral over the potential in that equation. In this procedure, the wave function becomes a sum of terms involving integrals of ever-increasing complexity, which appears at first to be a futile process. The procedure, however, is quite useful when the fraction of the wave scattered is small and the successive terms diminish in importance. Now, the wave function is written as a plane wave plus corrections for scattering. The first correction is equivalent to treating the wave function in the integral as a plane wave everywhere, that is, the interaction is weak. This concept is, of course, equivalent to that used in the impulse approximation to the classical trajectory calculation discussed in Chapter 2. Equation (3.55) can be approximated, in this case, as

$$\psi(\mathbf{R}) \simeq \exp\left[i\mathbf{K}\cdot\mathbf{R}\right] + \frac{m}{2\pi\hbar^2}\int \frac{\exp\left[iK\,|\,\mathbf{R}-\mathbf{R}'\,|\right]}{|\,\mathbf{R}-\mathbf{R}'\,|}\,V(\mathbf{R}')$$
$$\times \exp\left[i\mathbf{K}\cdot\mathbf{R}'\right]d^3R' \qquad (3.56)$$

which is referred to as the first Born approximation to the wave function.

To obtain the scattering amplitude, the wave function should be evaluated at the asymptotic limit where $R \to \infty$, or at least where R is large compared to the range of the interaction potential. When $R \gg R'$, the distance between the target and the detector becomes

$$|\mathbf{R} - \mathbf{R}'| = [R^2 + R'^2 - 2\mathbf{R} \cdot \mathbf{R}']^{1/2} \simeq R - \frac{\mathbf{R} \cdot \mathbf{R}'}{R} + \cdots \qquad (3.57)$$

Substituting the above expression into Eq. (3.56), we obtain

$$\psi(\mathbf{R}) \xrightarrow{R \to \infty} \exp i\mathbf{K} \cdot \mathbf{R} + \frac{e^{iKR}}{R}$$

$$\times \left[\frac{m}{2\pi\hbar^2} \int \exp[-i\mathbf{K}_f \cdot \mathbf{R}'] V(\mathbf{R}') \exp[i\mathbf{K}_0 \cdot \mathbf{R}] d^3 R' \right] \qquad (3.58)$$

where we have written $\mathbf{K}_f = K_0 \hat{R}$, and where $\hat{R}$ is in the direction of observation. The scattering amplitude is now easily identified by comparing Eq. (3.58) with the asymptotic form for $\psi(\mathbf{R})$ in Eq. (3.26),

$$f^{(1)}(\chi) = \frac{m}{2\pi\hbar^2} \int d^3 R \exp[-i\mathbf{K}_f \cdot \mathbf{R}] V(\mathbf{R}) \exp[i\mathbf{K}_0 \cdot \mathbf{R}] \qquad (3.59)$$

The superscript indicates that one iteration or substitution in Eq. (3.55) was kept.

The result in Eq. (3.59) is the first Born approximation to the scattering amplitude. Noting that the wave vector $\mathbf{K}_f$ times $\hbar$ is the final momentum of particles scattered in the direction of observation, we then have $\mathbf{K}_f - \mathbf{K}_0 = \Delta\mathbf{p}/\hbar$. If the deflections are small, then $(\Delta\mathbf{p}/\hbar) \cdot \mathbf{R} \simeq (\Delta\mathbf{p} \cdot \mathbf{b})/\hbar$, and the first Born approximation to the scattering amplitude is identical to our earlier result in Eq. (3.54). There are two significant differences, however, which make this expression more general. First, to obtain Eq. (3.59), we did not have to assume the scattering angles were small, nor did we need to assume a spherically symmetric potential, thus allowing us to use Eq. (3.59) for molecular collisions. Second, after the diminishing spherical component e^{iKR}/R is removed, the scattered wave is, to first order, a plane wave moving in the direction of the detector, $\hat{R}$. Therefore, the Born approximation is often written as

$$f^{(1)}(\chi) = \frac{m}{2\pi\hbar^2} \int d^3 R \, \psi_f^* \, V(R) \psi_0 \qquad 3.60)$$

ψ_0 and ψ_f indicating the initial and final plane waves. Equation (3.60) has the form of a transition probability (Chapter 4) between an initial, incident wave and a final, scattered wave. This form will be useful when considering inelastic effects.

Whereas the interferences approximation (the semiclassical method) is directly related to the classical scattering formalism, the first Born approximation is not. In the latter case, we may expect a very different energy and angular dependence for the scattering cross section although phase shifts in the integral are still simply related to classical quantities, like the classical action.

For the power-law potentials discussed earlier, the first Born approximation to the scattering amplitude can be integrated, yielding a cross section of the form

$$\sigma^{(1)}(\chi) = \left(\frac{\hbar}{\Delta p}\right)^{6-2n}\left(\frac{2m}{\hbar^2}\,g_n C_n\right)^2 \tag{3.61}$$

where $\Delta p = 2p_0 \sin \chi/2$, which at small angles becomes $\Delta p \simeq p_0 \chi$. For the above expression, g_n is evaluated in Appendix D and is finite only for $n \leq 3$. The angular dependence of the first-Born cross section differs from that of the classical impulse approximation except, fortuitously, for $n = 1$. Here the first-Born cross section also equals the exact classical and quantum-mechanical cross sections when the small angle limit is *not* taken. The cross section for this potential, the Rutherford cross section, is

$$\sigma^{(1)}(\chi) = \left(\frac{Z_A Z_B e^2}{4E}\right)^2 \frac{1}{\sin^4 \chi/2} \tag{3.62}$$

when $C_1 = \pm Z_A Z_B e^2$ is used for the coulomb potential. Whereas, in the classical impulse approximation, $\rho = \chi \sin \chi\, \sigma(\chi)$ was a quantity that depended only on $\tau = E\chi$ for any power n, such is the case in the Born approximation only for $n = 1$. The ρ, τ plots described earlier are not useful, therefore, in diffractive regions.

The integrated cross section for power laws are infinite in the first Born approximation, as in the classical calculation, except for $3 > n > 2$. However, for screened interactions, where the potential falls off rapidly, like potentials of the Yukawa form,

$$V = \frac{C_n}{R^n} \exp\left[-\beta R\right] \tag{3.63}$$

the integrated cross section is finite for $n \leq 3$ (Appendix D). For the $n = 1$ case the cross section has the form

$$\sigma^{(1)}(\chi) = \left(\frac{2m}{\hbar^2}\,C_1\right)^2 \frac{1}{[(\Delta p/\hbar)^2 + \beta^2]^2} \tag{3.64}$$

This yields the integrated cross section

$$\sigma^{(1)} = \frac{\pi C_1^2}{e_0} \frac{1}{[(E + e_0)/4]} \tag{3.65}$$

where $\varepsilon_0 = (1/2m)(\hbar\beta)^2$. In the above expression the screening parameter acts like a small, energy-transfer cutoff. That is, integrating the classical cross section for the coulomb interaction down to some small angle or minimum energy of transfer, T_0, yields

$$\sigma \simeq \frac{\pi C_1^2}{E} \frac{M_A}{M_A + M_B} \frac{1}{T_0} \qquad (3.66)$$

if $E \gg T_0$ [viz. Eq. (2.69)]. Comparing Eqs. (3.65) and (3.66) at high collision energies, we see that the screening equivalent to setting a low energy-transfer cutoff of the form $T_0 = e_0/(1 + \mu)$. This concept allows the extension of classical calculations to a large number of otherwise inappropriate situations, as we assumed in Chapter 2. A word of caution, however: The cutoff due to screening does *not* arise classically. That is, the classical, integrated cross section obtained using the screened coulomb potential would be infinite and the differential cross section would become infinite at small angles! The screening parameter only provides a cutoff via the diffraction effect, which the Born approximation estimates when the interaction is weak.

earlier remarks on the long-wavelength approximation. When the wavelength became large compared to the size of the colliding particles, the angular differential cross section was found to be independent of scattering angle as in Eq. (3.37). Power-law cross sections, having no size parameter, may violate this rule, as exemplified by the coulomb interaction for which the cross section has the same angular dependence at all energies. For screened potentials, however, the "extent" of the interaction is the inverse of the screening parameter, β, in Eq. (3.63). When $\lambda \gg \beta^{-1}$, Eq. (3.64) reduces to

$$\sigma^{(1)}(\chi) \xrightarrow{\;E \to 0\;} \left(\frac{2mC_1}{\hbar^2 \beta^2}\right)^2 = \left(\frac{C_1}{e_0}\right)^2$$

which is also independent of the CM scattering angle. The Born approximation as defined in Eq. (3.59) assumes that the *amount* of scattering by an object is small so that the total wave function remains, primarily, a plane wave. If, indeed, the phase shifts are all small at low energies the Born results will be applicable, but for some systems the low l value phase shifts are significant as the average interaction potential is large. This will be considered further in Chapter 5 when we compare the Born results to results of experiments.

Charge Distributions

Before proceeding to calculate interaction potentials in the next chapter, it is useful to briefly review classical and wave-mechanical concepts relating to bound particles. In this discussion we will confine our attention to a single electron bound to a nucleus as in a hydrogen atom. The same ideas, however, apply to all bound particles, for example, atoms bound together to form molecules or a single electron bound in the combined field of a nucleus and other electrons.

Assuming that the zero of potential energy occurs when the bound electron is infinitely far from the nucleus, then, as the force between the electron and the nucleus is attractive, the energy of a bound electron will be negative, as in the examples in Figures 3.2 and 3.4. For a given binding energy, a classical electron can be in an infinite number of possible orbits each corresponding to a different angular momentum as indicated in Figure 3.9. Each orbit repeats with a given period or frequency determined by the binding energy. If only the binding energy is specified, all orbits corresponding to this energy are equally probable. A quantity one would like to know is the average distribution of charge in an atom with a particular binding energy, or the likelihood of finding an electron in a given position. This can be obtained by noting the positions of the electron in each possible orbit and weighting them by the time spent in the vicinity of a given position. A simple and equivalent procedure is to assume that all positions and momenta that yield the correct total energy are equally likely. For an electron in an atom with binding energy ε_0, the probability density of possible momenta and positions so defined is described by a delta function

$$\rho(\mathbf{p}, \mathbf{r}) = C_{\varepsilon_0} \delta \left[\varepsilon_0 - \frac{\mathbf{p}^2}{2m_e} - V(\mathbf{r}) \right] \qquad (3.67)$$

where $V(\mathbf{r})$ is the effective potential energy of the electron. By using the delta function in Eq. (3.67), we assume the electron to be a point charge. If the momentum and position of the electron are such that the total energy of the particle equals ε_0, the probability density in Eq. (3.67) is large; if not, the probability density is zero. Alternatively, the electron itself may be considered diffuse, in which case ρ will be better described by a function that is not infinitely sharp, like a gaussian or a step function.

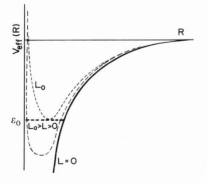

Figure 3.9. Effective potentials for an electron, $V(r) + L^2/2m_e r^2$ (dashed curves) for two L values and a given ε_0: L_0, circular orbit, R fixed; $L < L_0$, elliptic orbit of arbitrary orientation. Solid curve: coulomb potential $V(R) = -Ze^2/r$.

The constant C_{ε_0} in Eq. (3.67) is determined by the requirement that the probability of finding the particle somewhere is unity. Summing over all positions and momenta, this implies

$$1 = \int\int \rho(\mathbf{p}, \mathbf{r})\, d^3p\, d^3r.$$

Multiplying the probability density by the electronic charge $(-e)$ converts ρ into a charge density. That is, the electronic charge on the atom can be thought of statistically as a distributed charge, which, in Eq. (3.67), is a dynamic distribution in that it depends on both the position and momentum. To find the average, static charge density we average over all possible momenta. The static charge distribution so calculated is

$$\rho(\mathbf{r}) = \int \rho(\mathbf{p}, \mathbf{r})\, d^3p$$

$$= C_{\varepsilon_0} 4\pi(2m_e)^{3/2} [\varepsilon_0 - V(\mathbf{r})]^{\frac{1}{2}}; \qquad \varepsilon_0 - V(\mathbf{r}) \geq 0 \qquad (3.68)$$

$$\rho(\mathbf{r}) = 0; \qquad \varepsilon_0 - V(\mathbf{r}) < 0$$

As one would expect, the charge can only be found classically where the kinetic energy is positive. In quantum mechanics, the waves associated with bound particles are not so restricted, as was indicated in Figure 3.2 and 3.4. As it is also impossible in quantum mechanics to specify (measure) both the positions and momenta exactly, the spiked, infinitely narrow delta function should be replaced by a function with a width determined by the uncertainty principle. Such an approach yields a somewhat better static charge density and is equivalent to the Thomas–Fermi method for large atoms, which is discussed in Appendix H.

The exact quantum-mechanical charge density is obtained by solving the wave equation for an electron in a potential $V(\mathbf{r})$. As the square of the wave function is the probability density, the static charge distribution is

$$\rho(\mathbf{r}) = |\psi(\mathbf{r})|^2 \qquad (3.69)$$

The charge density for a particle in the absence of a potential (a plane wave) is uniform classically [viz. Eq. (3.68)] and quantum mechanically, as $\psi(\mathbf{r})$ is a plane wave. For a single electron on a fixed nucleus of charge Ze, the wave equation is

$$\left[-\frac{\hbar^2}{2m_e} \nabla_r^2 - \frac{Ze}{r} \right] \psi(\mathbf{r}) = \varepsilon\psi(\mathbf{r}) \qquad (3.70)$$

where m_e is the electron mass and $\mathbf{r}$ the position of the electron with respect to the nucleus. This equation is often used to also approximate the charge density of an electron in a many-electron atom in which the nucleus is shielded by the presence of the other electrons. When this is the case, we replace Z by Z' in Eq. (3.70) to represent the size of the effective charge seen by the electron. Solutions to Eq. (3.70) when $\varepsilon > 0$ describe scattering of the electron by a nucleus. For bound particles, $\varepsilon < 0$, solutions exist only for certain values of energy, as discussed earlier.

As the potential in Eq. (3.70) only depends on r the wave function can be separated, as before, $\psi(\mathbf{r}) = \mathscr{R}_l(\mathbf{r}) Y_{lm}(\theta, \phi)$. Solutions to the radial equation exist for $\varepsilon = \varepsilon_n = -(m_e/2)[(Ze^2)^2/\hbar^2 n^2]$, where n is an integer and $l \leq n - 1$. The first few radial functions, $\mathscr{R}_{nl}$, are

$$\mathscr{R}_{10} = \left(\frac{Z}{a_0}\right)^{3/2} \cdot \exp\left[-\frac{Zr}{a_0}\right] \qquad \text{for } \varepsilon = \varepsilon_1$$

$$\mathscr{R}_{20} = \left(\frac{Z}{2a_0}\right)^{3/2} \cdot 2\left(1 - \frac{Zr}{2a_0}\right) \exp\left[-\frac{Zr}{2a_0}\right] \qquad \text{for } \varepsilon = \varepsilon_2 \qquad (3.71)$$

$$\mathscr{R}_{21} = \left(\frac{Z}{2a_0}\right)^{3/2} \cdot \frac{2}{(3)^{1/2}}\left(\frac{Zr}{2a_0}\right) \exp\left[-\frac{Zr}{2a_0}\right]$$

which the reader should verify are solutions of the radial equation, Eq. (3.31). Here $a_0 = \hbar^2/m_e e^2$ is the Bohr radius. The mean radius of the ground-state hydrogen atom is $3a_0/2$. In Figure 3.10 we compare the radial charge distributions $\rho(r) = r^2 \int \rho(\mathbf{r}) \, d\hat{r}$ calculated using the above solutions to the wave equation with the classical distributions in Eq. (3.68). For the $n = 2$ state, an equal weighting of four quantum-mechanical states is used. The agreement between the distributions is quite good, except, of course, where the kinetic energy is negative. The mean radius of the classical charge distribution is $\frac{5}{8}|Ze^2/\varepsilon|$, which for the ground-state hydrogen atom is $5/4a_0$ slightly larger than the Bohr radius. (The extension of the above results to many-electron atoms and the labeling of such states is reviewed in Appendix G.)

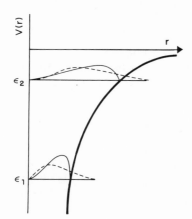

Figure 3.10. Radial charge distributions $\rho(r)$ calculated classically from Eq. (3.68) (solid lines) and quantum mechanically from Eq. (3.71) (dashed lines), for $n = 1$ and $n = 2$ states. Heavy solid curve, the coulomb potential $V = e^2/r$.

The static charge distribution discussed above will be useful for determining the effective interaction potentials between atoms. For inelastic collisions the momentum distribution, $\rho(\mathbf{p}) = \int \rho(\mathbf{r}, \mathbf{p}) d^3r$, of the electron is the more important quantity. That is, an incident atom or electron collides with the electrons in the target, transferring energy to them. The amount of energy transferred depends primarily on the electronic speeds and not their distribution in space. For an electron attached to a nucleus of charge Z the classical momentum distribution, from Eq. (3.67), is

$$\rho(\mathbf{p}) = \frac{8}{\pi^2} \frac{\bar{p}^5}{(p^2 + \bar{p}^2)^4} \tag{3.72}$$

where $\bar{p} \equiv (-2m\varepsilon)^{1/2}$, and ε is the binding energy. This, fortuitously, equals the hydrogenic distribution averaged over angular momentum. Therefore, completely classical-calculated inelastic cross sections can often be quite accurate. Although classical momentum and spatial distributions are useful, and show considerable similarity to the wave-mechanical distributions, we emphasize that they cannot predict the binding energies. That is, as we stated earlier, classical distributions exist for *all* binding energies.

Before leaving this section, it is useful to point out that all bound one-dimensional motion is periodic and the simplest example of periodic phenomenon is that of mass on a spring. Because the description of the collision itself is sufficiently complex, it is often useful both classically and quantum-mechanically to simplify the bound-state motions. This can be done by replacing the binding potential by the potential function of a spring with a spring constant that reproduces the periodicity. That is, a potential of the form

$$V(x) = V_0 + \tfrac{1}{2}k(x - x_0)^2 \tag{3.73}$$

describes a bound particle in one dimension with a fundamental frequency $\omega = (k/m)^{1/2}$, where m is the mass of the particle. A particle in an elliptical orbit can be described by two oscillators with different amplitudes, and so on. The solutions to the wave equation for a one-dimensional oscillator,

$$\left[-\frac{h^2}{2m} \frac{d^2}{dx^2} + V_0 + \tfrac{1}{2}k(x - x_0)^2 \right] \psi(x) = \varepsilon\psi(x) \tag{3.74}$$

have a set of equally spaced energy levels,

$$\varepsilon_n = (n + \tfrac{1}{2})\hbar\omega + V_0, \qquad n = 0, 1, 2, \ldots \tag{3.75}$$

which is essentially what is observed in the vibrational spectra of molecules. This energy spacing, however, is *not* much like the energy-level spacing of electrons bound in atoms. For interacting atoms the broad minimum seen in the potential is well approximated by a parabola in the vicinity of the minimum, but the coulomb potential for electron binding cannot be modeled so simply. However, as each classical electronic orbit can be separately approximated by one or more oscillators with a frequency determined by the binding energy of the orbit, the potential in Eq. (3.73) can be useful for describing bound electrons in some instances. A classical calculation treating the bound electron as an oscillator is carried out in Appendix C.

Vibrational and Rotational Levels of Molecules

We can employ the ideas developed above to describe the bound motion of atoms in molecules. A molecule consisting of N atoms has $3N$ degrees of freedom. Three degrees of freedom are needed to describe the CM translational motion, and three degrees to describe the overall rotation about the CM of the molecules, except for linear molecules for which there is no moment of inertia about the molecule axis. The remaining $3N - 6$ degrees of freedom (or $3N - 5$ for linear molecules) correspond to vibrational modes of the molecule; hence H_2O has three types of vibrational motions, as indicated in Figure 3.11, and O_2 has one. The linear CO_2 has four vibrational modes, two of which are degenerate, that is, they differ only in orientation. The vibrational motion of CO_2 indicated in Figure 3.11 was exploited in the CO_2 laser considered in Figure 1.5. An ammonia molecule exhibits six vibrational modes; methane, 9; etc.

The rotations and vibrations of molecules cannot be totally separated as the moments of inertia change with the molecular vibrational motion. However, as the two types of motion tend to be of very different frequency, an approximation separation is valid, i.e., many vibrations occur for low-lying states during one rotational period. Here we will use the simple diatomic molecule to exhibit this separation. The wave motion in the CM system can be broken into radial and angular components, as we saw in Eqs. (3.10) and (3.11) and again in Eq. (3.31), which correspond to the

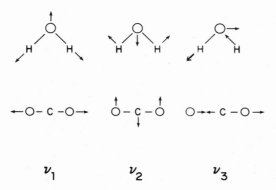

Figure 3.11. Vibrational modes for H_2O and CO_2: ν_1, symmetric stretch; ν_2, symmetric bending mode (doubly degenerate for CO_2, i.e., motion in and out of figure also, which for H_2O is a rotation); ν_3, antisymmetric stretch. Energy spacing for H_2O: 0.453 eV, 0.198 eV, 0.466 eV; for CO_2: 0.165 eV, 0.083 eV, 0.291 eV.

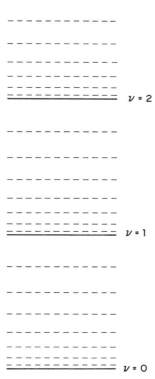

Figure 3.12. Schematic diagram of vibrational and rotational spacing in a diatomic molecule. Vibrational levels nearly equally spaced [Eq. (3.78)], rotational levels increase in spacing [Eq. (3.76)].

vibrational and rotational motions respectively. As before, we write the wave function in the form $\psi(\mathbf{R}) = \mathscr{R}(R)Y_{lm}(\Omega)$. As a first approximation, if the vibrational motion is rapid, we can assume that the angular (rotational) motion occurs with some fixed average separation between the atoms, $\bar{R}_M$. From Eq. (3.31) and following the rotational energy is

$$E_{rot} = \frac{L^2}{2m\bar{R}_M^2} \tag{3.76}$$

which has the form of the classical rotational energy with $m\bar{R}_M^2$ the moment of inertia. However, unlike the classical rotor for which L is a continuous variable, here, $L^2 = l(l + 1)\hbar^2$, yielding a series of rotational levels with increasing spacing as shown in Figure 3.12.

As the vibrational motion for bound particles will be centered about a minimum in the effective potential, one can always expand V_{eff} about $\bar{R}_M$, yielding, as in Eq. (3.73),

$$V_{eff}(R) \simeq E_{rot} + V(\bar{R}_M) + \tfrac{1}{2}k(R - \bar{R}_M)^2 + \cdots \tag{3.77}$$

where k, the "spring" constant, is the second derivative of $V_{eff}(R)$ evaluated at $\bar{R}_M$. Now the radial equation for $\mathscr{R}(R)$ takes on the form of Eq. (3.74), and the vibrational-energy-level spacing will be

$$E_{vib} = (n + \tfrac{1}{2})\hbar\omega \qquad (3.78)$$

where $\omega = (k/m)^{1/2}$. Knowing the potential form, we can determine $\bar{R}_M$ and k and hence the rotational (angular) and vibrational (radial) energy levels of the molecule. Or, conversely, the vibrational and rotational spectra can be used to determine $\bar{R}_M$ and k (the second derivation of V_{eff} at $\bar{R}_M$). The reader should recall that the angular levels of a one-electron atom are degenerate. That is, the binding energy depends on the index n only, not on l [cf. Eq. (3.71)]. This is due to the fact that the CM of the electron and the nucleus is essentially centered on the nucleus, that is, the moment of inertia is very nearly zero. Here the molecular levels separate, yielding a rather complex set of rotational and vibrational levels, which for fast collisions can often be approximated by a classical continuum. In the next chapter we will discuss methods for determining $V(R)$ which can be used to describe the interaction of colliding atoms or the motion of bound atoms.

Exercises

3.1. Verify that Eqs. (3.13) are solutions to Eq. (3.11) and that the sums $(j_l \pm in_l)$ form incoming and outgoing radial waves at large R when the time dependence is included. Verify also that such waves satisfy Eq. (3.18).

3.2. Find the asymptotic form for j_l and n_l as $R \to 0$ from Eq. (3.11) and verify that this is consistent with Eq. (3.16).

3.3. Find the solutions to Eq. (3.19) in the region $R > d$ when $\mathscr{S}(\mathbf{R})$ is a uniform source, $\mathscr{S}_0$, of radius d. Find $\psi(\mathbf{R})$ for a line source $\mathscr{S}$.

3.4. For the potential in Figure 3.2 find the solutions discussed by matching the functions and their slopes in each region at the boundaries of the region. Find the resonance condition for the case $V_3 > E > 0$ if $V_2 = 0$.

3.5. Compare the WKB solutions for the potential in Figure 3.2 to the ones obtained in Problem 3.4.

3.6. As the radial function $R\mathscr{R}_l(R)$ from Eq. (3.31) satisfies a simple one-dimensional wave equation like Eq. (3.22), use the oscillatory form of the WKB solution, Eq. (3.23b), to obtain an estimate for η_l defined in Eq. (3.34). This result is referred to as the JWKB phase shift. Show that this phase shift is equivalent to η^{sc} in Eq. (3.50b) and that χ^{sc} in Eq. (3.47) is equal to the classical deflection function $\chi(b)$ in Eq. (2.55).

3.7. Use the potential $V = -C_4/R^4 + C_8/R^8$ to determine the form of $\sigma^{sc}(\chi)$ in Eq. (3.51) for $\chi > \chi_r$ and $\chi < \chi_r$, where χ_r is the rainbow angle. Plot and compare to the classical calculation, $\sigma(\chi)$.

3.8. Generalize the concept of interfering, contributing cross sections described by Eq. (3.51) to calculate $\sigma^{sc}(\chi)$ for a molecular target consisting of two hard spheres. Assume they both have radius $\bar{r}$, are separated by $\bar{R} > 2\bar{r}$, and are aligned perpendicular to the incoming beam.

3.9. Compare the first-Born differential cross sections to the classical impulse approximation for the screened potentials of Eq. (3.63). Use the results in Appendix D. How well can Born cross sections fit the data in Figure 2.14? Discuss regions of validity.

3.10. Calculate the first-Born differential cross section for an incoming neutron colliding with a diatomic molecule aligned perpendicular to the incoming beam. Assume that the potential of interactions is a delta function (a hard sphere of radius zero in Problem 3.8) and the separation of the nuclei is $\bar{R}$.

3.11. Carry out the calculation in Problem 3.10 for arbitrary alignment of the molecule and calculate a rotationally averaged cross section.

3.12. Using Eq. (3.67), verify the results in Eqs. (3.68) and (3.72). Obtain the result in Eq. (3.72) for the ground-state hydrogen atom using the fourier transform of $\psi(\mathbf{r})$, Eq. (3.71), i.e.

$$\psi(\mathbf{p}) = \frac{1}{(2\pi)^{3/2}} \int \exp\left[(i\mathbf{p} \cdot \mathbf{r})/\hbar\right]\psi(\mathbf{r})\, d^3r$$

is the momentum wave function and $\rho(\mathbf{p}) = |\psi(\mathbf{p})|^2$.

3.13. Calculate the Boltzmann population of the lowest vibrational and rotational levels of H_2 and CO_2 at room temperature using appropriate parameters m, $\bar{R}_M$, and ω. How does this change at $2000°K$?

Suggested Reading

Much of the content of this chapter is well presented in a number of texts on molecular collisions or quantum mechanics, a few of which are listed below. The reader should be cautioned that many texts use Green function techniques to obtain the scattering amplitudes, a topic that we have not considered.

Introductory Material: Quantum Mechanics

H. W. MASSEY, *Atomic and Molecular Collisions*, Halsted Press, New York (1979), Chapters 1–2.

F. K. RICHTMEYER, E. H. KENNARD, and J. N. COOPER, *Introduction to Modern Physics*, 6th edn. McGraw-Hill, New York (1969), Chapters 11–14.

W. KAUZMAN, *Quantum Chemistry*, Academic Press, New York (1980), Part II.

P. TIPLER, *Foundations of Modern Physics*, Worth, New York (1969), Chapters 5–8.

Introductory Material: Scattering

D. RAPP, *Quantum Mechanics*, Holt, Rinehart and Winston (1970), Chapters 5, 7, 9, and 27.

M. S. CHILD, *Molecular Collision Theory*, Academic Press, New York (1974), Chapters 3–5.

J. B. Hasted, *Physics of Atomic Collisions*, 2nd edn., American Elsevier, New York (1972), Chapter 2.

B. H. Bransden, *Atomic Collision Theory*, W. A. Benjamin, New York, 1970, Chapters 1–3.

M. R. C. McDowell and J. P. Coleman, *Introduction to the Theory of Ion–Atom Collisions*, North-Holland, Amsterdam (1970), Chapter 2.

Semiclassical Scattering

M. R. C. McDowell and J. P. Coleman, *Introduction to the Theory of Ion–Atom Collisions*, North-Holland, Amsterdam (1970), Chapter 2.

K. W. Ford and J. A. Wheeler, *Ann. Phys.* 7, 259 (1959).

R. B. Bernstein, in *Molecular Beams (Advances in Chemical Physics)*, Vol. 10, ed. J. Ross, Wiley, New York (1966), Chapter 3.

Vibrational and Rotational States of Molecules

C. N. Banwell, *Fundamentals of Molecular Spectroscopy*, McGraw-Hill, New York (1966).

G. Herzberg, *Spectra of Diatomic Molecules*, 2nd edn., Van Nostrand-Reinhold, Princeton (1950), Chapter 3.

4

Interaction Potentials and Transition Probabilities

Introduction

Calculation of the angular differential cross section (for comparison with experiment or for use when experimental data are not available) requires a description of the interaction between the two colliding particles. This endeavor has kept physicists and chemists busy since the beginning of the century, the description being complicated by the fact that there are three different effects involved and, generally, many particles.

The primary effect is the coulombic force of repulsion or attraction of charged particles, which for many atomic systems is all that needs to be considered to obtain a reasonable description of the interactions. However, as there are generally a number of electrons and two nuclei involved, the net force may be rather complicated. The second effect is not so much a force, in the classical sense, as a property of electrons and nuclei at the atomic level. As any two *identical* atomic level particles (electrons or nuclei) cannot be distinguished by experiment, their calculated probability distributions should be identical. That is, in a theoretical description one should be able to interchange the position of any two electrons and obtain the same result for the probability amplitude to within a sign. This requirement restricts the possible distributions of electrons and is often referred to as the "exchange force." Such a "force" also occurs when the target and incident nuclei are identical, that is, they can be interchanged without changing the character of the collision, as when a proton and a hydrogen atom collide. Lastly, the electrons and nuclei have small intrinsic magnetic fields, the magnitudes of which are proportional to a quantity called the spin of the particle, as if the fields were created by a spinning charge. These magnetic fields interact with moving charges, e.g., orbiting electrons or colliding atoms, resulting in forces which generally, although not always, are smaller than the previous two.

To accurately describe the atomic behavior associated with the above effects, we resort to quantum mechanics. It was seen in our previous discussion (Chapter 3) that the wave motion of the heavy atomic centers could generally be treated classically, except for small-angle scattering. This is also true for very fast electrons only, as their light mass implies relatively longer wavelengths. Therefore it is useful in many instances to make the following separation: the motion of the bound electrons will be obtained from the quantum-mechanical wave equation, while the incident particle motion will be calculated classically. This separation, a special case of the Born–Oppenheimer separation, provides a starting point for understanding atomic-collision problems and has also been useful in interpreting molecular spectra.

Rather than calculate the net force on the collision centers, it is more convenient, as seen in our discussion of the deflection function, to obtain the net interaction potential. This interaction potential is the change in potential energy of the collision partners and clearly depends on the initial charge distributions of the colliding atoms. Each atom is known to have a set of states (charge distributions) characterized by a binding energy, total orbital angular momentum, and total spin. These states determine which inelastic energy losses may occur during the collision, that is, the possible final states of the colliding atoms. Associated with *each* pair of levels of the collision partners is an interaction potential, which makes the detailed solution of a collision problem almost impossibly complex. However, it also provides a continuous challenge to the imagination of the atomic and molecular physicist or chemist and is the source of the richness of effects associated with atomic and molecular phenomena. This complexity should not discourage the reader, as there are some obvious methods of simplification, which we will be discussing. The most obvious one, occurs when the potentials are all similar, suggesting that a common average potential can be used for all the states, an approximation we will resort to in a number of instances.

The actual interaction between collision partners is a dynamic interaction as at least one particle carries with it a cloud of electrons moving about the nucleus. Because these electrons are involved in the collision, this interaction is very different in *character* from the force between two bare nuclei. For two bare nuclei, the interaction at any given separation is the same whether the nuclei are standing still or moving (as long as the velocities are not near the speed of light and spin is ignored.) However, when two atoms collide, the electronic charge distributions change continuously. If the motion of the centers is slow, the electrons adjust to or follow this motion, and after the collision, essentially return to their initial configuration. This collision process is called adiabatic, as the electrons do not gain or lose energy. That is, even though the atoms may be deflected and change kinetic energies, their initial and final internal energies are the same, and the collision is elastic. If the atomic centers approach too rapidly, the electrons on each center cannot adjust to the changes occurring in the position of the other nuclei and electrons. Because of this, abrupt changes in the charge distribution (state) may occur, similar to those occurring when light is absorbed or emitted. The effective forces between the centers obviously must differ in these two speed regions, and therefore the force is velocity dependent, making the calculations at intermediate velocities rather complicated. In the latter case, although elastic collisions may occur, there is a high probability for inelastic collisions.

Rather than treating all collision velocities by a single theoretical method, the above discussion suggests we divide collisions into rough categories. Fast collisions are those for which the relative speeds of the particles are much larger than the average speeds of the bound electrons, v_e, i.e., $v \gg v_e$. For this case, the electron cloud can be thought of as frozen or static *during* the collision, except for those instances for which an abrupt change or transition occurs. The force between two atoms or an electron and an atom is, therefore, the force between *static* charge distributions.

At the other extreme (slow collisions, for which $v \ll v_e$), the electrons adjust continuously to the motion. For heavy particle interactions, when the two atomic centers are very far apart and standing still, a situation about which we can conjecture, the atoms can be thought of as distinct charge distributions. The interaction potential is simply the sum of the interactions between the charges in the two distributions and is calculated by procedures developed in classical electrostatics, as are the interactions of the frozen charge distributions discussed above. When the two atoms are brought together *slowly* to a new position, (Figure 4.1), their charge clouds distort, or are polarized, and the electrons redistribute themselves about the nuclei in a way determined by their initial (large R) atomic states. The net binding energy of the electrons changes in response to the distortion of the electron charge cloud, and the new charge distribution would be obtained from quantum mechanics as were the original separate, atomic charge dis-

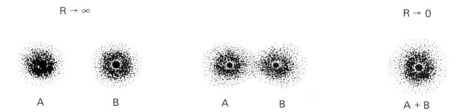

$R \to \infty$ $R \to 0$

A   B   A   B   A + B

Figure 4.1. Charge distributions on two nuclei at various internuclear separations, R.

tributions. The total potential energy change of the combined system is the sum of the increased nuclear repulsion and the change in electronic binding energy. This situation is continued until $R = 0$. For two atoms the coulomb force between the nuclei becomes infinite at $R = 0$ (i.e., the potential is infinitely steep), and the electrons distribute themselves as if a single atomic center of charge $(Z_A + Z_B)e$, a united atom, were formed. Of course, the situation $R = 0$ is never reached, particularly in low-energy collisions; however, the concept of a united atom is useful theoretically for estimating the potentials at small R values.

For the example in Figure 4.1, the force at large R is attractive as the electron distributions are distorted to lie between the nuclei. The binding energy increases, hence the potential decreases, with decreasing R until a minimum is passed and the force becomes repulsive. Such a potential with a minimum is typical for atoms in their lowest states. Adiabatic potentials can, in principle, be calculated for every orientation when molecules are involved, and such potentials are the basis for interpreting the spectra of diatomic and polyatomic molecules. Whereas fast incident electrons interact with an atom in much the same way as a bare nucleus of the same speed, slow collisions are quite different. In fact, when the incident electron speed is comparable to or less than the speeds of the bound electrons, the distinction between incident and bound electrons disappears in a close collision.

One problem with the above distinction between fast and slow collisions is that all electrons in the atoms do *not* have the same orbital speeds or periods. However, this can prove to be a distinct advantage. For instance, when a collision is fast with respect to the periods of the outermost electrons (outer-shell electrons), it may be adiabatic with respect to the inner-shell electrons. Therefore the inner-shell electrons return, after the collision, to their original state and only screen the interaction between the atoms. Alternatively, when considering inner-shell excitations that occur at much higher collision speeds, the outer-shell distributions can be considered frozen during the collision. For reference it is useful to remember that a proton with a speed equal to the speed of an electron in the ground-state orbit of hydrogen has an energy of about 25 keV, and, further, the orbital speed scales as Z', the effective nuclear charge seen by the bound electron.

Expanding on the notion, introduced earlier, of a collision time, τ_c, we define τ_c as

$$\tau_c \approx d/v \tag{4.1}$$

where d is the range of the interaction, roughly the size of the colliding atoms, $d = r_A + r_B$, introduced in Chapter 2. For large τ_c the collision will be nearly adiabatic (no energy change), whereas for short τ_c inelastic effects are likely. A rough estimate of the average energy transfer can be obtained using the collision time. The uncertainty principle states that for short times large uncertainties in energy can exist, the uncertainty being determined by

$$\Delta E \Delta \tau \geq h \qquad (4.2)$$

where h is Planck's constant. During the collison we are clearly unable, experimentally, to ascertain the charge distribution, and therefore the energy levels are uncertain by at least

$$\Delta E \sim h/\tau_{\rm c} \sim \left(\frac{h}{d}\right) v \qquad (4.3)$$

Equation (4.3) then gives an estimate of the average inelastic energy loss expected. It is seen to be proportional to the speed, which is consistent with the above discussion; i.e., as v goes to zero, the collisions become adiabatic.

For collisions between atoms or molecules in the two limiting cases— fast and slow collisions—charge distributions, and therefore interaction potentials, can be established at each value of R, the internuclear separation. In the following we obtain interaction potentials between two atomic systems, assuming that the initial charge distributions are known. Transition probabilities between the internal states of the colliding system are then estimated as a function of the speed, and inelastic cross sections are calculated from the transition probabilities. Those procedures appropriate for fast ion–atom collisions can also be used to approximate interaction potentials and transition probabilities in fast electron–atom collisions, where, in the discussion, the incident particle, A, is replaced by an electron. Slow electron collisions, however, are distinct in a number of ways, as is pointed out in the discussion. In much of the subsequent discussion we use a convenient set of units for atomic and molecular calculations appropriately called atomic units. These are defined such that $\hbar = 1$, $m_{\rm e} = 1$, and $e = 1$ and are described further in Appendix J.

Electrostatic Interaction between Atomic Charge Distributions

Consider two atoms, A and B, which have electronic charge distributions $\rho_A(\mathbf{r}_1)$ and $\rho_B(\mathbf{r}_2)$ and nuclei of charge $Z_A e$ and $Z_B e$, as shown in Figure 4.2, where $\mathbf{r}_1$ and $\mathbf{r}_2$ are electron position vectors about nuclei A and B respectively. The integrals, $\int \rho_A(\mathbf{r}_1) d^3 r_1 = N_A$ and $\int \rho_B(\mathbf{r}_2) d^3 r_2 = N_B$, give the total number of electrons on each of the atoms. For neutral atoms $Z_A = N_A$ and $Z_B = N_B$. The net force between the atoms is obtained from the coulombic potentials between all the constituents. For charge distributions which do not depend on R, i.e., frozen charge distributions, the potential is a sum of nuclear–nuclear, electron–nuclear, and electron–

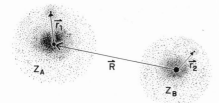

Figure 4.2. Coordinate system for Eq. (4.4); $\mathbf{r}_1$ and $\mathbf{r}_2$ are coordinates of the electron distributions on A and B respectively. Z_A and Z_B are the nuclear charges, and $\mathbf{R}$ is the internuclear separation.

electron interactions, $V_{AB}(R) = V_{NN} + \langle V_{eN} \rangle + \langle V_{ee} \rangle$. These have the form

$$V_{AB}(R) = \frac{Z_A Z_B e^2}{R} - Z_A e^2 \int \frac{\rho_B(\mathbf{r}_2)\,d^3 r_2}{|\mathbf{r}_2 - \mathbf{R}|} - Z_B e^2 \int \frac{\rho_A(\mathbf{r}_1)\,d^3 r_1}{|\mathbf{r}_1 + \mathbf{R}|}$$

$$+ e^2 \iint \frac{\rho_B(\mathbf{r}_2)\rho_A(\mathbf{r}_1)\,d^3 r_1 d^3 r_2}{|\mathbf{r}_1 - \mathbf{r}_2 + \mathbf{R}|} \tag{4.4}$$

where the brackets $\langle \ \rangle$, imply an averaging of the interaction over the charge distribution. The potential given in Eq. (4.4) is for two centers of charge, A and B, but is easily generalized to include more centers when molecules collide. The cases where ρ_A and ρ_B distort will be considered shortly. The integrals in Eq. (4.4) are treated extensively in any course on electricity and magnetism, and that presentation will not be repeated here. To obtain the general nature of $V_{AB}(R)$ for a particular pair of charge distributions, it is enough to consider the form of $V_{AB}(R)$ for the limiting cases: R very large and very small. Also, as the differential cross section is determined primarily by the shape of the potential near the distance of closest approach, it is often not necessary to know V_{AB} accurately at all R.

Defining the average radius of the atom as

$$\bar{r} = \int r\rho(\mathbf{r})\,d^3 r / N$$

we consider the centers far apart when $R \gg \bar{r}_1$, $R \gg \bar{r}_2$. Replacing r_1 and r_2 by zero in Eq. (4.4), we see that the long-range potential becomes

$$V_{AB}^L(R) \xrightarrow{R \to \infty} \frac{(Z_A - N_A)(Z_B - N_B)}{R} e^2 \tag{4.5}$$

which is zero unless the particles are ions. This does *not* mean that the interaction potentials between neutral atoms are zero at large R, but rather that they become zero faster than $1/R$. As $R \to 0$, the short-range expression from Eq. (4.4) (Problem 4.1) can be written as

$$V_{AB}^S(R) \xrightarrow{R \to 0} \frac{Z_A Z_B e^2}{R} - \frac{Z_A Z_B e^2}{a_{AB}} \tag{4.6}$$

where the first term is just the repulsion between the nuclei, and the second term, a constant, is the net change in potential energy of the electrons. This is written in terms of a characteristic length a_{AB}, a measure of the sizes of the charge distributions. For a bare ion, A, colliding with a neutral atom, B, having a hydrogenic charge distribution, we have

$$a_{AB} = a_0/Z \qquad (4.7)$$

where $Z = Z'_B$ is the effective nuclear charge of the atom, used to determine the charge distribution in Eq. (3.70). For two colliding atoms with identical hydrogenic charge distributions, i.e., $Z'_A = Z'_B$, Z in Eq. (4.7) is found from Eq. (4.4) to be $Z = \frac{8}{11}Z'_B$.

Separating the nuclei gradually from $R = 0$, we see that the nuclear repulsion will be partially screened by the intervening cloud of electrons. As the atomic charge density has an exponential behavior [e.g., Eq. (3.71)], it has become customary to approximate this potential by the screened coulomb form

$$V^S_{AB}(R) = \frac{Z_A Z_B}{R} e^2 \exp\left[-R/a_{AB}\right] \qquad (4.8)$$

which reduces to Eq. (4.6) as R approaches zero. Considerable effort has been expended in determining a_{AB} for many-electron atoms. Bohr first suggested using $a^B_{AB} = a_0(Z_A^{2/3} + Z_B^{2/3})^{-1/2}$, based on a Thomas–Fermi charge distribution, which Lindhard later showed should be multiplied by 0.8853. Based on a similar analysis, Firsov recommended using $a^F_{AB} = 0.8853 a_0(Z_A^{1/2} + Z_B^{1/2})^{-2/3}$, which differs from the Lindhard estimate by at most 10%. For light, singly ionized, ion–atom collisions the expression $a^s_{AB} = 2a_0(Z'_A + Z'_B)^{-1}$ has been used to obtain the screening constant. Here Z'_A and Z'_B are effective charges determined from the first ionization potential of the outer–shell electrons on A and B, e.g., $I_B = \frac{1}{2}Z_B'^2(e^2/a_0)$, where I_B is the ionization energy. When such screening lengths are used, potentials in the form of Eq. (4.8) can describe the static short-range interactions fairly accurately.

Examining the long-range force more carefully, we note that the denominators in Eq. (4.4) can be written

$$\frac{1}{|\mathbf{r} - \mathbf{R}|} = \frac{1}{R[1 + (r/R)^2 - 2(r/R)\cos\theta]^{1/2}} \qquad (4.9)$$

where θ is the angle between r and R. Expanding at large R, as in Eq. (3.57), but keeping higher-order terms, one finds

$$\frac{1}{|\mathbf{r} - \mathbf{R}|} = \frac{1}{R}\left\{1 + \cos\theta\left(\frac{r}{R}\right) + \frac{1}{2}\left(\frac{r}{R}\right)^2 (3\cos^2\theta - 1) + \cdots\right\} \qquad (4.10a)$$

We note that the angular terms are the same Legendre polynomials, Eq.

(3.15), used in the expansion of the scattering amplitude. The full expression can be written as

$$\frac{1}{|\mathbf{r} - \mathbf{R}|} = \frac{1}{R} \sum_{l=0}^{\infty} \left(\frac{r}{R}\right)^l P_l (\cos \theta) \tag{4.11}$$

where, at large R, each successive term is smaller as $|P_l| \leq 1$. We noted earlier that the multipole expansion of the scattering amplitude implied that, at successive impact parameters or l values, scattering occurred from different parts of a spherically averaged potential. Here the charge distributions are not necessarily spherical and the multipole moments reflect their shapes.

If there are no electrons on one of the centers, e.g., center A, the potential in Eq. (4.4) simplifies. Substituting Eq. (4.11) into the electron–nucleus interaction, we note that the long-range potential, when A is a bare nucleus or an electron, has the form

$$V_{AB}^L(R) = \frac{(Z_B - N_B)Z_A e^2}{R} - \frac{Z_A e^2}{R} \sum_{l=1}^{\infty} \frac{\overline{X_B^l}}{R^l} \tag{4.12}$$

where $\overline{X_B^l} = \int \rho_B(\mathbf{r}_2) r_2^l P_l (\cos \theta_2) d^3 r_2$ are the moments of the charge distribution about center $B, e\overline{X_B^1}$ is referred to as the dipole moment, often written μ_B, $e\overline{X_B^2}$ the quadrupole moment, etc. For a spherical charge distribution all moments but the zeroth vanish. (Although the atoms may initially have spherical charge distributions, they clearly will not remain so as they approach each other, a case we will treat shortly.) Molecules, which generally do not have spherical charge distributions, may have nonzero multipole moments, with the size of $\overline{X_B^l}$ depending on the orientation of the molecule. For the interaction of A with a diatomic molecule, the orientation in question is the angle between the axis of the molecule and the vector R between A and the molecular center (Figure 4.3).

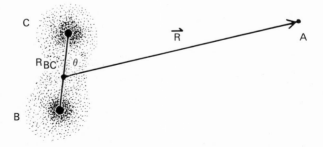

Figure 4.3. Coordinates for bare ion, A, interacting with a diatomic molecule, BC.

To complete the discussion, we return to the general case in which A has a charge distribution also. An expansion similar to that in Eq. (4.10a) is made for the electron–electron interaction in Eq. (4.4),

$$\frac{1}{|\mathbf{R} - (\mathbf{r}_2 - \mathbf{r}_1)|} = \frac{1}{R} \left\{ 1 + \frac{r_2 \cos \theta_2 - r_1 \cos \theta_1}{R} \right.$$

$$\left. + \frac{1}{2R^2} [3(r_2 \cos \theta_2 - r_1 \cos \theta_1)^2 - |\mathbf{r}_2 - \mathbf{r}_1|^2] + \cdots \right\}$$

(4.10b)

where $|\mathbf{r}_2 - \mathbf{r}_1|^2 = r_2^2 + r_1^2 - 2\mathbf{r}_1 \cdot \mathbf{r}_2$. The leading terms in the long-range potential in Eq. (4.4) can now be written, using Eqs. (4.10b) and Eq. (4.12), as

$$V_{AB}^L(R) = \frac{(Z_A - N_A)(Z_B - N_B)e^2}{R} \qquad \text{ion–ion}$$

$$- (Z_A - N_A)e^2 \frac{\overline{X_B^1}}{R^2} - (Z_B - N_B)e^2 \frac{\overline{X_A^1}}{R^2} \qquad \text{ion–dipole}$$

$$- (Z_A - N_A)e^2 \frac{\overline{X_B^2}}{R^3} - (Z_B - N_B)e^2 \frac{\overline{X_A^2}}{R^3} \qquad \text{ion–quadrupole}$$

$$- e^2 \frac{\overline{X_B^1 X_A^1}}{R^3} \qquad \text{dipole–dipole}$$

(4.13)

That is, the potential is the sum of the interaction of the multipole moments of the charge distributions for the two atoms, the zeroth moment being that of a point charge or ion. Here the dipole–dipole term has the form

$$e^2 \overline{X_B^1 X_A^1} \equiv \int \rho_A(\mathbf{r}_1) d^3 r_1 \int \rho_B(\mathbf{r}_2) d^3 r_2 (2z_1 z_2 - x_1 x_2 - y_1 y_2) \qquad (4.14)$$

where x_1, y_1, z_1 and x_2, y_2, z_2 are the components of $\mathbf{r}_1$ and $\mathbf{r}_2$ respectively. As before, the moments of spherical or spherically averaged charge distributions in Eq. (4.13) are all zero except for the ion–ion term. For neutral, molecular species the largest term in the long-range potential goes as $1/R^3$ and would depend on the relative orientation of the two molecules. Such an interaction depends on both A and B having permanent dipole moments. In the following section, we will allow the electronic charge distributions to distort. This produces additional contributions to the long- and short-range forces and is of particular interest when the charge distributions of A and B have no permanent moments.

Induced Moments: The Polarizability and the van der Waals Forces

When constructing an adiabatic potential, ρ_A and ρ_B depend on the separation R, and therefore the force is not the gradient of the potential in Eq. (4.4). To obtain the appropriate potential, one calculates the binding energy of the electrons in the combined system, $\varepsilon^{AB}(R)$. Now the net change in potential energy of the system, as described in the introduction to this chapter, is

$$V_{AB} = V_{NN} + [\varepsilon^{AB}(R) - \varepsilon^{AB}(\infty)] \tag{4.15}$$

where $\varepsilon^{AB}(\infty)$ is the initial binding energy of the electrons on the separated atoms at $R = \infty$. For the short-range behavior at $R = 0$, the nuclei can be imagined to be united, as stated earlier, and the total potential energy in Eq. (4.15) becomes

$$V_{AB}^S \xrightarrow{R \to 0} \frac{Z_A Z_B e^2}{R} + [\varepsilon^{AB}(0) - \varepsilon^A - \varepsilon^B] \tag{4.16}$$

where $\varepsilon^{AB}(0)$, ε^A, and ε^B are the binding energies of the combined system and the separated atoms respectively, i.e., $\varepsilon^{AB}(\infty) = \varepsilon^A + \varepsilon^B$. The term in brackets in Eq. (4.16) can be used now to obtain the screening constant a_{AB} in Eq. (4.6) for the *adiabatic* potential at small R. For example, $a_{AB} = \frac{2}{3}a_0$ for $H^+ + H$, where He^+ is the united atom.

The adiabatic behavior at long range can be estimated from Eq. (4.15) or by expanding the *force* between distorted charge distributions in powers of $1/R$ and constructing a corresponding potential. For the latter case, the procedure followed is the same as that used in the previous section. Now, however, the moments of the position vectors $\mathbf{r}_1$ and $\mathbf{r}_2$ are calculated using the distorted charge distributions on each center. These charge distributions can be estimated by treating the electrons as classical oscillators, a procedure outlined at the end of Chapter 3. The interaction of a charge with a classical oscillator is treated in Appendix C. Here we use only the simple result that, if the electron oscillator is placed in a constant field, like the field of the neighboring atom, the position vector is shifted by an amount proportional to the applied field, $\mathscr{E}$,

$$\mathbf{r}(t) = \mathbf{r}_0(t) - \frac{e\mathscr{E}}{m_e \omega^2} \tag{4.17}$$

In the above expression, $\mathbf{r}_0(t)$ is the initial position vector of the electron, ω is the frequency of the oscillator, and the nuclear position is assumed fixed. The field experienced by B due to A depends on the charge distribution on atom A. This distribution is in turn distorted by the field of B and, therefore, the effective potential is obtained iteratively, being careful with orders of $1/R$.

In this example (i.e., using classical oscillators rather than quantum-mechanical charge distributions), the moments of the charge distribution are time averages of various powers of the position vector of the electron. For an initially spherical charge distribution, we have $\langle \mathbf{r}_0(t) \rangle_t = 0$, where the brackets indicate a time average over the position vector of the oscillator. The field set up by atom A, $\mathscr{E}_A$, induces a net dipole moment in the charge distribution of oscillator B,

$$e\langle \mathbf{r}_2(t) \rangle_t = -\alpha_B \mathscr{E}_A \tag{4.18}$$

where, from Eq. (4.17), the quantity $\alpha_B = e^2/m_e \omega^2$ is the polarizability of oscillator B. The quantum-mechanical system can be thought of as made up of a collection of oscillators, each corresponding to a frequency ω_k, observed in the atomic spectrum, and each weighted by an oscillator strength f_k. The sum of these oscillator strengths,

$$\sum_k f_k = N_B \tag{4.19}$$

is set equal to the number of electrons in the atom. Now the total polarizability of the atom is a weighted sum of the polarizabilities of the individual oscillators:

$$\alpha_B = \frac{e^2}{m_e} \sum_k f_k/\omega_k^2 \tag{4.20}$$

Employing our earlier notation, we write the dipole moment along R induced in B by the field of the bare ion A or electron, using Eq. (4.18), in the form

$$(e\overline{X_B^1})_i = \alpha_B \left(\frac{Z_A e}{R^2} \right) \tag{4.21}$$

Since the long-range *force* would have the same form whether B had an intrinsic or induced dipole moment, the potential corresponding to the dipole moment in Eq. (4.21) is easily shown to be

$$V_{AB}^L = \frac{Z_A(Z_B - N_B)e^2}{R} - \frac{\alpha_B}{2} \frac{(Z_A e)^2}{R^4} + \cdots \tag{4.22}$$

In taking the gradient of V_{AB}^L, we now differentiate the induced moment, as well as the field. Therefore, Eq. (4.22) differs from Eq. (4.12) or Eq. (4.13) by the factor $1/2$ in the ion–dipole interaction term. Equation (4.22) indicates that the largest induced term in the potential at *large* R, due to distortions of the charge distribution, goes as $1/R^4$. This can be generalized to the case where B is not spherically symmetric. One writes the one-electron dipole moment as a permanent plus an induced moment

$$e\langle \mathbf{r}_2(t)\rangle = e\langle \mathbf{r}_{2_0}(t)\rangle_t + \mathbf{a}_B \cdot \mathscr{E}_A$$

and the *induced* part of V_{AB}^L,

$$(V_{AB}^L)_i = -\tfrac{1}{2}\mathscr{E}_A \cdot \mathbf{a} \cdot \mathscr{E}_A + \cdots \tag{4.23}$$

is added to Eq. (4.12). The polarization $\mathbf{a}$ in any direction depends on the distribution of charge in all other directions, and hence the polarizability is a tensor quantity. The reader should verify that the angular dependence of the polarizability interaction, when B is a diatomic molecule, has the form

$$(V_{AB}^L)_i = -\tfrac{1}{2}(\cos^2\theta\,\alpha_{zz} + \sin^2\theta\,\alpha_\perp)\,\frac{Z_A^2 e^4}{R^4}$$

where $\alpha_\perp$ and α_{zz} are the polarizabilities perpendicular to and along the axis of the molecule, $[\alpha = (\alpha_{zz} + 2\alpha_\perp)/2]$.

From Eq. (4.20), the polarizability of an atom is seen to be inversely proportional to how tightly the electron is bound via ω_j. For an atom involving a number of electrons N_j in each shell, j, we can very roughly estimate the average polarizability in atomic units as

$$\alpha_B \sim \sum_j N_{Bj} I_j^{-2} \tag{4.24}$$

where I_j is the energy in atomic units required to remove an electron from the jth shell. Values of polarizabilities for a number of atoms and molecules are given in Table 4.1 and compared to this estimate, which shows the general behavior only. It is seen that the closed-shell systems, like He and Ne, have lower polarizabilities than their neighboring open-shell atoms, and the loosely bound alkali metals and larger molecules have large values of α.

Table 4.1. Atomic and Molecular Polarizabilities[a]

Atom	$\alpha(a_0^3)$	Eq. (4.24)	Molecule	$\alpha(a_0^3)$	$\bar{X}^1(a_0)$
H	4.5	4.0	H_2	5.45	0
He	1.37	1.5	H_2O	9.8	0.72
Li	165.0	26.0	NH_3	14.6	0.58
N	7.6	4.6	O_2	10.6	0
Ne	2.5	2.4	CO	13.1	0.05
Na	165.0	29.0	CO_2	18.0	0
Ar	11.1	5.0	SO_2	25.5	0.64

[a] Atomic polarizabilities: see R. R. Teachout and R. T. Pack, *Atomic Data* **3**, 195 (1971) which contains results and references. Molecular polarizabilities and dipole moments: see J. B. Hasted, *Physics of Atomic Collisions*, 2nd edn., American Elsevier, New York (1972), p. 732, which contains primary references and results for other systems. See also: C. W. Allen, *Astrophysical Quantities*, 2nd edn., The Athlone Press, University of London (Oxford University Press, New York), p. 87 (1963). The estimate from Eq. (4.24) was obtained using removal energies from C. E. Moore, *Atomic Energy Levels*, Vol. 1 (1949) and Vol. 2 (1952), National Bureau of Standards, Washington, D.C.

Permanent dipole moments are also given in Table 4.1 for the non-symmetric molecules.

When center A is an atom or molecule rather than a bare ion, $\mathscr{E}_A$ in Eq. (4.18) depends on the distribution of the electrons on A. The induced dipole moment of B then interacts with the permanent moments of A and vice versa, yielding terms of higher order in $1/R$. The proliferation of terms should not lead the reader to despair. Except in rare circumstances one uses only the largest one or two terms in the expansion. If, in fact, a large number of terms are required to describe the potential accurately, the expansion method is *not* appropriate and the interaction energies should be calculated by an alternative scheme.

Even if A and B are both neutral *and* spherically symmetric, a long-range interaction will occur, as their charge distributions are symmetric in an average sense only. In fact the positions of the electrons fluctuate continuously. The dipole–dipole interaction for frozen charge distributions was a result of averaging an expression involving the coordinates r_1 and r_2 over *independent* charge distributions in Eq. (4.14). If the charge distributions fluctuate, then the instantaneous position of $r_2(t)$ is affected by the field $\mathscr{E}_A$, which in turn depends on $r_1(t)$. Therefore, the instantaneous charge distributions are not independent, producing a nonvanishing, average interaction between the induced dipoles. Note that, from the ion-induced dipole example, the instantaneous field of each oscillator is proportional to $1/R^3$, indicating that the induced dipole–induced dipole interaction is of order $1/R^6$,

$$(V_{AB}^L)_d = -C_{VW}/R^6 + \cdots \tag{4.25}$$

The result in Eq. (4.25) is the well-known van der Waals interaction used to describe differences in the thermodynamic behavior between a "real" gas and an ideal gas. It is the first nonvanishing term for neutral, symmetric systems and is called a dispersion interaction (hence the subscript d). That is, the interaction depends on the oscillator strengths of atoms A and B, which also determine the dispersion of light by these atoms. The van der Waals coefficient for spherical atoms has the form

$$C_{VW} = \frac{3}{2} \frac{e^4 h}{m_e^2} \sum_i \sum_j \frac{f_j^A f_i^B}{\omega_i \omega_j (\omega_i + \omega_j)} \tag{4.26}$$

where the sums are over the oscillator strengths for the two atoms. The quantity C_{VW} can be estimated from the polarizabilities using the expression

$$C_{VW} \sim \tfrac{3}{2} \alpha_A \alpha_B (\bar{I}_A^{-1} + \bar{I}_B^{-1})^{-1}$$

Here $\bar{I}_A^{-1}$ and $\bar{I}_B^{-1}$ are the averages of the inverse of the ionization potentials of centers A and B respectively. That is, $\bar{I}_A^{-1} = \sum N_{Ai}/I_{Ai}$.

These long-range forces can be combined with the repulsive short-range force discussed earlier to construct semiempirical potentials for all R. As an example, we consider two often-used expressions. The first, for neutral symmetric systems, is the Lennard-Jones six-twelve potential [Eq. (2.62)]

$$V_{AB} \sim V_M[(R_M/R)^{12} - 2(R_M/R)^6] \tag{4.27}$$

Here the constants R_M and V_M are determined by fitting to data and/or matching with theoretical expressions. Although the repulsive part of Eq. (4.27) is not a very accurate approximation to the short-range potentials discussed earlier, the power-law form is useful, as we have seen, in estimating cross sections determined from potentials with a minimum. The constants R_M and V_M are the positions and depth of the minimum which may be obtained, for instance, from the molecular spectra of bound molecules. As the long-range force gradually gives way to an exponentially repulsive potential at small R, an expression used in ion–atom collisions is

$$V_{AB} \sim \frac{Z_A Z_B e^2}{R} \exp\left(-\frac{R}{a_{AB}}\right) - \frac{\alpha_B}{2R^4}(Z_A e^2)^2 \left[1 - \exp\left(-\frac{R}{a'_{AB}}\right)\right]^n \tag{4.28}$$

where the constants a'_{AB} and n are chosen to reproduce the minimum in the potential with $n \geq 4$. There have been many attempts to describe interaction potentials with analytic forms involving a few parameters, and the reader is encouraged to look into the many excellent texts on quantum chemistry. A word of caution, however, is in order: Trying to describe a complex interaction over a broad range of internuclear separations using a couple of parameters is a hopeless procedure unless the accuracy required is not great. The best procedure, which was pointed out earlier, is to describe the potential over the small range of internuclear separations of primary importance for a particular energy regime. As a final reminder, the classical methods employed here are also limited by the fact that the constants, such as α and C_{VW}, are in fact quantum-mechanical entities and, therefore, can only be evaluated accurately from experiment or by a wave-mechanical description of the atoms and molecules. We consider this in the following section, where the potentials found are identical in form to those above, but expressions for calculating quantities like the oscillator strength are obtained also.

Wave-Mechanical Treatment of the Interactions

Before proceeding further, we briefly examine the wave-mechanical treatment of the induced interactions. We remind the reader that in the Born–Oppenheimer approximation the electrons are treated separately as moving in the field set up by the nuclei. The wave equation for electrons on fixed nuclear centers has the form

$$\left\{ -\frac{\hbar^2}{2m_e} \sum_i \nabla_i^2 + V_{eN} + V_{ee} \right\} \psi^{AB} = \varepsilon^{AB}(R)\psi^{AB} \tag{4.29}$$

where the sum is over all electrons in the combined system, and ψ^{AB} depends on the electron coordinates and the separation R. The quantity $\varepsilon^{AB}(R)$ is the total electronic binding energy of the combined system, and the term in brackets is the energy or Hamiltonian operator for the electrons, which we will write symbolically as H_e. After solving for $\varepsilon^{AB}(R)$ in Eq. (4.29) we can obtain the interaction potential from Eq. (4.15); that is, $V_{AB} = V_{NN} + [\varepsilon^{AB}(R) - \varepsilon^{AB}(\infty)]$.

When considering the long-range induced interaction between atoms A and B, it is useful to break H_e into three parts:

$$H_e = H_A + H_B + V_{AB}^e \tag{4.30}$$

In Eq. (4.30), H_A and H_B are the Hamiltonians for the separate atoms containing the interactions between the nuclei and electrons of each atom. V_{AB}^e is the electronic part of the interaction potential we have been discussing, *prior* to averaging over electronic charge distributions in Eq. (4.4). In the following, we again consider the case of a bare charge interacting with neutral atom B. For this situation, $H_e = H_B + V_{AB}^e$, and at large R, as $V_{AB}^e \to 0$, $\psi^{AB} \to \psi_j^B$. That is, the wave function of the combined system must become an atomic wave function placing the electrons on B in some state j. Therefore, ψ^{AB} itself should assume the labels of the atomic state of B with which it correlates at large R, i.e., we write ε_j^{AB} and ψ_j^{AB}. In like manner $\varepsilon_j^{AB}(R) \xrightarrow{R \to \infty} \varepsilon_j^B$, where the atomic wave function ψ_j^B is a solution of the atomic wave equation

$$H_B \psi_j^B = \varepsilon_j^B \psi_j^B \tag{4.31}$$

Treating V_{AB}^e as nonzero but small at large R, we know that the effect of the ion A is to distort the electrons on atom B from their original configuration or state. Therefore, the wave function ψ_j^{AB} becomes an admixture of the atomic states for the electrons on B,

$$\psi_j^{AB} = \sum_k c_k(R)\psi_k^B \tag{4.32}$$

where the $c_k(R)$ are coefficients that change as R decreases. Clearly, when $R \to \infty$, we must require that

$$\begin{aligned} c_k(R) &\to 0 \qquad k \neq j \\ c_j(R) &\to 1 \end{aligned} \tag{4.33}$$

to establish the conditions described above. Mathematically the expansion in Eq. (4.32) is equivalent to saying that the solutions of Eq. (4.31) for the electrons on B form a complete orthogonal set of functions. That is, any perturbed or distorted wave function for the electrons can be described by a sum of the unperturbed wave functions. By the orthogonality of the solutions to the Schrödinger equation, there is no net overlap of the wave functions of H_B when integrated over all positions of the electron, much like the orthogonality of the Legendre polynomials in Eq. (3.32). This condition is expressed as

$$\int \cdots \int \psi_j^{B*} \psi_k^B \, d^3 r_1 \, d^3 r_2 \cdots d^3 r_{N_B} = \delta_{jk} \qquad (4.34)$$

where $r_1, r_2, \ldots, r_{N_B}$ are the position vectors for all the electrons on B. Substituting ψ_j^{AB} as expressed in Eq. (4.32) into the Hamiltonian $H_B + V_{AB}^e$, and using Eq. (4.31), one obtains

$$\sum_k c_k(R)\{\varepsilon_k^B + V_{AB}^e\}\psi_k^B = \varepsilon_j^{AB} \sum_k c_k(R)\psi_k^B$$

An expression for any one of the coefficients, $c_k(R)$, is obtained by multiplying by ψ_i^{B*} on the left and integrating over all the electron coordinates, as in Eq. (4.34). This procedure yields

$$c_k(R)[\varepsilon_j^{AB}(R) - \varepsilon_k^B] = \sum_n c_n(R)\langle V_{AB}^e \rangle_{kn} \qquad (4.35)$$

where the averaging implied by the brackets $\langle \ \rangle$ means

$$\langle V_{AB}^e \rangle_{kn} = \int \cdots \int \psi_k^{B*} V_{AB}^e \psi_n^B \, d^3 r_1 \cdots d^3 r_{N_B} \qquad (4.36)$$

The differential equation, Eq. (4.29), with which we started, has been reduced in Eq. (4.35) to a coupled set of linear equations which can be solved by standard matrix methods. For example, $\varepsilon_j^{AB}(R)$ is determined exactly by diagonalizing the determinant $|\langle V_{AB}^e \rangle_{kn} + \varepsilon_k^B \delta_{kn}|$. [As there are an infinite number of states the set used in Eq. (4.32) has to be truncated.] To estimate ε_j^{AB}, we note that when $\langle V_{AB} \rangle_{kn}$ is small all the coefficients $c_k(R)$ are small, except of course $c_j(R)$ which remains approximately unity. Setting all the coefficients except c_j equal to zero in Eq. (4.35), we obtain a first estimate of ε_j^{AB}:

$$^1\varepsilon_j^{AB} = \varepsilon_j^B + \langle V_{AB}^e \rangle_{jj} \qquad (4.37)$$

This shows that the change in the binding energy, $\langle V_{AB}^e \rangle_{jj}$, is just the electronic contribution to the static potential we calculated classically! That is, replacing $\rho_B \rightarrow |\psi_j^B|^2$, which is the charge distribution of the initial state j, we see that the total interaction potential, $V_{AB} \approx V_{NN} + \langle V_{AB}^e \rangle_{jj}$, has the form of Eq. (4.4) if A is a bare charge.

The first-order estimate of ε_j^{AB} in Eq. (4.37) does not give any information on the effect of distortion of the wave function as the coefficients c_k, $k \neq j$, which account for the alteration of the charge cloud, are set to zero.

To account for the distortion of the charge cloud, improved estimates of the coefficients $c_k(R)$ are needed, and we approach this using an iterative scheme to solve the set of equations in Eq. (4.35). Assuming the correction to $\varepsilon_j^{AB}(R)$ due to distortion is small, and all coefficients c_k are small except c_j, we find a first estimate of the *coefficients* from Eq. (4.35):

$$c_k^1(R) = -V_{kj}/(\varepsilon_k - \varepsilon_j) \qquad k \neq j \qquad (4.38)$$

For convenience, we have dropped the labels B on the energies and have written $V_{kj} = \langle V_{AB}^e \rangle_{kj}$. Using the coefficients from Eq. (4.38) in Eq. (4.35), we find an improved estimate of ε_j^{AB},

$$\varepsilon_j^{AB}(R) = \varepsilon_j + V_{jj} - \sum_{k \neq j} \frac{|V_{kj}|^2}{\varepsilon_k - \varepsilon_j} \qquad (4.39)$$

Continuing this procedure, we can obtain better and better estimates of the coefficients and the total binding energy. Each improvement will depend on a higher power of V_{AB}^e and therefore should decrease in importance if the interaction is weak, as it is at large R.

The second term in Eq. (4.39) is the static interaction, and the third term can be related to the induced interactions discussed earlier. The latter term arises from the changes in the charge distribution, i.e., changes in the coefficients c_k. The correction to the binding energy can be calculated exactly, or at large R can be estimated by expanding V_{AB}^e in powers of $1/R$. Substituting the expansion in Eq. (4.10a) into V_{kj} yields

$$V_{kj} = \frac{1}{R^2} \left\langle \sum_{i=1}^{N_B} z_i \right\rangle_{kj} + \cdots$$

Here $z_i = r_i \cos \theta_i$, the coordinates of the electrons on B along the internuclear axis, are averaged over the wave function as in Eq. (4.36). Substituting this expression for V_{kj} into Eq. (4.39), we see that the largest term for the ion–atom interaction goes as $1/R^4$, as found in the previous section. Now, however, the polarizability of B in state j can be specified and, presumably, calculated. That is, using the above expression for V_{kj}, we obtain the polarizability

$$\alpha_B = 2e^2 \sum_{k \neq j} \frac{\left| \langle \sum_{i=1}^{N_B} z_i \rangle_{kj} \right|^2}{\varepsilon_k - \varepsilon_j} \qquad (4.40a)$$

Comparing Eq. (4.40a) with Eq. (4.20), we see that the oscillator strengths for an initial state j, which we introduced in an *ad hoc* manner before, have the form

$$f_{kj} = \frac{2m_e}{\hbar^2} (\varepsilon_k - \varepsilon_j) \left| \left\langle \sum_{i=1}^{N_B} z_i \right\rangle_{kj} \right|^2 \qquad (4.40b)$$

where the frequency of the oscillator is $\omega_{kj} = (\varepsilon_k - \varepsilon_j)/\hbar$. It is left as a problem for the reader to verify that the property of the oscillator strengths $\sum_k f_{kj} = N_B$ follows directly from Eq. (4.40b). The quantum-mechanical expression for the van der Waals coefficient, C_{VW}, can be obtained by generalizing the above discussion to the case in which both atoms have charge, a problem also left for the reader, Problem 4.9.

The perturbation or iteration method above yields the static and induced interactions as separate corrections to the binding energy of the bound electrons at large R. When the distortion in atom B due to center A is small, the effect is to mix in other possible charge distributions of center B. The amount of mixing is seen in Eq. (4.39) to be proportional to the square of the interaction potential and inversely proportional to the difference in binding energy between states. Therefore, states close in energy have the greatest effect on the charge distribution *if* the quantity $\langle V_{AB}^e \rangle_{kj}$ is not zero, a fact we will return to below. Further, included in the sum in Eq. (4.39) are states that are solutions to Eq. (4.31) for which B is ionized, as well as the bound states of B. Such states, in fact, form a continuum of energy levels, in which case the sums in Eq. (4.39) and Eq. (4.40a) should be changed to integrals. For simplicity of notation, the sums over the bound levels and integrals over the continuum levels were indicated jointly by $\sum_k$, as is customary.

Implicit in the perturbation approach was the fact that the corrections to ε_j^{AB} were small. At certain values of R for which either the energy spacing between two states is very small or the coupling interaction very large, the perturbation approach used above is not valid. In these regions $\psi_j^{AB}(R)$ often can be approximated by only two states, that is, the two closely spaced or strongly interacting states. If only two states are considered, the binding energies from Eq. (4.35) are easily solved for by diagonalizing a two-by-two determinant. This procedure yields

$$\varepsilon_{\pm}^{AB}(R) = \tfrac{1}{2}(\varepsilon_j + V_{jj} + \varepsilon_k + V_{kk})$$
$$\pm \tfrac{1}{2}[(\varepsilon_j + V_{jj} - \varepsilon_k - V_{kk})^2 + 4|V_{kj}|^2]^{1/2} \qquad (4.41)$$

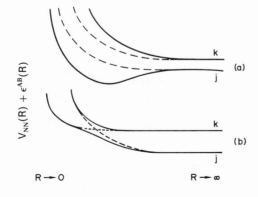

Figure 4.4. Hypothetical potential energy diagrams for two strongly interacting states; $V_{NN} +$ binding energy vs R; dashed curves are the static binding energy $(\varepsilon_j + V_{jj})$ and $(\varepsilon_k + V_{kk})$; solid curves are the adiabatic binding energy ε_+^{AB} and ε_-^{AB} from Eq. (4.41) when $|V_{kj}| \neq 0$.

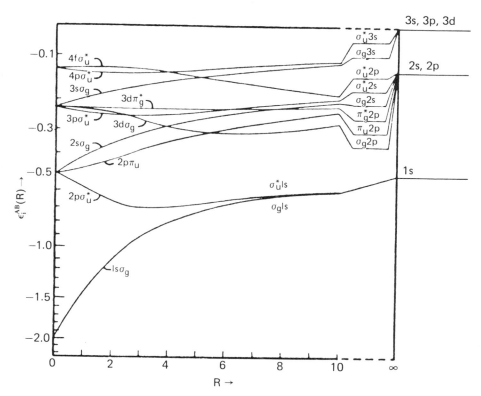

Figure 4.5. Correlation diagram for binding energies ε_i^{AB} of $H^+ + H$ or the H_2^+ molecular ion energies and separations in atomic units.

where the plus corresponds to the upper state, and the minus the lower state, either k or j. If the potential is again assumed to be weak, an expression for the polarizability involving only two states is recovered. Two cases for which the potential is not weak are indicated in Figure 4.4 in which are plotted hypothetical potential curves using Eq. (4.41). In the first example (a), the static estimates of the binding energies, $\varepsilon_j + V_{kk}$, are similar and in the second (b) they cross (become degenerate) at some value of R. In both cases, the effect of the distortion is to cause the adiabatic potential curves to separate from each other if, of course, V_{kj} is nonzero. The fact that potential curves associated with states of the same symmetry, that is, those for which V_{kj} is nonzero, do *not* cross allows one to correlate states of the separated atoms uniquely with the states of the united atom. In Figure 4.5 is given the correlation diagram for the binding energies of an $H^+ + H$ system or the H_2^+ molecular ion. The states at large separation are those associated with the H atom, and those at small separation are associated with He^+. We will consider such diagrams further and will discuss the symmetries and labeling of the states below.

State Identification and Molecular Symmetries

In Appendix G the labeling and ordering of the states of atoms is reviewed. This procedure is extended here to diatomic molecules, the labeling of which will be applied to the interaction potentials between atoms. The total angular momentum, which is a well-defined quantity in a molecule, is a combination of electronic and nuclear angular momentum. In the Born–Oppenheimer approximation these can be considered separately with the nuclear angular momentum being determined by the rotational state of the diatomic, discussed in Chapter 3. For a slow collision between atoms, the corresponding quantity is the collisional angular momentum of the nuclei, $L = mvb$. Similarly, diatomics are labeled according to their vibrational energy, which in a collision, an unbounded system, corresponds to the kinetic energy of radial motion, $m\dot{R}^2/2$. The remaining problem, therefore, is the labeling of the electronic states of the diatomic molecule.

As the electrons in a diatomic move in a field which is *not* spherically symmetric, the total electronic angular momentum is not a quantity that characterizes the state for all R. However, the nuclear potential is symmetric with respect to rotations about the internuclear axis, hence the component of angular momentum about this axis is a well-defined quantity, i.e., one that does not depend on R. Even for a single-electron diatomic molecule, there are two schemes for specifying this angular momentum and ordering the states according to energy. These schemes are based on the atomic states considered earlier. The molecular state can be imagined as being constructed by bringing two atoms together or by separating a single large atom into two nuclei. These were referred to as the separate- and united-atom limits of the molecule and were the basis for our constructing estimates for molecular potentials in the preceding sections.

In both the separate- and united-atom limit the weak electric field along the internuclear axis splits the degenerate, single-electron atomic states. This is referred to as a Stark effect. The substates of angular momentum, l, separate in energy due to this field according to the m_l values, with the lowest $|m_l|$ lying lowest in energy. In the notation of molecular spectroscopy, σ, π, δ, ... are used for $|m_l| = 0$, 1, 2, ... The single-electron state is also labeled according to the atomic state of origin at either $R = 0$ or $R \to \infty$. For instance, a $(2p)$ state of a Li^{+2} united ion having $|m_l| = 0$, 1 can be imagined to split into $(2p\sigma)$ and $(2p\pi)$ states of a HeH^{+2} diatomic ion. The corresponding notation in the separate-atom limit is: a $He^+(2p)$ state is split by the field of a proton into a $(\sigma 2p)_{He}$ or $(\pi 2p)_{He}$ state of

HeH^{+2}, where the subscript labels which center the electron was on at infinite separation.

The problem remains to correlate the states created at large R with those at small R. As R varies, the binding energy changes, according to our earlier discussion, but the $|m_l|$ label remains unchanged. The correct correlation is obtained by matching states of the same $|m_l|$ at large and small R in such a way that the potentials associated with states of the same symmetry (here $|m_l|$) do not cross [viz. Eq. (4.41)]. Such a correlation diagram is shown in Figure 4.6 where the states are ordered according to their binding energy, $\varepsilon_j^{AB}(R)$. These binding energies are drawn schematically as simple lines to indicate the general nature of the change in energy as R

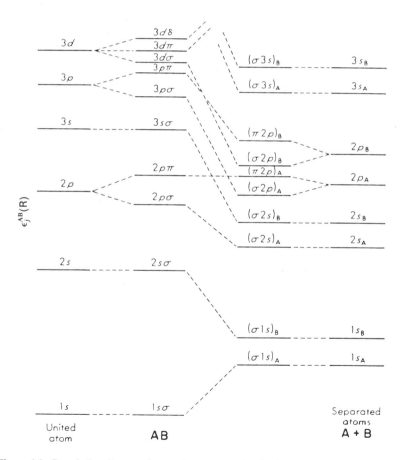

Figure 4.6. Correlation diagram for an electron associated with two centers of charge Z'_A and Z'_B. Vertical axis, electronic binding energy; horizontal axis, R.

changes. Interaction potentials are obtained by adding the nuclear repulsion, as in Eq. (4.15).

If the nuclei are identical, then an additional symmetry is present, as the potential experienced by the electrons does not change when *all* the coordinates are reversed. The states themselves, therefore, can be either symmetric (gerade, g) or antisymmetric (ungerade, u) with respect to this reversal (i.e., the probability densities are unchanged). The molecular orbital energies for H_2^+ were shown in Figure 4.5. Interactions for symmetric systems are discussed in the following section.

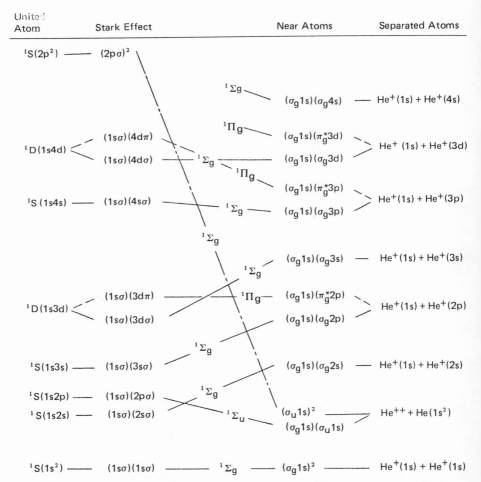

Figure 4.7. Correlation diagram for He_2^{+2} gerade states and lowest ungerade state based on orbitals in Figure 4.5, modified so orbitals of the same symmetry don't cross. Dot-dash line: diabatic state associated with "promoted" electrons. Vertical axis, electronic binding energy; horizontal axis, R.

A building-up principle, similar to that in Appendix G, is now used to construct multielectron states, a procedure referred to as the molecular orbital method. In this method, the total angular momentum along the axis of the diatomic molecule is a well-defined quantity, labeled $\Sigma, \Pi, \Delta, \ldots$ according to $|\sum_i (m_l)_i| = 0, 1, 2, \ldots.$ The Π, Δ, etc. states are doubly degenerate as $\sum_i (m_l)_i$ can be positive or negative. In addition, as the potential is symmetric with respect to reflection about any plane containing the internuclear axis, the nondegenerate Σ states are labeled either Σ^- or Σ^+, indicating whether the wave function changes sign or does not change sign under such a reflection. As an example, the lowest state of HeH at small R is designated $(1s\sigma)^2(2s\sigma)$ $^2\Sigma^+$, where the superscript 2 indicates a spin doublet. The lowest state at large R is $(\sigma 1s)^2_{He}(\sigma 1s)_H$ $^2\Sigma^+$. As there will exist a number of states of the same symmetry, it has become customary to label the ground state by X and subsequent states of higher energy by A, B, C, to indicate the order. States differing in symmetry from the ground state are ordered in energy with the labels $a, b, c, \cdots$.

The correlation diagram for the lowest gerade states of the He_2^{+2} molecular ion is drawn schematically in Figure 4.7 from the molecular orbitals in Figure 4.5. Included also is the lowest ungerade state, which plays a role in the He^{++} on He collisions. Like the atomic case, the orbital labels in a multielectron molecule give a useful, approximate picture of the molecule. As the correlation between the *motions* of the individual electrons is often weak, Lichten and Fano have shown that the molecular-orbital labels are quite useful in understanding transitions occurring during the collision. For instance, in a fast collision the individual electrons may follow, with high probability, the dashed curve in Figure 4.7, often referred to as a diabatic curve, which correlates with a highly excited state of the united atom and crosses many states of the same symmetry. Since there are two orbitals at large R for each orbital of the same symmetry at small R, many such crossings will occur, a fact referred to as electron promotion. These crossings will be considered further in our discussion of transition probabilities.

The Exchange Interaction

Perturbation methods are useful for calculating the binding energy when the electronic charge distribution of one center does not overlap significantly with that of the other center. As R decreases, the probability of finding an electron from B in the vicinity of A increases. In the polarizability calculation above, this is described by the more weakly bound and continuum states in the sum. These have broad charge distributions that overlap significantly with center A. In estimating the long-range interaction, the coefficient of such states is small. At very small R the overlap between the electrons on A and on B is complete as the two centers merge. At intermediate separations, when A begins to penetrate the charge cloud of B significantly, the successive corrections in the perturbation approach become significant; that is, the coefficients c_k, $k \neq j$, are not small. At intermediate R, it is therefore advantageous to include in the approximation to ε_j^{AB} wave functions for which the electrons *are* on center A.

From the energy level diagram for $H^+ + Ar$ in Figure 4.8, the state closest in energy to the ground state is the lowest state of $H + Ar^+$. Based on our previous discussion, this state will play an important role in determining the distortion of the charge cloud when the interaction matrix V_{jk} between the two states is significant. In treating such a state, extra care must be taken because the wave function which places one electron on H and the rest on Ar^+ is not orthogonal to the wave functions for which all the electrons are on Ar. The integral of the form of Eq. (4.34) involving the ground states of $H + Ar^+$ and $Ar + H^+$, written symbolically as $\langle O_A | O_B \rangle$, is a measure of the overlap of these charge distributions. Both the overlap integral and the average interaction potential, $\langle O_A | V_{AB}^e | O_B \rangle$, between the states decrease exponentially with increasing R because of the exponential nature of the atomic wave functions, e.g., Eq. (3.71). That is,

$$\langle O_A | O_B \rangle \sim \text{const} \cdot \exp[-\tfrac{1}{2}(Z_A' + Z_B')R/a_0] \qquad (4.42a)$$

where Z_A' and Z_B' are effective charges for the *extra* electron being on H and Ar respectively; in this example $Z_A' = 1$. Including such a state in the expansion of ψ_j^{AB} would not change our previous results at large R, as the coefficient associated with this state would also decrease exponentially with increasing R. Therefore it was correct to ignore it when calculating long-range potentials even though the energy difference between the states is small. However, at intermediate R, when R approaches the size of the atomic radii, this state will be the dominant contribution to the distortion of the charge cloud.

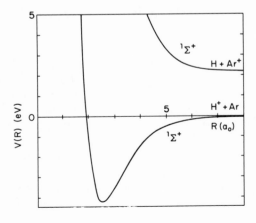

Figure 4.8. ArH^+ potentials ($V_{NN} + \varepsilon^{AB}$) associated with lowest states of $H^+ + Ar(^1S)$ and $H(^2S) + Ar^+(^2P)$. [From approximation of R. E. Johnson, C. E. Carlson, and J. W. Boring, *Chem. Phys. Lett.* **16**, 119 (1972).]

Approximating ψ_j^{AB} by two states at intermediate R, which place an electron on different centers, physically implies the electron is shared by the centers. This, of course, is the nature of the covalent bond used to describe molecular structure. When the ion and atom are identical, as in the $H^+ + H$ system, the two states of interest, one with the electron on center A and the other with the electron on center B, have identical energies and the electron spends an equal amount of time on each center. The resulting interaction energy is referred to as an exchange interaction as two nuclei can be interchanged without affecting the physical situation. For systems like $H^+ + Ar$, which have ground states close in energy, the interaction at intermediate R is like an exchange interaction.

In describing the interaction of a proton and a hydrogen atom, an approximate electronic wave function for the ground state at intermediate R can be written as a sum of states placing the electron on center A or B,

$$\psi_0^{H_2^+} \simeq c_A(R)\psi_0(\mathbf{r}_{1A}) + c_B(R)\psi_0(\mathbf{r}_{1B}) \qquad (4.42b)$$

where ψ_0 is the ground-state atomic wave function for hydrogen. Based on the above discussion, the probability that the electron is associated with either center is the same, $|c_A|^2 = |c_B|^2$. Substituting $\psi_0^{H_2^+}$ into the Schrödinger equation, we obtain a set of two, coupled, linear equations. This is just the strong-coupling situation which led to Eq. (4.41), but now the atomic states are identical and the overlap is nonzero. Diagonalizing the two-by-two matrix gives $\varepsilon_0^{AB}(R)$. For ground-state hydrogen two energies are obtained, as ψ_0 is real, these have the form

$$\varepsilon_\pm^{AB} \simeq \varepsilon_0 + \frac{\langle O_A | V_{AB}^e(r_{1B}) | O_A \rangle \pm \langle O_A | V_{AB}^e(r_{1B}) | O_B \rangle}{1 \pm \langle O_A | O_B \rangle} \qquad (4.43a)$$

where the plus and minus solutions correspond to $c_B = \pm c_A$. Note that when the overlap of the wave function is negligible, the correction in Eq. (4.43a) to ε_0 is again the static interaction of Eq. (4.4) between an ion and an atom. When the overlap is *not* small, two states of different symmetry and different in binding energy associated with the combined molecule evolve from the two degenerate atomic states. The state lower in energy (note that $\langle V_{AB}^e \rangle$ is negative) is symmetric with respect to an interchange of A and B (i.e., $c_A = c_B$) and the upper state is antisymmetric (i.e., $c_A = -c_B$) as the sign of the wave function in Eq. (4.42b) changes on switching centers A and B. The difference in energy between these two states, $\Delta\varepsilon^{AB}$, is referred to as the exchange energy. When the overlap is small but nonnegligible, the exchange energy can be written using Eq. (4.43a) as

$$\Delta\varepsilon^{AB} \sim 2[\langle O_A | V_{AB}^e | O_B \rangle - \langle O_A | V_{AB}^e | O_A \rangle \langle O_A | O_B \rangle] \qquad (4.43b)$$

Physically it is clear that the symmetric and antisymmetric ground states of H_2^+ should differ in energy. In Figure 4.9 a one-dimensional plot of the

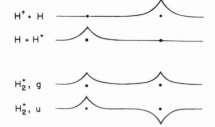

$H^+ + H$

$H + H^+$

H_2^+, g

H_2^+, u

Figure 4.9. Schematic diagram of wave functions for H_2^+. Upper: atomic states on either center. Lower: molecular states; g (gerade, symmetric), sum of atomic states; u (ungerade, antisymmetric), difference between atomic states.

exponential-like wave functions indicates that the electron density between the two nuclei will be larger for the symmetric wave function leading to a lower binding energy. In fact, the electrons are excluded from the midpoint in the antisymmetric state.

Firsov has suggested an alternate, and more accurate, method for calculating $\Delta\varepsilon^{AB}$, which is easily extended to larger systems. This method is based on the flow of electrons between the two centers, yielding a general exponential form for the exchange energy, like the overlap in Eq. (4.42a),

$$\Delta\varepsilon^{AB} \sim V_0(\beta R)^n \exp[-\beta R] \tag{4.43c}$$

where $\beta = Z'/a_0 \simeq (2I)^{1/2}$; I is the binding energy of the exchanged or shared electron in atomic units. The constant V_0 and the power of R are determined from the atomic-orbital wave functions. Whereas the result of the polarization of the electron cloud was to add an attractive contribution to V_{AB} at very large R, the effect of the exchange interaction is to split the degenerate ground states, adding a large repulsive part to one and a large attractive part to the other. Potential curves for the lowest two states of H_2^+ are given in Figure 4.10; these curves include the short-range repulsive force. It is clear from this diagram that half the collisions of an ion with its parent atom will involve a repulsive force, and half an attractive force, at

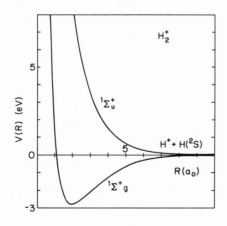

Figure 4.10. H_2^+ potentials associated with $H^+ + H(1s)$ at $R \to \infty$: $V(R) = e^2/R + \varepsilon^{AB}(R)$. [From tabulations of D. R. Bates and R. H. G. Reid, in *Advances in Atomic and Molecular Physics*, Vol. 4, ed. D. R. Bates, Academic Press, New York (1968), p. 25.]

intermediate R. At very large R, the much weaker long-range polarization force is very nearly identical for both states. Because of the exchange interaction, the potential of the lowest state of H_2^+ has a form that can support bound states of the molecular ion, which are observed spectroscopically. Such states cannot be described by potentials using the polarizability as in Eq. (4.28).

In nonsymmetric systems, like the interaction of protons with argon, the interaction potentials at intermediate R have a character similar to the symmetric, resonant interaction in H_2^+. For this nearly resonant case, an exchange-like interaction energy $\Delta\varepsilon^{AB}$ splits the states, which, as we saw earlier, is generally the case for two close-lying, interacting states [e.g., Eq. (4.41)]. This energy splitting has the same general form as that for the resonant interaction in Eq. (4.43c). Now, as in Eq. (4.42a), $\beta \simeq \frac{1}{2}(Z_A' + Z_B')/a_0$, or $\beta \simeq 1/(2)^{1/2}[I_A^{1/2} + I_B^{1/2}]$, I_A and I_B being the ionization potentials in atomic units for the extra electron on center A and center B respectively. The constant and the power of R again are determined from the atomic-orbital wave functions. This screening constant β is often used to estimate the short-range screening length a_{AB} in Eq. (4.8), i.e., $\beta^{-1} \simeq a_{AB}$. In the $(ArH)^+$ system, unlike the H_2^+ system, the collision is *either* H^+ + Ar or Ar^+ + H and, therefore, proceeds along one potential or the whole unless a transition occurs. This will be considered shortly.

The above procedures can be extended to symmetric or nonsymmetric systems for which more than one electron is involved, for example, He^{++} + He or H^+ + H^-. For interactions between identical neutral atoms, like H + H and He + He, the nuclei again can be swapped without affecting the electronic Hamiltonian. The ground-state wave function of H_2 will therefore reflect this symmetry. The Heitler–London (valence-bond) method uses

$$\psi_0^{AB} \simeq C_{AB}[\psi_0(r_{1A})\psi_0(r_{2B}) \pm \psi_0(r_{1B})\psi_0(r_{2A})] \tag{4.44}$$

where in the first term electron 1 is on A and electron 2 on B, and in the second term they are interchanged. In using Eq. (4.44) to calculate binding energies, the resulting exchange interaction, referred to as the Heitler–London interaction, again splits the degenerate levels as seen in Figure 4.11.

Whereas the two lowest states of H_2^+ have the same weight, in H_2 the relative weights of the upper and lower state are 3 to 1. This arises because the *total* electronic wave function must be antisymmetric with respect to the interchange of any two electrons, a basic principle of quantum mechanics (the Pauli principle) which we have not as yet invoked. The total electronic wave function is made up of the spatial wave function and the spin wave function which indicates the relative orientation of the *intrinsic* magnetic moments of the electrons. The spin of the electron, S, has a value of $\frac{1}{2}$, yielding a total intrinsic angular momentum $[S(S + 1)^{1/2}]\hbar$, e.g., Eq. (3.12). The angular momentum has two relative orientations, "up" $(+\frac{1}{2})$ and "down" $(-\frac{1}{2})$, with spin wave functions written symbolically as α and β

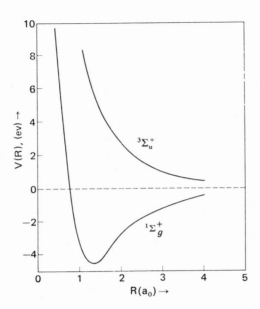

Figure 4.11. Potential curves for the lowest two states of H_2: $V_{NN} + \varepsilon^{AB}$ vs R.

respectively. The spin wave functions for *two* electrons are symmetric and antisymmetric products of α and β,

$$\alpha_1\alpha_2, \; \beta_1\beta_2, \; \frac{1}{(2)^{1/2}}(\alpha_1\beta_2 + \beta_1\alpha_2), \; \frac{1}{(2)^{1/2}}(\alpha_1\beta_2 - \beta_1\alpha_2) \qquad (4.45)$$

where the subscripts indicate a particular electron. These are shown schematically in Figure 4.12. The first three, referred to as a triplet, are all symmetric with respect to interchanging 1 and 2, and all have $S = 1$. The other antisymmetric function, a singlet, has a net zero spin $S = 0$. To construct an antisymmetric wave function for the electrons, the symmetric triplet functions of Eq. (4.45) must be combined with the antisymmetric spatial wave function of Eq. (4.44), resulting in a repulsive state having three different spin orientations. The antisymmetric singlet wave function is combined with the symmetric spatial wave function to form an attractive ground state with a single spin orientation. This is the ground state of H_2 which is, of course, a bound state. This pair of potentials, associated with H + H, therefore occurs with a three-to-one weighting (i.e., three out of four elastic collisions between two hydrogen atoms occur along a repulsive potential).

To complete the discussion of exchange effects, we note that in using the Born–Oppenheimer separation we have made an artificial distinction between the two identical charge centers in H_2^+ and H_2. The full molecular wave function should be antisymmetric with respect to an interchange of the protons (half-integer spin particles, fermions) as well as being anti-

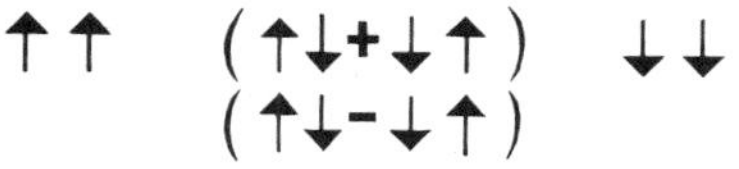

$\uparrow\uparrow$ $\left(\dfrac{\uparrow\downarrow+\downarrow\uparrow}{\uparrow\downarrow-\downarrow\uparrow}\right)$ $\downarrow\downarrow$

Figure 4.12. Spin orientations for two electrons: three symmetric and one antisymmetric.

symmetric with respect to electron exchange. For He_2^+ and He_2 the full molecular wave function should be symmetric with respect to interchange of the nuclei (bosons, integer spin particles). Such symmetry properties affect the weighting of the nuclear wave function governing the vibrational and rotational motion when combined with the nuclear spin functions. In the same way that the electronic spin functions fix the Σ_u to Σ_g ratio in H_2 at 3 to 1, the nuclear spin functions for H_2 fix the ratio of odd rotational states (antisymmetric) to even rotational states (symmetric) at 3 to 1. However, if one of the protons in H_2 is replaced by a deuteron (proton + neutron) the nuclear symmetry is broken and the even and odd rotational states are weighted equally. For our purpose, this nuclear symmetry is manifested only in low-energy and/or large-angle collisions.

The interaction potential for an incident electron is assumed to have the same form as that for a bare nucleus in a fast collision. As we saw above, however, when an electron was attached to an incident nucleus, the electronic wave function was antisymmetrized with respect to all the electrons in the incident-plus-target particle system. Similarly, the interaction between an electron and a target atom is modified by electron exchange. In fact, the incident electron and the target, $e + B$, can be treated as an ionized state of the B^- negative ion which is fully antisymmetrized among all the electrons. As a detector cannot distinguish between a scattering of the incident electron and a capture of this electron by B with an ejection of an electron from B, this exchange effect will be manifest in slow collisions.

In the above discussion the words "exchange interaction" are seen to describe a number of different effects involving identical particles. In the following discussion, exchange interaction will refer to interactions of the type described in Eq. (4.43b) and the Heitler–London interaction. These reflect the fact that the electrons "see" identical centers of charge and, therefore, are shared equally. Charge-exchange collisions involving this interaction, as well as collisions involving identical particle interchange (e.g., protons in H_2^+ and electrons in $e + B$), will be considered shortly. In the following section we describe the calculation of transition probabilities used in inelastic-collision calculations.

Transition Probabilities: The Impact Parameter Method

Integrated inelastic cross sections can be calculated semiclassically using a quantum-mechanical expression for the impact parameter transition probability $P_{AB \to j}(b)$ in Eqs. (2.24 and 2.25). These transition probabilities can also be used to describe angular differential cross sections for inelastic processes by identifying a scattering angle with each b. Extensive use of the impact parameter method has been made in describing inelastic energy loss in crystalline materials. Unlike single-particle experiments, in which our instruments cannot control impact parameters, multiple collisions in crystalline solids or large molecules are spatially interrelated. That is, in the short-wavelength limit, a first collision at a given impact parameter will determine the impact parameters of the subsequent collisions and, hence, the transition probabilities. In this section we approach the calculation of these transition probabilities using methods closely related to those employed above in determining the interaction potentials. Alternate approaches have been developed particularly for treating many-electron systems. Firsov, for example, describes the inelastic energy loss at each impact parameter as due to the "drag" force produced by the exchange of indistinguishable electrons between the incident and target particles during the collision. Lindhard, similarly, describes the inelastic loss via the force on the incident particle produced by the distorted electron cloud, a model we will consider further in Chapter 5. In a subsequent section of this chapter we will consider inelastic cross-section calculations via the Born approximation, a fully wave-mechanical model.

The interaction energies in the preceding sections are not only useful in calculating trajectories, but also allow us to understand inelastic transitions that occur when atoms collide. In heavy-particle collisions, the interaction potential maps the change in internal energy of the electrons and nuclei as they approach. Inelastic effects occur as transitions between two states where the energy difference of the transition *during* the collision is determined from the potential curves. That is, a transition can be thought of, roughly, as taking place between two potential curves, each corresponding at large R to a given state of the colliding species. These transitions are usually nonradiative, implying that the energy gain or loss due to the transition comes from a slowing or speeding up of the nuclei. Quantum-mechanically, such changes in internal energy are allowed, as the nuclear kinetic energy has rather significant uncertainties during the collision [cf. Eq. (4.3)]. The net inelastic energy change after the collision is, of course, the difference between the binding energies of the final and initial states of the *separated atoms*.

As the electrons and nuclei must exchange energy, transitions involving small energy changes are generally favored. Because the binding energies of the electrons change during the collision, the relative spacings between the states change and, hence, the transition probabilities also vary, implying that there will be favored regions of R for transitions. In Figure 4.13 three typical examples of relative spacings are shown for potential curves. In

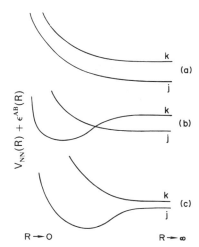

Figure 4.13. Hypothetical potential energy diagrams for two states: $V_{NN} + \varepsilon^{AB}$ vs R. Case (a): separation changes little, weak interaction between states; case (b): curve crossing between states; case (c): increasing separation at intermediate R due to strong interaction between states.

example a, the spacing changes little; therefore transitions are favored at that point in the trajectory at which the interaction is strongest and the motion is slowest. This generally occurs at the distance of closest approach of the two nuclei, where the radial velocity is zero. This situation pertains to fast collisions in which the static-potential curves have roughly the same shape for each state, and the transition probabilities are small. In example b, transitions are favored where the energy spacing of the states is the smallest, that is, in the region where they cross. This produces a relatively small, abrupt change in the nuclear motion *at* the transition point. However, as the atoms separate, the forces on the centers are very different for the two potentials involved, resulting in a much larger gradual change in net kinetic energy of the nuclei. For case c, in which two states closely spaced at $R \to \infty$ separate significantly at smaller R, there again will be a most-likely transition region. At intermediate and small R the significant spacing between the potentials diminishes the likelihood of a transition, and at very large R the interactions are too weak to cause a transtion. In the region where the curves begin to separate, the combination of the size of the interaction between the colliding particles and separation in the levels is optimal for transitions to occur. Often *adiabatic* curves separate, as shown in Figure 4.4a, *because* the states involved mix significantly. For $H^+ + Ar$ and $Ar + H^+$ ground-state curves Figure 4.8, the separation between states is a result of the strong exchange-like interaction. This interaction produces not only an optimum transition region, but also transition probabilities between the two states that may be quite large.

To calculate transition probabilities, we introduce a time dependence by assuming that the separation R is a known function of time, $R(t)$, that is, the trajectory is given. The wave equation for the electronic motion is now written

$$\left\{ H_e[\mathbf{r}, \mathbf{R}(t)] - i\hbar \frac{\partial}{\partial t_r} \right\} \psi^{AB}(t) = 0 \tag{4.46}$$

where the subscript r indicates that the time derivative is taken with the electronic coordinates fixed. The solution to Eq. (4.46) can be found in the same manner that the adiabatic equation (4.29) was solved, the only difference being that the nuclei are allowed to move. As in Eq. (4.32), the wave function is expanded in terms of the states of the separated atom which, for a bare charge interacting with an atom, we write as

$$\psi_0^{AB}(t) = \sum_k C_{0k}(t) \psi_k^B \exp\left[-i \frac{\varepsilon_k^B}{\hbar} t \right] \tag{4.47}$$

In Eq. (4.47), the time-dependent factors discussed in Chapter 3 [Eq. (3.5)] are included. The coefficients C_{0k} are functions of time and, therefore, implicit functions of R, whereas, in the time-independent problem, R was a parameter. The subscript zero now indicates the state of atom B at the start of the collision, i.e., at $t \rightarrow \infty$,

$$\begin{aligned} C_{00}(-\infty) &= 1 \\ C_{0k}(-\infty) &= 0, \qquad k \neq 0 \end{aligned} \tag{4.48}$$

As in the time-independent case, the coefficients C_{0k} are obtained by substitution of the expansion in Eq. (4.47) into Eq. (4.46) and using the orthogonality property, Eq. (4.34), of the atomic wave functions. Now, however, the coupled linear equations are linear differential equations for each coefficient,

$$i\hbar \frac{\partial C_{0j}}{\partial t} = \sum_k C_{0k} V_{jk} \exp[i\omega_{jk} t] \tag{4.49}$$

where again $\omega_{jk} = (\varepsilon_j^B - \varepsilon_k^B)/\hbar$.

Using a perturbation method like that employed earlier, i.e., the V_{jk} are small, an approximate solution for the C_{0k} is obtained by assuming these coefficients change very little from their initial values in Eq. (4.48). With this assumption Eq. (4.49) simplifies to

$$i\hbar \frac{\partial}{\partial t} C_{0j}^1 = V_{j0} \exp[i\omega_{j0} t], \qquad j \neq 0 \tag{4.50}$$

yielding a first-order estimate for the coefficients

$$C_{0k}^1(\infty) = \frac{1}{i\hbar} \int_{-\infty}^{\infty} V_{k0} \exp[i\omega_{k0} t] dt \tag{4.51}$$

The absolute value squared of the coefficients, $C_{0k}(\infty)$, is the likelihood of finding the electrons in the state k *after* the collision. Therefore, the first-order estimate of the transition probability into a particular final state, which we label f, is written

$$P^1_{0 \to f}(b) = |C^1_{0f}(\infty)|^2 \tag{4.52}$$

When the nuclei were moved infinitely slowly (the adiabatic problem discussed earlier), the wave function returned to the same state at large R. Here, however, the electrons have a finite probability of being in a new state if the integral in Eq. (4.51), which is the fourier transfrom of the interaction, is nonzero. It is now explicit that the transition probability is determined, as discussed earlier, by the size of the interaction V_{f0} and the size of the inelastic energy difference $\hbar\omega_{f0}$. Often there are situations for which the coupling interaction is simply zero, in which case no transition occurs. This generally can be determined directly from the nature of the two states involved, in which case we say a selection rule for transitions has been found.

To evaluate the integral in Eq. (4.51) the time dependence of $R(t)$ has to be specified. Because each state has a separate interaction potential, it is appropriate to use an average trajectory in the transition probability calculation. One such trajectory is obtained using the average instantaneous binding energy of the electrons at each time step. This reasonable but involved procedure for determining the average trajectory is generally not warranted, and at low energies is not even correct. For many collisions simple straight-line trajectories or, for close collisions, trajectories obtained from a simple screened coulomb interaction are adequate for estimating $P^1_{0 \to f}$. With the straight-line trajectory ($R^2 = b^2 + Z^2$ and $Z = vt$) Eq. (4.51) becomes

$$C^1_{0f}(\infty) = \frac{1}{i\hbar v} \int_{-\infty}^{\infty} V_{f0}(R) \exp\left[+ \frac{i\omega_{f0}}{v} Z \right] dZ \tag{4.53}$$

The oscillating exponential in Eq. (4.53) causes cancellation in the integration. The amount of cancellation, or rapidity of the oscillations, is seen to be determined by the speed v and the energy separation, $\hbar\omega_{f0}$. Therefore, for a given velocity, smaller changes in internal energy yield larger transition probabilities, a fact we guessed earlier. A change in velocity for a fixed ω_{f0} affects both the oscillating exponentials and the $1/v$ in front of the integral. As $v \to 0$, the exponential term oscillates infinitely rapidly for even small increments in Z, and the transition probability goes to zero in spite of the $1/v$ dependence in front of the integral. Such a collision is adiabatic, and no transitions between the states occur unless $\omega_{f0} = 0$ (i.e., the states are degenerate). For very fast collisions, as $v \to \infty$, the exponential does not oscillate but the $1/v$ factor causes the coefficient to go to zero.

Here the transition probabilities are small *not* because of the size of the energy change, but because the collision time is short. For very fast collisions, the energy difference between the states does not play an important role and the net disturbance, $\int_{-\infty}^{\infty} V_{f0}\, dt$, becomes small compared to $\hbar$. The reader cannot help but notice the similarity in form between the phase shifts calculated in the impulse approximation in Eq. (3.52), which determine the elastic scattering amplitude, and the first-order transition probabilities calculated here at high energies. This similarity should not be too surprising. In wave mechanics, deflections are not the result of classical impulses to the nuclei, but rather of relative changes in phase of contributing waves. These phases can be estimated as a ratio of the classical action to the quantum-mechanical unit of action. Similarly, the coefficients in Eq. (4.53) indicate that changes in the electronic motion are due to shifts in phase, which, for fast collisions, are estimated from the ratio of the change in an electronic action, $\int_{-\infty}^{\infty} V_{f0}\, dt$, to $\hbar$.

As the transition probabilities go to zero both at high and low velocities, a maximum transition probability exists at some intermediate velocity for fixed ω_{f0}. The discussion can be made concrete by considering an interaction of the form $V_{f0} = V_0 \exp[-\beta R]$. From Appendix D, the first-order transition probability based on Eq. (4.53) is

$$P^1_{0 \to f}(b) = \left(\frac{2V_0 b}{\hbar v}\right)^2 \left(\frac{\beta}{\beta'}\right)^2 K_1^2(b\beta') \tag{4.54}$$

where K_1 is a modified Bessel function and $\beta' = [\beta^2 + (\omega_{f0}/v)^2]^{1/2}$. $P^1_{0 \to f}$ in Eq. (4.54) has the asymptotic forms

$$P^1_{0 \to f}(b) \to \begin{cases} \text{const} \cdot v \exp\left[\dfrac{-2b\omega_{f0}}{v}\right], & \text{as } v \to 0 \\[3mm] \text{const} \cdot \dfrac{1}{v^2}, & \text{as } v \to \infty \end{cases}$$

going through a maximum located, except for very small b, at $v \sim \omega_{f0}/\beta$.

The transition probabilities in Eq. (4.54) can be used to estimate the integrated inelastic cross section of Eq. (2.24),

$$\sigma^1_{0 \to f}(v) = 2\pi \int_0^{\infty} P^1_{0 \to f}(b) b\, db$$

$$= \frac{4\pi}{3} \left(\frac{2V_0}{\hbar \omega_{f0} \beta}\right)^2 \frac{(\omega_{f0}/v\beta)^2}{[1 + (\omega_{f0}/v\beta)^2]^3} \tag{4.55a}$$

which goes through a maximum at $v_{\max} = (2)^{1/2} |\omega_{f0}|/\beta$. Since β^{-1} indicates the size of the interaction region, then $(v\beta)^{-1} = \tau_c$ is a collision time. Therefore, the location of the maximum, the so-called Massey criterion, has the *form* of an uncertainty principle, as discussed in the introduction to this chapter. That is, the maximum in the cross section occurs when the allowed uncertainty in energy during the collision roughly matches the inelastic energy loss. Further, in this approximation, the sign of inelastic energy loss, $Q = \omega_{f0}\hbar$, is not important, i.e., endothermic and exothermic interactions are equally likely. The asymptotic behavior of the inelastic cross section given in Eq. (4.55a) is

$$\sigma^1_{0 \to f}(v) \to \begin{cases} \text{const} \cdot \dfrac{1}{v^2}, & \text{as } v \to \infty \\[2mm] \text{const} \cdot v^4, & \text{as } v \to 0 \end{cases} \tag{4.55b}$$

Although the behavior at large v obtained from this method is reasonably accurate for the potential chosen, the low velocity dependence is not very reliable, as seen in Figure 4.14.

In calculating the first-order transition probabilities above, we assume that the level spacing remains constant throughout the collision. To allow

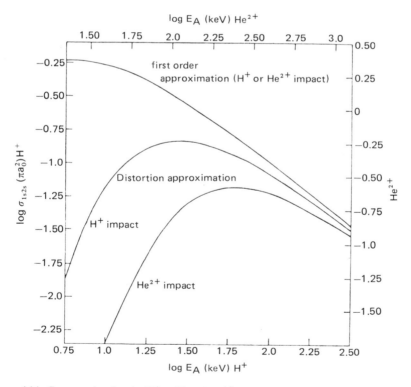

Figure 4.14. Cross section for the $H^+ + H$ and $He^{+2} + H$ $1s \to 2s$ transition calculated using the first-order transition probability in Eq. (4.53) and a better approximation, the distortion approximation in Eq. (4.56): [From D. R. Bates, *Proc. Phys. Soc (London)* **73**, 227 (1959).]

for relative changes in this spacing, we recall that the diagonal interactions, V_{jj}, determine the first-order improvement in the potential curves in Eqs. (4.37). Keeping the diagonal energy terms solving Eq. (4.49), we can partially account for the distortion of the electron charge cloud. The improved coefficients, so calculated, are

$$C_{0f}^{(d)}(\infty) = \frac{1}{i\hbar} \int_{-\infty}^{\infty} V_{f0} \, dt \exp\left[\frac{i}{\hbar} \int_{-\infty}^{t} (\varepsilon_f + V_{ff} - \varepsilon_0 - V_{00}) \, dt'\right] \qquad (4.56)$$

when the superscript d implies the "distortion" approximation. If V_{ff} and V_{00} are similar, as in Figure 4.13a, then the result in Eq. (4.53) is recovered. When this is not the case, the distortion modifies the cross section significantly at low energies, as seen in Figure 4.14 on the $1s \rightarrow 2s$ transition in hydrogen.

In heavy-particle collisions, the *approximate* binding energies $\varepsilon_f + V_{ff}$ and $\varepsilon_0 + V_{00}$ may become degenerate at a point R_x. That is, the corresponding potential curves cross at R_x, as in Figure 4.11b, in which case the phase in the integrand of Eq. (4.56) has stationery points. By applying the stationary phase approximation to the integral in Eq. (4.56), the reader should verify that the transition probability becomes

$$P_{0 \rightarrow f}^{(d)} \simeq 4 \frac{\tau_x}{\tau_{0f}} \sin^2\left[\frac{1}{2\hbar} \int_{-t_x}^{t_x} (\varepsilon_f + V_{ff} - \varepsilon_0 - V_{00}) \, dt + \frac{\pi}{4}\right] \qquad (4.57)$$

where t_x and $-t_x$ indicate the crossings on the incoming and outgoing passes. Unlike the first-order estimate, the transition probability in Eq. (4.57) has an oscillatory part. This interference term arises because the crossing point is passed twice if $R_x > R_0$, the distance of closest approach. In Eq. (4.57) τ_x is the transit time for each crossing region,

$$\tau_x = \frac{\delta R_x}{|dR/dt|_{R_x}}$$

where the extent of the interaction region, δR_x, is

$$\delta R_x = \left| 2\left[V_{0f} \Big/ \frac{d}{dR}(V_{ff} - V_{00}) \right] \right|_{R_x}$$

Finally, the quantity $\tau_{0f} = [\pi |V_{0f}|/\hbar]_{R_x}^{-1}$ is a characteristic transition time. The above expression is quite accurate when the transition region is well separated from R_0 and the transition probability is small. It is usually sufficient to use a straight-line trajectory for evaluating the phase in Eq. (4.57), although for collisions involving significant deflections a common average potential in the crossing region has been employed frequently. The reader has probably noticed that the argument of the sine function in Eq. (4.57) is the impulse approximation to the difference in the semiclassical phase shifts of the two potential curves between the crossing points. This is just the shift in phase between a trajectory for which the transition occurs on the approach and one for which the transition occurs as the particles recede, as indicated in Figure 4.15.

When the transition probability between two states is large, the coefficients $C_{0k}(t)$ and $C_{00}(t)$ in Eq. (4.49) cannot be estimated separately. As in the adiabatic problem, we can solve for $C_{00}(t)$ and $C_{0f}(t)$ by integrating the two-state coupled equations. The coefficients for all other weakly coupled states can then be obtained, if necessary, using the perturbation methods. Such two-state equations have been studied extensively both for crossing and noncrossing potential curves. Solutions must be obtained numerically, but for a number of special cases analytic approximations to the transition probabilities have been found. For the case in Figures 4.13b and 4.4b in which the diabatic potentials cross each other, Landau and Zener and, independently, Stueckleberg, estimated the transition probability by solving the equations analytically in the crossing region. Defining

$$p_{0f}^{LZS} = 1 - \exp\left[-\tau_x/\tau_{0f}\right] \tag{4.58}$$

to be the probability of a transition at each crossing, we see that the system has a probability $p_{0f}^{LZS}(1 - p_{0f}^{LZS})$ of following each of the trajectories indicated in Figure 4.15. The net probability is found to be

$$P_{0 \to f}^{LZS} = 4p_{0f}^{LZS}(1 - p_{0f}^{LZS})\sin^2\left[\Delta\eta_{0f} + \gamma_{0f}\right] \tag{4.59}$$

where $\Delta\eta_{0f}$ is the difference in the semiclassical, radial phase shifts for the trajectories and γ_{0f} is the additional phase factor that occurs in interference

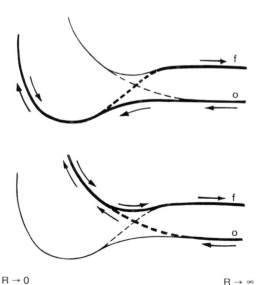

Figure 4.15. Effective potentials for the two trajectories leading to a transition when the diabatic states cross. Solid lines: adiabatic levels; dashed lines: diabatic; arrows indicate path of colliding particles.

$R \to 0$ $R \to \infty$

problems [e.g., Eq. (3.51)] due to differences in signs of the contributing amplitudes. In the semiclassical approximation, the phase shift is

$$\Delta\eta_{0f}(b) = \int_{R_+}^{R_x} p^+(R)dR - \int_{R_-}^{R_x} p^-(R)dR \qquad (4.60)$$

where the plus indicates the potential determined from the electronic binding energy of the upper state and the minus the lower state in Eq. (4.41). The quantities R_+ and R_- are the distance of closest approach for these potentials. With $\gamma_{0f} = \pi/4$, Eq. (4.59) reduces to the result in Eq. (4.57) when V_{0f} is small.

For case (a) shown in Figure 4.4, in which the static potentials are similar but the coupling potential is large, direct numerical intergration of the time-dependent equation indicates that a most-likely transition region exists. Approximating the lowest two states of the $H^+ + Ar$ collision by a constant energy separation to represent the static interaction potential, and an exponential coupling term to represent the exchange-like mixing of the two closely spaced states, we calculate the probability of being in the upper state as a function of time. The result is shown in Figure 4.16 for a collision with the system initially in the lower state. Transitions are seen to occur, as predicted earlier, in a well-defined region of R for this case also. The effective transition region occurs at that separation for which $|V_{0f}| \approx \frac{1}{2}|\varepsilon_f + V_{ff} - \varepsilon_0 - V_{00}|$, labeled R_x in Figure 4.16 and the extent of the region is,

$$\delta R_x = \frac{1}{2}\left|\frac{\varepsilon_f + V_{ff} - \varepsilon_0 - V_{00}}{dV_{0f}/dR}\right|_{R_x}$$

In these expressions, the roles of the coupling potential and the energy difference, here nearly constant, are reversed from the curve crossing case. The transition probability at the first passage is seen in Figure 4.16 to be a simple function of velocity. On exiting, however, the interference effects produce a final transition probability which oscillates significantly as a function of velocity and, of course, impact parameter.

A number of approximate expressions for the transition probability

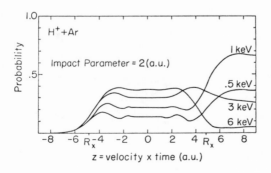

z = velocity x time (a.u.)

Figure 4.16. The probability of being in the excited state of ArH^+ (viz. Figure 4.8) during the collision. Results were obtained by numerically integrating the two-state impact-parameter equation. R_x indicates the location of the transition region using the criterion $|V_{0f}| \approx \frac{1}{2}|\varepsilon_f + V_{ff} - \varepsilon_0 - V_{00}|_{R_x}$. [From approximation of R. E. Johnson, C. E. Carlson, and J. W. Boring, *Chem. Phys. Lett.* **16**, 119 (1972)].

have been used for the case in which the static interaction potentials are nearly parallel and the coupling significant. These have the form

$$P_{0 \to f}(b) = 2\bar{P}_{0f} \sin^2 \left[\frac{1}{\hbar} \int_{-\infty}^{\infty} V_{0f} \, dt \right] \tag{4.61a}$$

The Rosen and Zener approximation to the average transition probability is

$$\bar{P}_{0f}^{RZ}(b) = \frac{1}{2} \frac{| \int_{-\infty}^{\infty} V_{0f} \exp\left[(i/\hbar) \int_{-\infty}^{t} (\varepsilon_f + V_{ff} - \varepsilon_0 - V_{00}) \, dt \right] |^2}{| \int_{-\infty}^{\infty} V_{0f} \, dt |^2} \tag{4.61b}$$

At high energies and short collision times, Eqs. (4.61a) and (4.61b) reduce to the distorted wave form of $P_{0 \to f}$ from Eq. (4.56). Also, the argument of the sine function approximates $\Delta \eta_{0f}$ of Eq. (4.60) when the effective crossing point, R_x, is large. Demkov considered an exponential coupling potential, $V_{0f} = V_0 \exp(-\beta R)$, and obtained a somewhat simpler expression for $\bar{P}_{0f}$:

$$\bar{P}_{0f}^{D} = \tfrac{1}{2} \operatorname{sech}^2 \{ (\pi/2\hbar\beta)[(\varepsilon_f + V_{ff} - \varepsilon_0 - V_{00})/\dot{R}]_{R_x} \} \tag{4.61c}$$

This expression has been used extensively to describe nonsymmetric charge exchange collisions in the intermediate velocity region and is quite accurate for $b \ll R_x$, where R_x is the location of the transition region discussed above.

If the static potentials are identical and the differences in binding energies ε_0 and ε_f are very small, then the average probability of a transition in Eqs. (4.61b) and (4.61c) is 1/2. This is the limiting case, discussed earlier, of symmetric resonant charge transfer, where, in fact, the initial and final states are indistinguishable and the coupling potential is the exchange energy $\Delta \varepsilon^{AB}$ of Eq. (4.43). For this symmetric resonant case, the two-state equations from Eq. (4.49) can be solved exactly, and the reader should verify that the transition probability is simply

$$P_{ct}(b) = \sin^2 \left[\frac{1}{\hbar} \int_{-\infty}^{\infty} \frac{\Delta \varepsilon^{AB}}{2} \, dt \right] \tag{4.62}$$

Now the crossing or transition region can be thought of as occurring at infinity where the states are degenerate. The form for the symmetric resonant charge-transfer probability in Eq. (4.62) has an interesting physical interpretation. The electron can be thought of as switching between the two colliding centers during the collision, with a frequency equal to the exchange energy divided by $\hbar$. The probability of a charge transfer then depends on the collision time over the exchange period, τ_c/τ_{ex}, which is, essentially, the form of the argument of the sine function in Eq. (4.62). We also note that, for slow collisions, the symmetric resonant charge-transfer probability does not go to zero. That is, even for adiabatic collisions, the electron can end up on the other center as the nuclei are indistinguishable.

This is true only for the symmetric resonant case. For accidental resonant collisions, in which $\omega_{f0} = 0$, the interaction terms V_{ff} and V_{00} are *not* equal, causing the transition probability in Eq. (4.56) or Eq. (4.61b) to go to zero at low velocities. In Appendix I, the transition probabilities in Eqs. (4.59), (4.61a) and (4.62) are used to obtain approximate analytic expressions for inelastic cross sections.

In this discussion we have been somewhat cavalier in applying the above methods to high-energy collisions which involve charge exchange, or, for that matter, any other rearrangement. In a rearrangement collision the masses of the colliding species change during the collision; this must be accounted for in the impact parameter method. Although the mass of the electron is small, if the incident ion is moving at a speed comparable to or greater than the orbital speed of the outer electrons, the kinetic energy gain by the captured electron may be comparable to or greater than other inelastic effects. This will modify the charge-exchange probabilities and will be discussed in the following section.

An alternative approach to that given above for calculating transition probabilities in heavy-particle collisions is to begin by expanding the time-dependent wave function $\psi_j^{AB}(t)$ in terms of the adiabatic wave functions $\psi_k^{AB}(\mathbf{r}, \mathbf{R})$, from Eq. (4.29). Substituting the wave function, with $\mathbf{R} = \mathbf{R}(t)$,

$$\psi_0^{AB}(t) = \sum_k a_{0k}(t)\psi_k^{AB}(\mathbf{r}, \mathbf{R}) \exp\left[-\frac{i}{\hbar} \int_{-\infty}^{t} \varepsilon_k^{AB}(R)\,dt \right] \qquad (4.63)$$

into the wave equation, Eq. (4.46), we obtain the first-order estimates of a_{0k}:

$$a_{0f}^1(\infty) = \frac{1}{i\hbar} \int_{-\infty}^{\infty} \mathcal{V}_{f0} \exp\left[\frac{i}{\hbar} \int_{-\infty}^{t} (\varepsilon_f^{AB} - \varepsilon_0^{AB})\,dt' \right] dt \qquad (4.64)$$

where

$$\mathcal{V}_{f0} = \int \psi_f^{AB*} \left(-i\hbar \frac{\partial}{\partial t_r} \right) \psi_0^{AB}\, d^3r$$

The phase factor in Eq. (4.64) now depends on the differences in the *adiabatic* binding energies, which is simply an extension of the distorted wave result in Eq. (4.56). The coupling interaction $\mathcal{V}_{f0}$ is very different in form, however, as it depends on the rate of change of the adiabatic states. If the collision is slow, $\psi_0^{AB}(t)$ remains the adiabatic wave function $\psi_0^{AB}(\mathbf{r}, \mathbf{R})$ throughout the collision. Transitions are induced by the rate of change of the wave functions rather than by a potential. The time derivatives in the equation above can be written as derivatives in $\mathbf{R}$ once the trajectories are specified, i.e., $\partial/\partial t_r = \mathbf{v}(\partial/\partial \mathbf{R}_r)$. This expression involves changes in length, radial motion, and rotation of the internuclear axis (angular motion), referred to in the literature as radial and rotational coupling respectively. The radial coupling clearly will not affect the angular momentum of the state. Hence, for collisions between atoms $\Sigma \rightarrow \Sigma$, $\Pi \rightarrow \Pi$, etc., transitions are allowed between the adiabatic states during the collision. The rotational coupling, however, causes transitions between states differing by one unit of angular momentum, $\Sigma \rightarrow \Pi$, $\Pi \rightarrow \Sigma$, $\Pi \rightarrow \Delta$, etc. These are referred to as the selection rules for the collisions, which are transparent in this expansion.

The static interaction potentials used in the exponent in Eq. (4.56) are a limiting case for determining the relative separation of the potential curves during the collision. The adiabatic levels, obtained by solving the full electronic Hamiltonian, H_e, are difficult to obtain, and the effort is often not warranted, as the wave functions are subsequently used to approximate the time-dependent electronic wave function. Other sets of wave functions and corresponding potential curves, referred to as diabatic, are often more appropriate for describing collisions at intermediate and low energies. One such set, discussed earlier, is constructed from the molecular orbitals of the quasi-molecule, in which the electron–electron interactions in Eq. (4.29) are averaged. That is, during the collision the electrons are assumed to behave independently, reacting to the field of the nuclei and the *average* field of the other electrons. Such states have been proven to be extremely useful, as shown by Lichten and Fano, for locating the transition regions (crossing points) of even very complex atomic systems. Although the collision time may be long, in the sense that the electrons may make a couple of orbits and the electronic cloud distorts significantly, it is often not long enough for the electrons to adjust completely to the details of the motion of all the other electrons, hence the appropriateness of the molecular orbital set of wave functions. Whereas adiabatic potential curves of the same symmetry do not cross (e.g., Figure 4.4b), these diabatic potentials do cross (e.g., Figure 4.7), and their crossing points indicate the likely transition regions. Other sets of diabatic wave functions can be imagined, with the atomic basis set, leading to the static interaction, being the appropriate set to use at high energies.

Rearrangement Collisions: Charge Exchange

In a rearrangement collision, one or more particles are transferred between the colliding species. If two identical particles (i.e., electrons or nuclei) are exchanged, the result is indistinguishable from collisions in which no interchange occurs. Such collisions, therefore, are not classified as rearrangement collisions, but are clearly related to the following discussion and will be considered shortly. We classify as rearrangement collisions cases in which the masses of the scattered and target particles change during the collision. We accounted for this kinematically in Chapter 2 when considering collisions of the form $A + BC \rightarrow AC + B$. Electron capture or charge exchange is a special case for which C represents one or more electrons transferred between the particles. For heavy-particle rearrangement, the velocities change when the particle is transferred. In the present formulation these velocities occur in the time derivative, which tracks the motion of the three heavy particles,

$$\frac{\partial}{\partial t_r} = v_A \frac{\partial}{\partial R_A}\bigg|_r + v_B \frac{\partial}{\partial R_B}\bigg|_r + v_C \frac{\partial}{\partial R_C}\bigg|_r$$

An electron transfer also affects the velocities slightly, but the modifications take a different form as the electron motion is described by wave mechanics. In the problem considered in the previous section, those electrons moving initially with B have, in addition to their internal motion, an overall translational velocity in the CM system, v_B. This motion should be described by a plane wave but was completely neglected in Eq. (4.47). When the plane-wave motion of the electrons is added to Eq. (4.47), the time-dependent wave function becomes

$$\psi_0^{AB}(t) = \sum_k C_{0k}(t)\psi_k^B(r_{jB}) \exp\left[-i\frac{\varepsilon_k^B}{\hbar}t\right]$$
$$\times \exp\left[\frac{i}{\hbar}\sum_{j=1}^{N_B}\left(m_e v_B \cdot r_{jB} - \frac{m_e v_B^2}{2}t\right)\right] \quad (4.65)$$

The last factor gives for the translational motion of the electrons, e.g., Eq. (3.1), with r_{jB} measured from the CM of the system. That this is the asymptotic form for the electronic wave functions can be verified by substituting Eq. (4.65) into Eq. (4.46), where R is large, taking care with the kinetic energy derivatives for the electrons. Using Eq. (4.65) in the electronic wave equation, Eq. (4.46), we obtain the same time-dependent equations, Eq. (4.49), for the coefficients, as the additional factor is common to all the states considered. Therefore nothing changes in calculating excitation or ionization. However, when describing charge transfer, the final state of interest places at least one electron from B on A moving with CM velocity v_A after the collision. The wave function for this state will involve different electron translation factors accounting for the change in momentum of the electrons.

A final, charge-exchange wave function could be described from sums of the ionized states associated with B which overlap center A. Alternatively, one can include, in the description of $\psi_{AB}(t)$, states placing electrons on A (as in the discussion of $H^+ + H$ and $H^+ + Ar$). In a manner similar to that used to estimate the exchange energy, we can approximate the time-dependent wave function by the two states, initial and final, involved in the charge exchange with the appropriate translation factors. This substitution will be left as a problem for the reader. Here we point out that the two-state time-dependent equations are exactly solvable for the symmetric resonant collision, yielding a transition probability for charge transfer identical in form to that in Eq. (4.62). Now, however, the exchange integrals in $\Delta\varepsilon^{AB}$ [viz. Eq. (4.43b)] contain the momentum change factor $\exp(im_e \mathbf{v} \cdot \mathbf{r}/\hbar)$. This translation factor has no effect when $v \to 0$; however, for fast collisions the effect of the momentum exchange is considerable. For $H^+ + H$, the overlap integrals are exactly integrable, yielding

$$P_{ct} \xrightarrow{v \to \infty} \frac{64\pi(b/a_0)^3}{(v/v_0)^7} \exp\left(-\frac{bv}{a_0 v_0}\right) \qquad (4.66a)$$

for charge exchange between the ground states, where v_0 is the velocity of the electron in the lowest Bohr orbit. Now the charge transfer cross section goes as

$$\sigma_{ct} \xrightarrow{v \to \infty} \text{const } v^{-12} \qquad (4.66b)$$

which differs markedly from the behavior of the excitation and ionization cross sections at high energy [e.g., Eq. (4.55b)] and decreases rapidly with increasing velocity at $v \gg v_0$.

These ideas can be applied to identical particle exchange also. That is, in a collision involving two atoms or an electron colliding with an atom, experiment cannot distinguish between the incident and target electrons after the collision. This is accounted for in the antisymmetrization of the wave function that we discussed earlier. However, the wave function describing the collision must also include the initial motion of the electrons, or the translation factors. An exchange during the collision involves a change in momentum for at least two electrons and, therefore, is also a low-velocity or large-angle effect. In resonant charge-transfer collisions (e.g., $H^+ + H$) or molecular collisions (e.g., $A + AB$) exchange of identical heavy particles must also be considered. For example, experiment cannot distinguish between a charge exchange, $H^+ + H \to H + H^+$, in which the final neutral is detected at a particular CM angle, and an elastic collision, $H^+ + H \to H^+ + H$, in which the target neutral is knocked into the same angle. In a semiclassical calculation, the two identical processes, each associated with a particular trajectory, produce interference effects in addition to those described in Chapter 3. The momentum transfer difference between these processes means that such interference effects become important only at low collision energies and/or large scattering angles. In the following section, the general nature of semiclassical, inelastic, differential cross sections is described.

Semiclassical Approximation for Inelastic Differential Cross Sections

The above discussion covered only transition probabilities and, hence, intergrated cross sections. We postponed discussion of the semiclassical scattering amplitude and differential cross sections for inelastic collisions until both the potential curves and transition probabilities were considered. By analogy with the results for elastic scattering in Eq. (3.35), the inelastic scattering amplitude is written

$$f_{0 \to f}(\chi) = \frac{1}{i(k_0 k_f)}^{1/2} \sum_{l=0}^{\infty} (l + \tfrac{1}{2}) P_l \{ \mathscr{A}_{0 \to f}(l) \exp[i\eta_f^l + i\eta_0^l] - \delta_{0f} \} \qquad (4.67)$$

In this expression $\mathscr{A}_{0 \to f}(l)$ is the transition amplitude, which for simple elastic scattering is unity. The phase factor is now a sum of two phase factors associated with the initial and final states of the system, and the δ_{0f} indicates that the subtraction at zero degrees, discussed in Chapter 3, only applies for elastic scattering. In the short-wavelength approximation the sum is again replaced by an integral, and for very long wavelengths only the $l = 0$ contributes. The integrated inelastic cross section for short wavelengths has the classical impact parameter form of Eq. (2.24), using Eq. (4.67) in Eq. (3.29) and the orthogonality of the Legendre polynomials:

$$\sigma_{0 \to f} = 2\pi \int_0^{\infty} b \, db \, \frac{p_f}{p_0} P_{0 \to f}(b) \qquad (4.68)$$

Identifying l with an impact parameter, $b \simeq (l + \tfrac{1}{2})/(k_0, k_f)^{1/2}$, we see that the transition probability is

$$P_{0 \to f}^{(b)} = |\mathscr{A}_{0 \to f}(b)|^2 \qquad (4.69)$$

and the momentum ratio, p_f/p_0, accounts for the flux change as described in Eq. (3.29). If the semiclassical approximations apply, then $p_f \sim p_0$, which we will assume for the remainder of this section.

Although the inelastic amplitudes $\mathscr{A}_{0 \to f}$ in Eq. (4.67) should be calculated directly from the full wave equation, it is clear from Eq. (4.69) that, to within a phase factor, these amplitudes can be approximated by the amplitudes $C_{0f}(\infty)$ or $a_{0f}(\infty)$ of Eqs. (4.47) and (4.63) calculated using the impact parameter method. When this identification is made, η_0 and η_f are the semiclassical phase shifts in Eq. (3.50b) determined from the interaction potentials of the initial and final states used in the approximation to $\psi_0^{AB}(t)$. That is, if the adiabatic wave functions are employed, as in Eq. (4.63), these phase shifts are determined from the adiabatic binding energies. If the atomic states are used, as in Eq. (4.47), then the phase shifts are determined using the static interaction potentials, and similarly for other diabatic sets of states.

To calculate the semiclassical, angular differential cross section, the stationary-phase approximation is applied to the scattering amplitudes. Since $\mathscr{A}_{0 \to f}$ can be written quite generally as the product of a slowly varying function of b and an oscillatory function, the stationary phase approximation yields

$$\sigma^{sc}_{0 \to f}(\chi) = \left| \sum_{\substack{q, \text{ all} \\ \text{stationary} \\ \text{points}}} [\bar{P}^{(q)}_{0f} \, \sigma^{(q)}(\chi)]^{1/2} \exp [iA^{(q)}/\hbar + i\gamma^{(q)}] \right|^2 \quad (4.70)$$

In Eq. (4.70) q labels stationary points, each one associated with a trajectory leading to a transition $0 \to f$; $\bar{P}^{(q)}_{0f}$ is the transition probability associated with that trajectory with a corresponding classical differential cross section $\sigma^{(q)}(\chi)$ and change in classical action $A^{(q)}$. The extra phase factor $\gamma^{(q)}$ is determined as before, viz. Eq. (3.51).

By way of understanding the result in Eq. (4.70), we refer to a couple of cases. If $C^{(1)}_{0f}$ of Eq. (4.51) is used to approximate the transition amplitudes, then $\mathscr{A}_{0 \to f}$ is a slowly varying, nonoscillatory function of b. Therefore, the only phase factor in the integrand of the scattering amplitude of Eq. (4.67), besides the angular factor, is $(\eta_0 + \eta_f)$, implying that the system is in the state 0 on the way in and f on the way out. In essence, the transition occurs at the distance of closest approach. On the other hand, if a transition region exists at some point $R_x > R_0$, then $\mathscr{A}_{0 \to f}$ has an oscillatory factor, as expressed in Eq. (4.57), (4.59), or (4.61). When this is added to and subtracted from $(\eta_0 + \eta_f)$, two phase factors are derived, one for each of the trajectories indicated in Figure 4.15, each having a transition at R_x. For the curve-crossing case, $\bar{P}^{(q)}_{0f} = p^{LZS}_{0f}(1 - p^{LZS}_{0f})$ for each trajectory, and the cross section becomes

$$\sigma^{sc}_{0 \to f}(\chi) \simeq p^{LZS}_{0f}(1 - p^{LZS}_{0f})\{\sigma^{(1)}(\chi) + \sigma^{(2)}(\chi) - 2(\sigma^{(1)}_{(\chi)}\sigma^{(2)}_{(\chi)})^{1/2}$$
$$\times \cos [(A^{(1)} - A^{(2)})/\hbar + \gamma^{(1)} - \gamma^{(2)}]\} \quad (4.71)$$

where 1 and 2 label quantities associated with the two trajectories in Figure 4.15. For the Rosen–Zener and Demkov transition probabilities, $\sigma^{sc}_{0f}(\chi)$ has the same form, with $p^{LZS}_{0f}(1 - p^{LZS}_{0f})$ of Eq. (4.59) being replaced by $(\frac{1}{2}\bar{P}_{0 \to f})$ of Eq. (4.61). Lastly, for symmetric resonant charge transfer, the "crossing" is at $R \to \infty$ and $p_{0f} = \frac{1}{2}$, giving

$$\sigma^{sc}_{\pm}(\chi) = \frac{1}{4} \left\{ \sigma^{(g)}(\chi) + \sigma^{(u)}(\chi) \pm 2(\sigma^{(g)}(\chi)\sigma^{(u)}(\chi))^{1/2} \cos \left[\frac{A^{(g)} - A^{(u)}}{\hbar} \right] \right\} \quad (4.72)$$

where (g) and (u) are the labels for the symmetric and antisymmetric states (cf. Figure 4.10). The $+$ in Eq. (4.72) corresponds to elastic collisions and the $-$ to charge-transfer collisions. In Eqs. (4.71) and (4.72) a single trajectory was associated with each reaction path. Additional interference effects, such as the rainbow scattering discussed in Chapter 3, and nuclear symmetry, discussed in the preceding section, will complicate this expression somewhat as there are additional contributing trajectories. In Figure 4.17

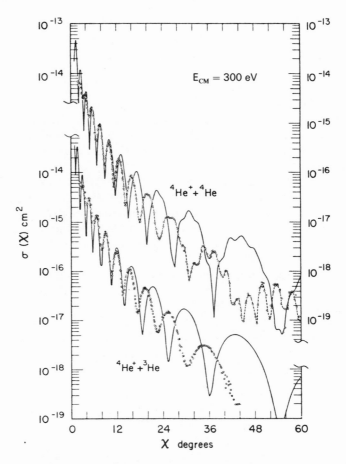

Figure 4.17. $He^+ + He$ charge-exchange cross section. Solid line: a semiclassical calculation; dots: experiment; $^4He^+ + ^3He$: indicates oscillatory pattern from Eq. (4.72); $^4He^+ + ^4He$: nuclear symmetry oscillation are superimposed at large scattering angle. [From W. Aberth, D. C. Lorents, R. P. Marchi, and F. T. Smith, *Phys. Rev. Lett.* **14**, 776 (1965).]

differential cross sections for $He^+ + He \rightarrow He + He^+$ are shown in which the oscillatory pattern suggested by Eq. (4.72) is evident. At large angles in the 600 eV data, a secondary interference pattern is seen as the nuclei are identical.

At high collision energies deflections are determined primarily by the common repulsive core of the static interaction potential. At these energies the $\sigma^{(q)}(\chi)$ are all roughly equivalent and the classical actions can be determined by the impulse approximation. The semiclassical cross section in Eq.

(4.10) simply reduces to

$$\sigma^{sc}_{0 \to f}(\chi) \sim P_{0 \to f}\, \sigma(\chi) \qquad (4.73)$$

where $P^{(b)}_{0 \to f}$ is the impact parameter transition probability and $\sigma(\chi)$ is a common, average differential cross section. This is a result we surmised in Chapter 2. In this form, the impact parameter transition probability is directly related to the differential cross section via a single average classical deflection function, $\chi(b)$.

The result in Eq. (4.70) can be applied to cases for which more than one crossing occurs and more than two states are involved, as long as the crossings are well separated and each trajectory is associated with a well-defined stationary point. If the number of stationary points becomes large, the scattering amplitude should be estimated from the integral form of Eq. (4.67) or via the Born approximation described in the following section.

Transitions in the Born Approximation

In the Born approximation described in Chapter 3, the nuclear motion was treated by wave mechanics and the interaction was assumed to be weak. The first Born result for elastic collisions in Eq. (3.60) showed that the scattering amplitude could be calculated from the initial and final plane-wave functions of the scattered particle. Generalizing this, we write the Born approximation for inelastic scattering as

$$f_{0 \to f} = \frac{m}{2\pi\hbar^2} \iint \psi^*_f(\mathbf{R}, \mathbf{r}) V(\mathbf{R}, \mathbf{r}) \psi_0(\mathbf{R}, \mathbf{r})\, d^3R\, d^3r \qquad (4.74)$$

The integral over $\mathbf{r}$ in Eq. (4.74) represents integrals over the coordinates of the electrons (or other composite particles) of A and B, R is the separation of A and B, and the potential is the total interaction potential before averaging over the charge distributions. If we neglect identical particle exchange, the initial and final wave functions are plane waves (for the overall motion) times the wave functions describing the initial and final atomic states in question, $\psi_f(\mathbf{r})$ and $\psi_0(\mathbf{r})$. In terms of the coupling potentials considered earlier, Eq. (4.74) can be written

$$f_{0 \to f} = \frac{m}{2\pi\hbar^2} \int \exp[-i\mathbf{K}_f \cdot \mathbf{R}][V_{f0}(\mathbf{R}) + V_{NN}(R)\delta_{0f}] \exp[i\mathbf{K}_0 \cdot \mathbf{R}]\, d^3R$$

$$(4.75)$$

where the integrations over the coordinates of the bound electrons have been performed to obtain $V_{f0}(\mathbf{R})$. As the scattering is *not* elastic, $K_f \neq K_0$;

the initial and final momenta in Eq. (4.75) are related by the inelastic energy loss $\hbar^2 K_f^2/2m = \hbar^2 K_0^2/2m - Q$, where $Q = \varepsilon_f - \varepsilon_0 = \hbar\omega_{f0}$ and $\mathbf{K}_f$ is in the direction of the detector. For elastic scattering, $Q = 0$, Eq. (4.75) becomes indentical to Eq. (3.59) and the interaction is the static potential.

The reader should verify (Problem 4.13) that Eq. (4.75) can be derived from the semiclassical expression in Eq. (4.67) using Eq. (4.53) for the transition amplitudes $\mathscr{A}_{0f}$. Therefore, the *integrated*, inelastic cross section obtained from Eq. (4.75) is equivalent to that calculated from the impact parameter transition probabilities if the inelastic energy loss is small compared to the CM energy. This can be demonstrated directly, following the method of Chapter 3, if the identification $(\mathbf{K}_f - \mathbf{K}_0) \cdot \mathbf{R} = (\Delta\mathbf{p} \cdot \mathbf{b})/\hbar$ is used to define the impact parameter. Now, direct integration of the scattering amplitude and the integrated cross section is carried out in a manner similar to that for the elastic cross section, except that $K_f \neq K_0$. For V_{f0} an exponential, e.g., Eq. (4.54), interaction, the form for the differential cross section is similar to that in Eq. (3.64), and the integrated cross section is

$$\sigma_{0 \to f} = \frac{2\pi m}{E} \left(\frac{2V_0 \beta}{\hbar^2}\right)^2 \int_{(\Delta p_l)^2}^{(\Delta p_u)^2} \frac{d(\Delta p)^2}{[(\Delta p/\hbar)^2 + \beta^2]^4} \tag{4.76}$$

In the limit $|Q| \ll \hbar^2 K_0^2/2m = E$, $\Delta p_l \sim Q/v$, $\Delta p_u \gg \hbar\beta$, and $K_0 \simeq K_f$, Eq. (4.76) now yields a result identical to that in Eq. (4.55). The advantage of starting with the Born approximation is that the angular differential cross section is obtained directly *and* the form of the potential can be quite general. When inelastic effects are studied, some caution has to be exercised in using first-order methods. Although a transition to any state may occur with a small probability, the sum total of such transitions might not be small. In such cases the above expressions would need to be modified.

We briefly mention another form for the Born scattering amplitude which has a useful physical interpretation. The Fourier transform of the potential for an incident bare ion or electron A interacting with target atom B can be used (see Problem 4.14) to write the scattering amplitude in Eq. (4.75) as

$$f^{(1)}_{0 \to f}(\chi) = \frac{2mZ_A e^2}{(\Delta p)^2} \left[Z_B \delta_{f0} - \sum_{j=1}^{N_B} \int \psi_f^* \exp\left[-i(\Delta\mathbf{p} \cdot \mathbf{r}_{jB})/\hbar\right] \psi_0 \, d^3r \right] \tag{4.77}$$

where ψ_f and ψ_0 are the final and initial electronic wave functions and d^3r implies an integration over all the electronic coordinates as usual. For an elastic collision, $\psi_f = \psi_0$, a result like that in Eq. (3.64) is found in which the static charge distribution $|\psi_0|^2$ determines the interaction, V_{00}. The first term in Eq. (4.77) accounts for the coulomb repulsion of the nuclei and the second term describes the impulse imparted to the electrons which may cause a change in state. Although the interaction is split into a sum of interactions with the individual target particles, we note the momentum transfer, Δp, involved is that for the whole target system. Momentum is *not* conserved separately between the incident particle and the individual particles making up the target as it was in the classical BEA discussed in Chapter 2.

The relationship between the Born and BEA approximations is very instructive and has been considered by a number of authors. Bethe pointed out that the contributions to the cross section could be understood quite easily if one considers large and small momentum transfers separately, a fact we use to simplify the calculation of the cross section. If the impulse received by the electrons is large, then after the collision the electron is very nearly a free particle not influenced by its nucleus. Now ψ_f can be represented by a plane wave with momentum $\hbar k$ and the differential cross section can be written, using Eq. (4.77), as

$$d\sigma^{(1)}(\mathbf{k}, \Delta p) \simeq \frac{8\pi\hbar^3}{v^2}\left(\frac{Z_A e^2}{\Delta p^2}\right)^2 \sum_i N_{B_i}\rho_i(\hbar\mathbf{k} - \Delta\mathbf{p})\, d^3k\, \Delta p\, d\Delta p \quad (4.78)$$

In the above expression, Δp is related to the CM angle χ, as before, and ρ_i is the momentum distribution of the electrons in the ith orbital of the target atom, i.e., the absolute value squared of the Fourier transform of the orbital wave function as described at the end of Chapter 3. When the target electron is initially at rest, $\rho_i(\mathbf{p}) = \delta(\mathbf{p})$, the energy transfer is $Q = \hbar^2 k^2/2m_e$, and we can simply integrate over Δp and the orientation of $\mathbf{k}$. The resulting energy-transfer cross section,

$$d\sigma^{(1)}(Q) \simeq \frac{2\pi}{m_e v^2} N_B\left(\frac{Z_A e^2}{Q}\right)^2 dQ \quad (4.79)$$

is identical to the BEA cross section of Chapter 2. This is, of course, the expected result for close collisions or large momentum transfers. It should also be evident that Eq. (4.78) is equivalent to a classical BEA approximation in which an initial speed distribution for the electrons, $\rho(\mathbf{p})$, is included.

If the momentum transfer is small, on the other hand, the Born approximation yields quite different results. (This limit could also be approximated by considering a collision with a classical oscillator, as in Appendix C.) If $|\Delta p/\hbar|$ is much less than the atomic size, then the exponential in Eq. (4.77) can be expanded. The differential cross section for a neutral target now becomes

$$d\sigma^{(1)}_{0 \to f}(\Delta p) \simeq \frac{8\pi}{v^2}\left(\frac{Z_A e^2}{\hbar\Delta p}\right)^2 \left|\left\langle \sum_{j=1}^{Z_B} z_j \right\rangle_{0f}\right|^2 \Delta p\, d\Delta p \quad (4.80)$$

where z_j is in the direction of Δp and $\Delta p\, d\Delta p = \hbar^2 K_0 K_f d\cos\chi$. This is the long-range (distance collision) dipole approximation for an inelastic collision. The dipole terms in brackets are just those used to determine the polarizability in Eq. (4.40a). The polarization interaction is, of course, an adiabatic effect in which the electronic charge cloud distorts and returns to its initial state after the collision. The result above emphasizes that, for fast collisions, the system does not always restore. The lower limit on the momentum transfer, approximately Q/v, is just the adiabatic cutoff postulated in Chapter 2. This quantity decreases with increasing velocity, implying that distant collisions become increasingly important at high collision energies. Therefore, the BEA approximation to the total collision cross section may be a poor approximation at high collision energies.

A useful comparison between the Born and BEA approximations can be made when considering the ionization cross section. Wave functions for ionized electrons in the vicinity of a charged particle are very much like scattered waves. They are plane-wave-like for large momentum and/or at large distances from the nuclei. Close to the nuclei they obviously are affected by the coulomb potential. Writing the energy transfer as $Q = \hbar^2 k^2/2m_e - \varepsilon_0$ and changing the CM scattering angle to a momentum transfer, one can compare the cross section $d^2\sigma/dQ\,d\Delta p$ in both approximations for target hydrogen. In this comparison we use the same initial velocity distribution for Born and BEA calculations, and the exact ionization states for ψ_f in Eq. (4.77). The results shown in Figure 4.18a for an incident ion differ significantly for the two approximations at small Δp and small Q. The corresponding total ionization cross sections, therefore, also will differ unless the lower limits on the integration in Q and Δp are large. If, on the other hand, the incident particle is a neutral, the screening of the interaction provides a built-in cutoff at small Δp (i.e., large b) and the two approximations in Figure 4.18b show considerable agreement. In the latter case the BEA cross section could be considered to be reasonably accurate.

To obtain the BEA results in Figure 4.18b, the Born cross section for the electron–hydrogen atom collision was used to represent the binary encounter between the electron and incident particle. The above comparison is based, therefore, on the premise that the cross section for an electron interacting elastically with a neutral atom is the same classically and quantum mechanically. This is a special property of the coulomb interaction (i.e., an electron colliding with an ion) but is not the case for the screened interaction, as is seen by comparing the results in Eq. (3.64) with the cross section deduced from Eq. (2.59). The following conclusion, however, holds: For incident neutrals having weak long-range interactions, the BEA is reasonably accurate.

The reader should remember that, when comparing various methods, one often forgets they are, after all, only approximate. That is, the Born and BEA methods, as well as the impact parameter and semiclassical methods,

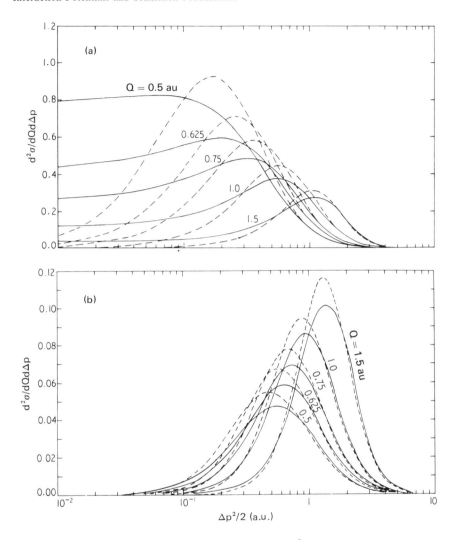

Figure 4.18. Double differential cross section for ionization, $d^2\sigma/dQ\,d\Delta p$, vs momentum transfer Δp at various energy transfers, Q. Dashed lines, BEA; solid lines, Born, (a) $H^+ + H$; (b) $H + H$. This quantity is often referred to as the generalized oscillator strength for the collision. [From J. H. Harberger, R. E. Johnson, and J. W. Boring, *J. Phys. B* **6**, 1040 (1973).]

are all approximate solutions to the collision problem and care must be taken to use each expression where appropriate. In the following chapter we discuss the results of experiments and evaluation of cross sections and related quantities using the methods developed in Chapters 2–4.

Detailed Balance

Before discussing and comparing results, we briefly consider a property of the cross sections which can be quite usefully exploited in some cases. The equations of motion, both quantum-mechanical and classical, are such that

$$| f_{0 \to f}(\chi; p_0)|^2 = | f_{f \to 0}(\chi; p_f)|^2 \quad \text{or} \quad P_{0 \to f}(b; p_0) = P_{f \to 0}(b; p_f)$$

(4.81)

which is due to the time reversal symmetry of the collision process. Using Eq. (3.29) and Eq. (4.81) the differential cross sections for the forward and reverse reactions can be related. Further, integrating over angle [or impact parameter in Eq. (4.68)], and using Eq. (4.81), the integrated inelastic cross sections are related by

$$p_0^2 \sigma_{0 \to f}(p_0) = p_f^2 \sigma_{f \to 0}(p_f)$$

(4.82)

In the semiclassical region $p_0 \approx p_f$ and therefore, by Eq. (4.82), the forward and reverse reactions have the same cross section. We had stated earlier that in this region endothermic and exothermic processes behaved similarly. At low velocities, near threshold for the endothermic process, the forward and reverse reactions can differ markedly but are simply related by Eq. (4.82). This relationship is a statement of the principle of detail balance used when describing equilibrium in statistical mechanics. Based on the notions of statistical mechanics Eq. (4.82) can be extended to cases where there are a number of equivalent initial states, η_0, and/or final states, η_f, (e.g., spin or angular momentum states)

$$p_0^2 \eta_0 \, \sigma_{0 \to f}(p_0) = p_f^2 \eta_f \, \sigma_{f \to 0}(p_f)$$

(4.83)

Such a relationship allows one to determine, for instance, deexcitation cross sections from data on excitation cross sections and provides a constraint when calculating cross sections by approximate methods.

Exercises

4.1 Using the static potential of Eq. (4.4) for the interaction between A and B, obtain the expressions in Eqs. (4.5) and (4.6). Using the ground-state wave functions of H for $p(\mathbf{r})$, obtain the result for a_{AB} in Eq. (4.7); determine Z for $H^+ + H$ and $H + H$.

4.2 For H + He and Ne + Ne, compare graphically the screened coulomb potential of Eq. (4.8) using various screening lengths a_{AB}^B, $a_{AB}^L = 0.8853 a_{AB}^B$, a_{AB}^F, and a_{AB}^s, all discussed below Eq. (4.8) and in Appendix H. Calculate the screened potential also from the adiabatic expression for a_{AB} using Eq. (4.16).

4.3. Consider a simple ionic-bonded molecule (B^+C^-) interacting with a charged particle A^+. If the equilibrium internuclear separation of the molecule is $2.5\,a_0$ and the binding force is characterized by a frequency $v_0 = 10^{12}\ sec^{-1}$, compare the ion–dipole and ion–induced dipole potentials graphically.

4.4. Using the ground-state hydrogenic wave function, calculate $V_{NN} + V_{00}$, the static interaction, for $H^+ + H$.

4.5. For $H^+ + H$, calculate $\langle O_A | O_B \rangle$ and $\langle O_A | V_{AB}^e | O_B \rangle$ using the ground-state hydrogen wave function. Show that they depend exponentially on R at large R. Evaluate and plot $V_{NN} + \varepsilon_{\pm}^{AB}$ in Eq. (4.43a). [*Note :* The integration can be carried out by expansion, as in Eq. (4.10a), or by using the prolate spheroidal coordinates, $\zeta = (r_{1A} + r_{1B})/2R$, $\eta = (r_{1A} - r_{1B})/2R$ and ϕ.] Compare to the polarizability interaction with $\alpha_H = 4.5a_0^3$ and to the results in Figure 4.10.

4.6. Consider a hydrogen atom in a small electric field (e.g., that produced by an approaching atom). Use the perturbation method employed in solving Eq. (4.35) to obtain the splitting of the degenerate m_l values. Employ the wave functions of Chapter 3.

4.7. For the system $A^{++} + B \rightarrow A^+ + B^+$, a series of curve crossings occurs. The potential V_{00} for $A^{++} + B$ goes as $V_{00} \sim -(\alpha^2/2)(Z^2/R^4)$, with $Z = 2$, whereas the potential V_{ff} for $A^+ + B^+$ goes as $V_{ff} \sim 1/R$, in atomic units. For $He^{++} + Ne$, plot the potentials and find the curve crossings for the first few states of He^+: $\alpha_{Ne} \simeq 13a_0^3$ and $I_{Ne} = 21.56$ eV.

4.8. Given the expression for the oscillator strength in Eq. (4.40b), show that $\sum_k f_{k0} = N_B$. Use the fact that $(\varepsilon_k - \varepsilon_0)\langle \sum_i^{N_B} z_i \rangle_{k0} = \langle [H_B, \sum_i^{N_B} z_i] \rangle_{k0}$, invoking Eq. (4.31), where $[H, z]$ means $Hz - zH$.

4.9. For $H + H$, use London's method to evaluate the coefficient C_{VW} from the first-order corrections to the wave functions for ψ^{AB}. This is obtained by writing $\psi^{AB}(R) = \sum_i \sum_j C_{ij}(R)\psi_i^A \psi_j^B$ and substituting into $H_e = H_A + H_B + V_{AB}^e$, a procedure similar to that used to obtain the set of equations in Eq. (4.35). Finally, employ the expansion for V_{AB}^e from Eq. (4.10b).

4.10. Calculate the transition probability in Eq. (4.53) if $V_{f0}(R)$ is a power law (Appendix D), $V_{0f} \propto R^{-n}$. What are the asymptotic limits to the cross section σ_{f0} at high and low v?

4.11. Use the stationary-phase approximation of Appendix F to obtain Eq. (4.57) from Eq. (4.56).

4.12. Solve the two-state, time-dependent equations for the transition probabilities in a symmetric resonant collision. Verify Eq. (4.62) for the charge-transfer probability and obtain the high-energy limit for the $H^+ + H$ collision using the wave functions in Eq. (3.71).

4.13. Use the semiclassical scattering amplitude in Eq. (4.67) to derive the Born expression in Eq. (4.75) when the phase shifts are small. Assume $\mathscr{A}_{0 \rightarrow f} = C_{0f}^{(1)}$ in Eq. (4.53) and define the impact parameter from $(K_f - K_0) \cdot R = \Delta p \cdot b/\hbar$.

4.14. Evaluate the fourier transform of the coulomb potential, $\int d^3R \exp[i\mathbf{K} \cdot \mathbf{R}]/R$. Use this to write the scattering amplitudes in Eq. (4.75) in the form of Eq. (4.77), when the incident particle is a bare nucleus.

4.15. Verify that Eq. (4.76) is equivalent to Eq. (4.55a) if $|Q| \ll E$.

Suggested Reading

As in Chapters 2 and 3, most of this material is contained in texts on atomic and molecular collisions, quantum mechanics, or quantum chemistry. Specific examples are given below; references for transitions, inelastic cross sections and charge exchange overlap considerably and the divisions of titles are somewhat arbitrary.

Intermolecular Potentials

J. O. HIRSCHFELDER, ed., *Advances in Chemical Physics*, Vol. 12, Wiley, New York (1967), Chapters 1–4.

J. O. HIRSCHFELDER, F. CURTISS, and R. B. BIRD, *Molecular Theory of Gases and Liquids*, Wiley, New York (1964), Chapter 2.

I. M. TORRENS, *Interatomic Potentials*, Academic Press, New York (1972), Chapters 1, 3, and 4.

J. GOODISMAN, *Diatomic Interaction Potential Theory*, Vol. 1, Academic Press, New York (1973), Chapters 1 and 2.

Exchange Energy

O. FIRSOV, *Zh. Eksp. Teor. Fiz.* **21**, 1001 (1951), and texts above.

W. HEITLER and Z. LONDON, *Z. Physik* **45**, 455 (1927).

Molecular Orbitals and Correlation Diagrams

W. KAUZMAN, *Quantum Chemistry*, Academic Press, New York (1960), Part III.

G. HERZBERG, *Spectra of Diatomic Molecules*, Van Nostrand-Reinhold, Princeton (1950), Chapter 6.

W. LITCHEN, *Phys. Rev.* **164**, 131 (1967); U. Fano and W. Litchen, *Phys. Rev. Lett.* **14**, 627 (1965).

Q. C. KESSEL, E. POLLACK, and W. W. SMITH, in *Collision Spectroscopy*, ed. R. G. Cooks, Plenum Press (1978), Chapter 3.

Stationary Perturbation Theory

A. DALGARNO, in *Quantum Theory, I. Elements*, ed. D. R. Bates, Academic Press, New York (1961), Chapter 5; any other quantum mechanics or quantum chemistry text.

Time-Dependent Perturbation Theory and Impact-Parameter Transition Probabilities

D. R. BATES, in *Quantum Theory, I. Elements*, ed. D. R. Bates, Academic Press, New York (1961), Chapter 8.

E. G. G. STUECKLEBERG, *Helv. Phys. Acta*, **5**, 369 (1932); L. Landau, *Phys. Z. USSR 2*, 46 (1932); C. Zener, *Proc. Roy. Soc. (London)* **137**, 696 (1932).

N. ROSEN and C. ZENER, *Phys. Rev.* **40**, 502 (1932); D. R. Bates, *Discuss. Faraday Soc.* **33**, 7 (1962).

Yu. Demkov, *Sov. Phys. JETP*, **18** 138 (1964); R. E. Olson, *Phys. Rev. A* **6**, 1822 (1972).

M. R. C. McDowell and J. P. Coleman, *Introduction to the Theory of Ion–Atom Collisions*, North-Holland, Amsterdam (1970), Chapters 4 and 8.

M. S. Child, *Molecular Collision Theory*, Academic Press, New York (1974), Chapter 8.

B. H. Bransden, *Atomic Collision Theory*, W. A. Benjamin, New York (1970), Chapters 8 and 9.

Born Approximation: Transitions

D. R. Bates, in *Atomic and Molecular Processes*, ed. D. R. Bates, Academic Press, New York, p. 549 (1962).

H. A. Bethe and R. Jackiw, *Intermediate Quantum Mechanics*, 2nd edn., W. A. Benjamin, New York (1968), Part II.

N. F. Mott and H. S. W. Massey, *The Theory of Atomic Collisions*, 3rd edn., Oxford University Press, London (1965), Chapter 5.

Semiclassical, Inelastic Differential Cross Sections

F. T. Smith, in *Physics of the One- and Two-Electron Atoms*, eds. F. Bopp and H. Kleinpoppen, North-Holland, Amsterdam (1969), p. 755. (F. T. Smith, D. C. Lorents, R. E. Olson, and co-workers have published a series of papers on semiclassical methods for differential cross sections in *Physical Review* from 1967 to the present.)

J. B. Delos and W. R. Thornson, *Phys. Rev. A*, **6** 720, 728 (1972).

T. A. Green and M. E. Riley, *Phys. Rev. A*, **8** 2938 (1973).

T. A. Green, *Phys. Rev. A*, **23** 532 (1981).

E. E. Nikitin, in *Advances in Quantum Chemistry*, *Vol. 5*, ed. P. Löwden, Academic Press, New York (1970), Chapter 4.

R. N. Porter and L. M. Raff, in *Dynamics of Molecular Collisions*, Vol. 2, ed. W. H. Miller, Plenum Press, New York (1976), Chapter 2.

L. D. Landau and E. M. Lifshitz, *Quantum Mechanics*, trans. J. B. Sykes and J. S. Bell, 3rd. edn., Pergamon Press, New York, (1977), Chapters VII and XVIII.

M. S. Child in *Advances in Atomic and Molecular Physics*, *Vol. 14*, ed. D. R. Bates, p. 225.

Charge Exchange

R. A. Mapleton, *Theory of Charge Exchange*, Wiley New York (1972).

D. R. Bates, *Proc. Roy. Soc.* (London), **A247** 294 (1958).

D. R. Bates and R. McCarroll, *Proc. Roy. Soc.* (London), **A245** 175 (1958).

D. R. Bates and R. McCarroll, *Phil. Mag. Supp.*, **11** 39 (1962).

T. A. Green and R. E. Johnson, *Phys. Rev.*, **152** 9 (1966).

R. E. Olsen and A. Salop, *Phys. Rev. A*, **16** 531 (1977).

5

Cross Sections and Rate Constants: Results

Introduction

In the previous three chapters basic methods for describing atomic and molecular collisions were developed. Although results and calculations were occasionally shown, the emphasis was to obtain the tools needed to understand which processes occur with what probabilities when atoms and molecules collide. In this chapter some experimental results and computations are presented and described in terms of the ideas already developed. The selection of results shown is not at all intended to be comprehensive. It is also somewhat arbitrary, as many excellent experiments have been performed on large numbers of systems and those chosen are not necessarily the "best" results or even the most recent. Those presented are ones with which I was familiar, which elucidate certain ideas already considered and/or will be used in Chapter 6 to discuss macroscopic phenomena. Again I refer the reader to not only the many excellent texts, but also reference tables of atomic and molecular collision results.

The chapter first contains a discussion of those collisions which give information on the interaction potentials both at short and long range. Following this, various inelastic atomic processes are discussed. In the last section molecular processes and reaction rates are considered.

Total Differential Cross Sections and Interaction Potentials

The interest in total differential cross sections is related to the notion that a single, average potential can be used to describe collisions, which is the basis of the impact-parameter approximation. From the discussion in

Chapter 4 it is clear that any such potential would have to be velocity dependent in order to describe collisions over a large range in angle and energy. However, this velocity dependence will be negligible if the distances of closest approach of interest correspond to scattering from the repulsive core of the atoms. Also, at very low energies, when transitions are unlikely, the average potential corresponds to a single potential associated with the adiabatic ground state of the collision complex. Static potentials, therefore, can be very useful; in fact, even inelastic molecular reactions often proceed along a single potential surface, a topic we will return to in the final section.

Classical trajectory calculations, using such potentials, are valid outside of the diffraction region, that is, where $\Delta p > \hbar/a$, with a, as usual, an atomic dimension. This region corresponds to scattering angles $\chi \sim \Delta p/p > \chi_c \sim \hbar/pa$. For light incident ions at thermal energies, χ_c is of the order of tens of degrees; at eV energies, tenths of degrees; and at keV energies, hundredths of degrees. Since angular differential cross-section experiments can have resolutions of better than a few tenths of a degree, if necessary, the classical trajectory method can be applied to a broad range of experimentally available angles and energies for incident ions and atoms. However, as the largest contributions to the integrated cross section come from angles less than χ_c, angular differential cross-section measurements generally cannot be used to construct integrated cross sections, except at the lowest velocities.

A set of data accumulated from a number of experiments is shown in Figure 5.1 on a ρ vs τ plot (Eq. 2.60). Two distinct types of experiments are involved. At large τ (small impact parameters), the results shown are total differential cross-section measurements in which all scattered particles are collected independent of state. At low τ (large impact parameters), the data exhibited are from elastic scattering experiments. There is an intermediate region in τ (or angle and energy) in which inelastic processes occur with high probability, particularly charge exchange, but the detection efficiencies of neutral and ionized particles are very different. From Figure 5.1 it is seen that summing the cross sections for all processes results in a nearly monotonically decreasing ρ vs τ plot over hugh changes in τ, if one ignores, for a time, the small-scale structure seen at low τ.

At high τ, the ρ vs τ plot is characteristic of a simple coulomb repulsion. The appropriate value of $Z_A Z_B$ would be 20 for this combination of collision partners. The observed slope appears to agree better with smaller effective charges because the tightly bound $(1s)^2$ shell of Ne is not penetrated except at the very highest τ shown. When Eq. (2.58) is used for a coulomb interaction, $\tau \sim C_1/R_0$, the distance of closest approach would be $R_0 \sim 0.3a_0$ at $\tau \sim 10^5$ eV-deg, with $C_1 = (Z_A Z_B)$ a.u. $= 20$ a.u. Therefore the penetration radius is larger than the mean radius of the $(1s)^2$ shell of Ne. It is also clear that the slope of the ρ vs τ plot changes rapidly with decreasing τ, implying an increase in the screening at larger impact pa-

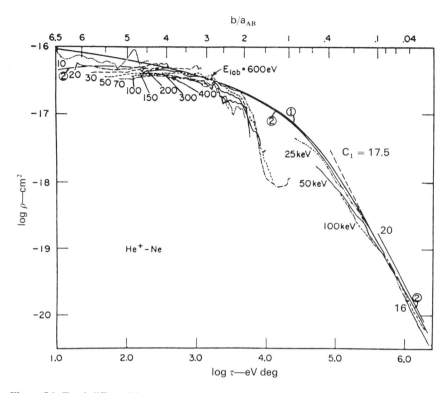

Figure 5.1. Total differential cross section, log ρ vs log τ, for the He$^+$ + Ne collision. Calculations: (1) one-term screened coulomb; (2) two-term screened coulomb with polarization attraction. $C_1 = 16$, 17.5, 20 are parameters for a coulomb potential, C_1/R, in atomic units. Experiments: low-τ elastic scattering; high-τ total differential cross section. Upper axis: scaled impact parameter, b/a_{AB} (see text). [From F. T. Smith, R. P. Marchi, W. Aberth, and D. C. Lorents, *Phys. Rev.* **161**, 31 (1967).]

rameters. Smith and co-workers used the static screened coulomb potential at Eq. (4.8), $V = (C_1/R)\exp(-R/a_{AB})$, to calculate ρ vs τ classically. They were able to fit the data in Figure 5.1 with $C_1 = 17.5 \pm 1$ and $a_{AB} = 0.68 \pm 0.04a_0$. The scaled impact parameter b/a_{AB} is indicated at the top of the figure and the derived screening constant is compared to models from Chapter 4 in Table 5.1. To account for different screening in the two shells of Ne, a two-term function is used having the right asymptotic limit $R \to 0$ ($\tau \to \infty$), $V = 2\{8\exp(-R/a_2) + 2\,\exp(-R/a_1)\}/R$. The classical cross section agrees with the above data when $a_1 \cong 0.07a_0$ and $a_2 \cong 0.70a_0$. These screening lengths are close in magnitude to the mean radius of the two shells of Ne. Lastly, use of the power-law inversion from Chapter 2 can reproduce these potentials quite well (Problem 5.1).

The difference between the screened potentials and the data at low τ

Table 5.1. Screening Length a_{AB} for He$^+$ + Ne in Atomic Units

Experiment	$(Z_A^{2/3} + Z_B^{2/3})^{-1/2}$	$(2)^{1/2}(Z_A' + Z_B')^{-1}$	$Z_A Z_B(\varepsilon^A + \varepsilon^B - \varepsilon_{(0)}^{AB})^{-1}$
$a_{AB}(a_0)$			
0.68 ± 0.04^a	0.40^b	0.77^b	0.5^b

^{a} F. T. Smith, R. P. Marchi, W. Aberth, and D. C. Lorentz, *Phys Rev.* **161**, 31 (1967).
^{b} From Chapter 4; $Z_A' = \sqrt{(2I_A)}$ in a. u.; ε_{AB} are Hartree–Fock binding energies Eq. (4.16).

can be accounted for by adding a small attractive part as indicated in curve 2. Screened coulomb forms, which allow for shell effects, combined with attractive, appropriate interactions can generally be used to describe the ρ vs τ data for colliding atoms. For some systems with very weak attractive components (e.g., collisions between closed-shell systems), the more slowly decaying, Born–Mayer form $V = V_0 \exp(-R/a)$ gives a better parametrized

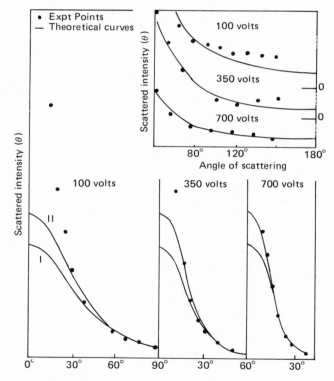

Figure 5.2. Total differential cross sections for e + He. Dots, experiment; curve, Born calculation using static interaction potential; I, based on a hydrogenic wave function, $Z' = 1.69$; II, based on Hartree–Frock approximation. [From N. F. Mott and H. S. W. Massey, *The Theory of Atomic Collisions*, 3rd edn., Oxford University Press, Oxford (1967), p. 458.]

fit to ρ at small τ. For incident, fast electrons, use of the static potential of Eq. (4.4) in the Born approximation yields total differential cross sections (Figure 5.2) in reasonable agreement with experiment except at small angles. The discrepancy at small angles is not as much a reflection on the accuracy of the potential as it is on the Born approximation. It arises because the diffraction region ($\chi < \chi_c$) is larger than that for incident heavy particles at any given velocity, due to the lighter mass of the electron. In the following discussion we consider the attractive part of the potential for heavy-particle collisions in more detail.

The above discussion confirms that static repulsive potentials can be extracted from total differential cross-section experiments. For collisions involving attractive potentials, interference phenomena occur even for $\chi > \chi_c$, as indicated by the fluctuations at low τ seen in Figure 5.1. At low velocities, where the thermal averaging is significant, these structures are often missed if the resolution is not sufficient. In Figure 5.3 the classical elastic cross section corresponding to the long-range van der Waals interaction is observed for K + Xe. For a $1/R^6$ potential, $\sigma_{AB}(\theta) \propto \theta^{-7/3}$, based on Eq. (2.61a). This corresponds to the slope observed in Figure 5.3. At even smaller angles this classical cross section gives way to a gaussian angular dependence resulting in a maximum at $\theta = 0°$ for the differential cross section. That is, the classical singularity disappears, as seen in Figure 5.3. This was shown to be the case for a screened coulomb potential in the Born approximation [Eq. (3.64)]. Here, however, it is seen to be the case also for power-law potentials, indicating a failure of the Born treatment in the diffraction region. The importance of the small θ behavior is clear. The total, integrated cross section exists (i.e., is not infinite) for realistic long-range interactions, except, of course, for the coulomb interaction. In fact the expression for $\sigma_{AB}(\chi)$ at $\chi = 0$ using Eq. (3.35) is simply related to σ_{AB} in Eq. (3.38), a result referred to as the optical theorem [Eq. (3.41)]. Classical cross sections, therefore, can only be used to calculate σ_{AB} if a cutoff in the vicinity of χ_c is assumed.

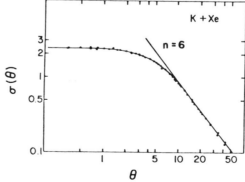

Figure 5.3. Elastic differential cross section versus laboratory angle θ for K + Xe using crossed thermal beams. Slope corresponding to classical cross section for $V_{AB} \propto R^{-6}$ indicated. [From R. Belbing and H. Pauly, *Z. Phys.* **179**, 16 (1964).]

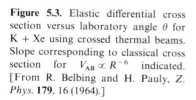

At very low energies and high resolution (reduced thermal averaging and increased angular resolution) the oscillatory structure in the differential cross section appears. Multiplying the differential cross section for the Na + Xe collision by $\chi^{7/3}$ to account for the classical angular dependence, we can clearly observe the rainbow enhancements in Figure 5.4. Two sets of oscillations are observed: a long-wavelength set, corresponding to the primary rainbow structure and described reasonably well by semiclassical methods (viz. Figure 3.8), and a faster set of quantum oscillations. Semiclassical phase shifts were calculated from the Lennard-Jones 6–8 potential [Eq. (2.62)] and used in the quantal expression for the scattering amplitude [Eq. (3.35)]. A "best" fit was obtained with $C_6 = 1.25 \times 10^3$ and $C_8 = 0.81 \times 10^5$ a.u. corresponding to a well depth of 0.013 eV at a separation of $9.3a_0$. Some thermal averaging was included in the comparison with experiment.

In collisions between ions and atoms, attractive potentials are a result of the target-atom polarizability at very large R and the much stronger electron-exchange interaction at intermediate R which produces the minimum. The effect of inelastic channels on the total differential cross sections can be seen quite clearly from experiments on $H^+ + Kr$ at higher energies. The close-lying charge-exchange channel plays a significant role at the energies considered in Figure 5.5 and has a very different potential form from that of the ground state (e.g., see Figure 4.8 for $H^+ + Ar$). The location of the rainbow maximum observed at low energies due to the attractive ground-state potential, V_0, is indicated by τ_r in Figure 5.5. The maximum in

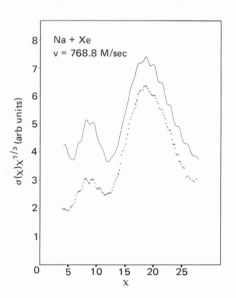

Figure 5.4. Elastic scattering cross section times $\chi^{7/3}$ vs χ. Low-frequency rainbow oscillations and high-frequency quantal oscillations. Solid line, calculation using semiclassical phase shifts and thermal averaging. [From P. Barwig, U. Buck, E. Hundhausen, and H. Pauly, Z. Phys. **196**, 343, (1966).]

Figure 5.5. Total differential cross section in terms of ρ vs τ for $H^+ + Kr$. τ_R indicates rainbow angle for elastic scattering from the ground-state potential at low velocities. Solid lines: (a) calculation using $\sigma_{AB}(\chi) \sim (1 - p_{of})\sigma_0(\chi) + P_{of}\sigma_f(\chi)$ and (b) using $\sigma_{AB}(\chi)$ calculated from $V_{AB} \sim (1 - p_{of})V_0 + p_{of}V_f$ (see text). [From: R. E. Johnson, C. E. Carlson, and J. W. Boring, *Chem. Phys. Lett.* **16**, 119 (1972).]

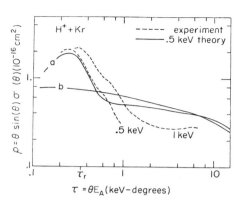

the observed cross sections is also located close to this τ_r, although transitions to the repulsive $H + Kr^+$ state, V_f, occur both on the inbound and outbound passes of the transition region. The average potential, which has the form $\bar{V}_{AB}(R) \sim (1 - p_{of})V_0 + p_{of}V_f$, where p_{of} is a transition probability on one pass through the strong-coupling region (e.g., see Figure 4.15), would have a smaller effective well depth than V_0 alone. Therefore τ_r [Eq. (2.63)] would be reduced, and hence the $\sigma_{AB}(\theta)$ calculated using $\bar{V}_{AB}$ indicates no rainbow in Figure 5.5. However, the averaged total differential cross section, elastic plus charge exchange, Eq. (4.71), $\sigma_{AB}(\chi) \sim (1 - p_{of})\sigma_0(\chi) + p_{of}\sigma_f(\chi)$ agrees with experiment. This implies that the total differential cross section is an average of cross sections each corresponding to a well-defined trajectory, as we emphasized in Chapters 3 and 4. Therefore, the use of an average potential, though convenient, is suspect in that region where inelastic processes are important but the distance of closest approach does not correspond to a strong repulsive interaction. This is the region, $10^3 \gtrsim \tau \gtrsim 5 \times 10^4$, for the results in Figure 5.1. Of course, for many practical situations the percentage error involved in using a single velocity-dependent (or static) interaction is small compared to other uncertainties.

To describe the elastic cross section only, optical model potentials, which add an imaginary part to $V(R)$, can be used to account for the expected drop in the *elastic* cross section (e.g., Figure 5.1) due to onset of the inelastic processes. Implicit in such models, however, is the fact that the threshold in τ for the decrease in the elastic channel is the same as the threshold for the increase in the inelastic processes. This is seen in a subsequent section not to be the case unless, of course, the states involved have identical potentials. In the following section we discuss the relationship between the integrated cross sections and the interaction potentials and continue the discussion of inelastic processes in subsequent sections.

Before proceeding, we note that multiple scattering experiments provide an additional test of our ability to obtain a single average interaction potential. For comparison with experiment, Eq. (1.2) is solved in the limit of small-angle scattering to obtain the beam spread after passage through a thin target. In such experiments the universal scaling laws used by Lindhard and co-workers tend to be quite useful for describing the data in a unified manner. In Figure 5.6 the lateral spread of the transmitted beam at half the maximum intensity, $\Delta r_{1/2}$, is plotted versus the thickness of the material, X, both as scaled quantities: $\Delta \tilde{r}_{1/2} = \Delta r_{1/2} (a_{AB}^L)^2 (E_A a_{AB}^L/2Z_A Z_B e^2) n_B$ and $\tilde{X} = \pi (a_{AB}^L)^2 n_B X$, where a_{AB}^L is the Thomas–Fermi screening parameter used by Lindhard (Chapter 4). The experimental results are compared to values computed using two potential functions discussed in Appendix H; also shown are the power-law potentials which best describe the data at various scaled thicknesses. It is seen that the thickness change corresponds, roughly, to a change in the important distances of closest approach. Hence, changing thickness allows one to test the average interaction potential in the screening region at different internuclear separations. A higher-resolution plot of such data indicates a dependence at small $\tilde{X}$ on Z_A and Z_B that does not obey the simple scaling laws, as was the case for small τ above. Further, in molecular or solid targets the spatial correlation of nearest neighbors modifies such results in a predictable way.

Integrated Total Cross Sections and the Diffusion Cross Section

The shadows cast by atoms exposed to a beam, the integrated total cross section, are an edge effect determined by the behavior of the potential at large R. Although the classical cross sections are infinite, the quantum-mechanical cross sections are finite for most realistic potentials (viz. the Born and Massey–Mohr approximations of Chapter 3). This was confirmed, at low collision energies, by the experimental results at small angles discussed in the previous section. To calculate the total cross section [viz. Eq. (3.38)] the radial phase shifts are required. As this cross section is dominated by distant collisions, we use the impulse approximation to the phase shifts from Eq. (3.52) and power-law potentials to obtain (Appendix B)

$$\eta(b) \sim -\frac{C_n a_n}{(n-1)\hbar v} \frac{1}{b^{n-1}} \tag{5.1}$$

where a_n is the constant in Eq. (2.58). This form of the radial phase shift reproduces the classical impulse CM scattering angle of Eq. (2.58) when used in Eq. (3.47), i.e., $\chi = (2/K)(\partial \eta/\partial b)$. In the Massey–Mohr approximation to the quantal cross section in Eq. (3.40), the cutoff impact parameter, b, for determining the size of the cross section is that impact parameter beyond which the radial phase shift is less than some small

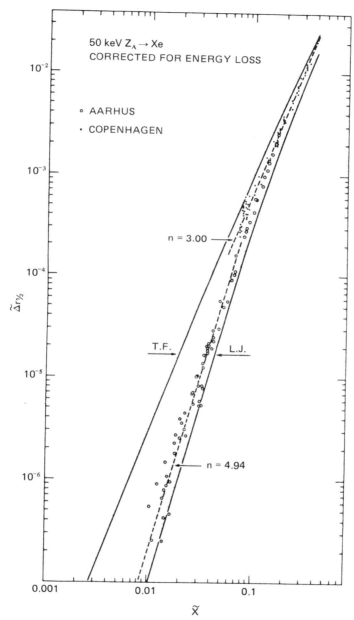

50 keV $Z_A \rightarrow$ Xe
CORRECTED FOR ENERGY LOSS

○ AARHUS

· COPENHAGEN

n = 3.00

$\tilde{\Delta}r_{1/2}$

T.F.   L.J.

n = 4.94

Figure 5.6. The radial spread in the transmitted beam at half the maximum intensity, $\Delta r_{1/2}$, vs the target "thickness," X, in scaled units (see text) for a variety of ions (Z_A) on Xe. Points and circles, data; dashed lines, power-law-potential calculations; solid lines, calculations using Thomas–Fermi (T.F.) and Lenz–Jensen (L.J.) potentials (Appendix H). [From H. Knudsen, F. Besenbacher, J. Heinemeier, and P. Hvelpund, *Phys. Rev. A* **13**, 2095 (1976).]

number, ε, i.e., $\eta(b_\varepsilon) = \varepsilon$. With this definition, using Eq. (5.1) in Eq. (3.40), we find $\sigma_{AB} \propto v^{-2/(n-1)}$. This velocity dependence is confirmed by the low-resolution cross-section measurements shown in Figure 5.7, indicating a van der Waals interaction, $n = 6$. Conversely, a cross-section measurement can determine the long-range dependence of the potential, with the velocity dependence yielding the approximate power law and the magnitude, the value of C_n.

Cross-section measurements like those above are only possible at energies for which the detector resolution, θ_d, is much less than θ_c (lab value of X_c). When this is not the case, a partially integrated cross section is measured, as pointed out in Chapter 2 (Figure 2.2),

$$\sigma_{AB}(\theta_A > \theta_d) = 2\pi \int_{\theta_d}^{\infty} \sigma_{AB}(\theta) \sin \theta \, d\theta$$

Such measurements have been used by Amdur and Jordan, and others to extract potential information outside the diffraction region. These results are in agreement with potentials obtained from differential measurements. As accurate integrated cross sections for heavy-particle collisions cannot be measured at all energies, one has to rely on calculations of these cross sections using potentials determined from differential cross-section measurements. At very low energies, σ_{AB} is generally a slowly varying function of energy which is often approximated by a constant.

Because of their much lighter mass, the long-wavelength scattering region is more easily studied using incident electrons, and rather interesting effects are found. In the limit of very long wavelengths, the cross section in Eq. (3.38) depends only on the $l = 0$ contribution, [viz. Eq. (3.37)]

$$\sigma_{AB} \simeq \frac{4\pi}{K^2} \sin^2 \eta_0 \quad \text{for} \quad K^{-1} > a_{AB} \tag{5.2}$$

which, if the phase shift is small, is consistent with the Born approximation. On the other hand, for certain collision pairs it is possible that η_0 may attain a value $n\pi$. Now, $\sigma_{AB} \sim 0$ in Eq. (5.2), rather, it depends on the $l = 1$ contribution, η_1. This is referred to as the Ramsauer–Townsend effect and is

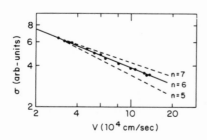

Figure 5.7. Integrated cross section at low resolution versus velocity for $K + N_2$ indicating R^{-6} dependence of long-range potential [From: H. Pauly, *Z. Naturforsch. Teil A* **15**, 277 (1960).]

seen in Figure 5.8 for electrons on Ar, Kr, and Xe. Comparing the radii of the outer-shell electrons for the targets shown in Figure 5.8 (He, $0.93a_0$; Ne, $0.97a_0$; and Ar, $1.3a_0$) to the wavelength of the incident electrons, we see that pure $l = 0$ (s-wave) scattering would be expected to occur at energies below 16 eV for He, 14eV for Ne, and 8 eV for Ar. However, because the target charge Z_B, hence the potential strength, increases considerably in going from He to Ar, a Ramsauer–Townsend is seen for Ar. In fact, at 0.54 eV, the lowest two phase shifts are $\eta_0 \sim 3\pi$ and $\eta_1 \sim \pi$ for target Ar. On the other hand, the cross section for target helium, with a low Z_B, is seen to behave like the Born cross section of Eq. (3.65) as the interaction is weak. For atoms with large radii, s-wave scattering would occur only at extremely low energies, and the above effect is not observed (e.g., for potassium the mean radius is $\sim 6.1a_0$ and pure s-wave scattering begins below 0.4 eV).

In very low energy heavy-particle collisions the thermal distribution of particle energies in the beam affects the extraction of a cross section. However, nozzle beams, discussed briefly in Chapter 1, can be used to reduce this spread in energy considerably. When the energy spread in the incident beam is small, the integrated, heavy-particle cross sections are found to be oscillatory at low collision energies. Very small angle scattering from both the long-range tail of the potential (the diffraction region) and the repulsive wall produce this interference (see Figure 2.15). This, in a sense, is an extension of the rainbow interference to zero-degree scattering, and is referred to as a forward glory, by analogy with the glory phenomena in light scattering. To describe this effect, we note that the integration in impact parameter in Eq. (3.39a) is well approximated by the Massey–Mohr method only if the phase shift $\eta(b)$ changes monotonically and rapidly with b. When this is the case, the integrand, $\sin^2 \eta(b)$, oscillates rapidly with increasing b and can be replaced by its average value. If, however, this radial phase shift,

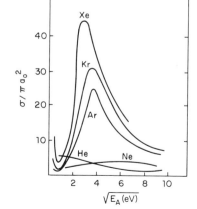

Figure 5.8. Integrated cross section versus $(E_A)^{1/2}$ for electrons incident on rare-gas atoms. [From H. S. W. Massey and E. H. S. Burhop, *Electronic and Ionic Impact Phenomena*, Oxford University Press (1968), pp. 25–26.]

i.e., not the total phase shift of Eq. (3.45), has a stationary point, then $\sin^2 \eta(b)$ cannot be approximated by its average value at all b. In the vicinity of this stationary point, which for total cross section corresponds to scattering near zero degrees, constructive or destructive interference will occur. The reader can use the stationary-phase method developed in Appendix F to obtain the cross section,

$$\sigma_{AB} \simeq 2\pi b_\varepsilon^2 \left[1 + \frac{2\varepsilon^2}{n-2} \right]$$
$$- \{4\pi b[\pi/(\partial^2 \eta/\partial b^2)]^{1/2} \cos[(2\eta + \pi)/4]\}_{b=b_s} \qquad (5.3)$$

where b_s is the stationary point of $\eta(b)$. The first term is the Massey–Mohr approximation to σ_{AB}, Eq. (3.40). The oscillatory part is clearly visible in Figure 5.9 for the K + Kr collision and is superimposed on the dominant van der Waals cross section, $\sigma_{AB} \propto v^{-7/3}$. As η depends roughly on v^{-1} [viz. Eq. (3.52)], the peaks are spaced equally in v^{-1} and can be indexed, a

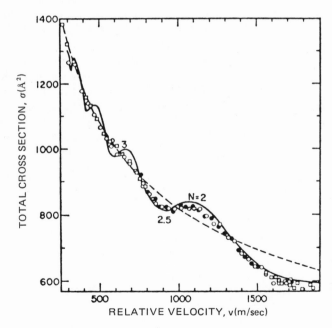

Figure 5.9. Integrated cross section velocity for the K + Kr collision indicating the glory oscillations. Points, experiments; dashed line, average cross section, as in Massey–Mohr approximation; solid line, cross section, including oscillations as in Eq. (5.3). [From E. W. Rothe, *J. Chem. Phys.* **39**, 493 (1963); see also R. B. Bernstein and J. T. Muckerman, *Low Energy Molecular Beam Scattering*, in *Intermolecular Forces* (*Advances in Chemical Physics*), Vol. 12, ed. J. O. Hirschfelder, Wiley, New York (1967), p. 466.]

procedure used in interpreting many interference phenomena. The oscillatory spacing is determined by $V_M R_M$ where V_M and R_M are the position and depth of the well minimum. For the Lennard-Jones 6-12 potential, for instance, [Eq. (4.27)]

$$\eta(b_s) \propto \frac{1}{\hbar v} V_M R_M \qquad (5.4)$$

and Bernstein has shown that the extracted $V_M R_M$ is nearly independent of the form chosen for the potential. Because these oscillations are related to the potential-well parameters, it should not be surprising that the number of glory maxima has been shown to be equal to the number of bound states of the potential. Therefore, the oscillatory pattern and the overall energy dependence can be used to characterize the primary features of the potential well rather accurately. For the data in Figure 5.8, $V_M = 0.0088$ eV and $R_M = 5.1$ Å using a 6-12 potential.

Considerable effort was made in the experiments discussed above to reduce the thermal spread in energy to make the oscillatory pattern manifest. However, in many applications there is a thermal distribution in energy associated with low-energy colliding particles. For instance, in Chapter 1 the diffusion coefficient had the form $D = kT/M_A v_A$, where v_A is a thermally averaged collision frequency. Quantities such as v_A are temperature dependent, as are the reaction cross sections to be discussed in the last section of this chapter. A simple estimate of v_A is $v_A \sim n_B \bar{v}\sigma_{AB}$, where $\bar{\sigma}_{AB}$ is a low-energy, averaged cross section (e.g., hard sphere), n_B is the target number density, and $\bar{v}$ is the mean thermal speed of the colliding particles. From the Maxwell–Boltzmann distribution (Problem 2.3), $\bar{v} = (8kT/\pi m)^{1/2}$: If A and B are at the same temperature and σ_{AB} is a constant, then $v_A \propto T^{1/2}$, which is about right for many systems.

To calculate v_A more carefully we note that resistance to diffusion is a result of collisions that change the momentum of the particle. Writing the change in the forward-directed momentum during a collision as $(\Delta \mathbf{p}_A)_{\parallel} = mv(1 - \cos \chi)$ and extending Eq. (2.3), we write the momentum transfer per unit path length as

$$2\pi \int (\Delta \mathbf{p}_A)_{\parallel} n_B \sigma_{AB}(\chi) d\cos \chi$$

Averaging this, as in Eq. (2.13b), over the thermal distribution of velocities, we determine the collision frequency v_A from

$$M_A v_A = 2\pi \int f(\mathbf{v}) d^3 v \int (\Delta \mathbf{p}_A)_{\parallel} n_B \sigma_{AB}(\chi) d\cos \chi \qquad (5.5)$$

where $f(\mathbf{v})$ is the Maxwell–Boltzmann distribution of relative velocities if A and B are at the same temperature. The angular integration in Eq. (5.5) generally is performed first, and the quantity so obtained is referred to as the diffusion cross section:

$$\sigma_d \equiv 2\pi \int_{-1}^{1} \sigma_{AB}(\chi)(1 - \cos \chi)\, d \cos \chi \tag{5.6}$$

Better thermodynamic models of this transport process (e.g., the Chapman and Enskog theory) indicate that $M_A v_A$ in the expression for the diffusion coefficient should be replaced by

$$M_A v_A = m(n_B + n_A) \int f(\mathbf{v})\, d^3 v\, v \left(\frac{mv^2}{3kT}\right) \sigma_d(v) \tag{5.7}$$

which involves the same diffusion cross section defined above. The expression for D should also be multiplied by $(1 + \varepsilon_0)$, where ε_0 is a quantity that is system dependent but generally small. If $\sigma_d(v)$ is independent of velocity, then *both* results indicate $v_A \propto T^{1/2}$, as predicted by our first simple argument.

Using the scattering amplitude of Chapter 3, Eq. (3.35), and integrating over angles, we see that the diffusion cross section in Eq. (5.6) becomes

$$\sigma_d = \frac{4\pi}{K^2} \sum_{l=0}^{\infty} (l + 1) \sin^2 (\eta_{l+1} - \eta_l) \tag{5.8}$$

This can be written in terms of impact parameters, replacing the sum by an integral, as was σ_{AB} in Eq. (3.39), yielding

$$\sigma_d \simeq 2\pi \int_{0}^{\infty} P_d(b) b\, db \tag{5.9}$$

with $P_d = 2 \sin^2 [\Delta\eta(b)]$. Writing $\Delta\eta(b)$ as $\Delta\eta(b) \simeq (1/K)(\partial\eta/\partial b)$ and using the definition of the deflection function in Eq. (3.4), we obtain $\Delta\eta(b) \simeq \chi(b)/2$, or $P_d = 1 - \cos [\chi(b)]$. This is the result we would obtain using the classical differential cross section, Eq. (2.44), directly in Eq. (5.6). That is, unlike the elastic scattering cross section, the quantal [Eq. (5.8)] and classical diffusion cross sections are equivalent! This, of course, is a result of the fact that the small-angle collisions (diffractions) are excluded in both sums by the angular factor $(1 - \cos \chi)$. Hence, σ_d is closely related to the classical concept of the cross section described in Chapter 2. In the power-law approximation

$$\sigma_d \propto E^{-2/n} \tag{5.10}$$

σ_d exhibiting a very different velocity dependence compared to σ_{AB}. For an exponential interaction potential both quantities exhibit a roughly logarithmic velocity dependence, although with very different constants, as does the Born calculation of σ_d (Problem 5.8).

When power-law cross sections are used, the collision frequency and diffusion coefficient have the following temperature dependence: $v_A \propto (kT)^{1/2-2/n}$ and $D \propto (kT)^{1/2+2/n}$. At low temperatures the diffusion coefficient for neutrals should vary as $D \propto (kT)^{5/6}$. At high temperatures the repulsive wall will determine the temperature dependence, which for an exponential potential varies slightly more rapidly than $T^{1/2}$. Measured high-temperature coefficients are often found to go as $T^{2/3}$ (actually $T^{0.69}$ is closer), which is part of the rationale for using a $1/R^{12}$ repulsive wall in the Lennard-Jones potential [viz. Eq. (4.27)].

At higher than thermal energies the diffusion cross section is of interest as it also describes the average elastic energy loss of a moving atom with a stationary target. This is the nuclear part of the stopping power of Eq. (2.71). That is, $S_n = \int T(d\sigma/dT)\,dT = (\gamma E_A/2)\sigma_d$, and hence S_n is a "classical" quantity. For heavy atoms Lindhard has obtained a general expression for S_n (Appendix H) which is frequently used in the literature and shows reasonable agreement with experiment over a broad range of energies.

The motion of ions through a material subjected to an electric field is controlled by collisions with the material atoms, as discussed in Chapter 1. At low field strengths the quantity of interest is the ion mobility, which is closely related to the diffusion coefficient, $\kappa = qD/kT$. Measurements of ion mobilities, therefore, also provide a measure of the diffusion cross section or the collision frequency. From our previous discussion $\kappa \propto T^{2/n-1/2}$, which would be temperature independent at low temperatures, where the long-range polarizability interaction dominates. Measuring κ, therefore, provides a direct measure of the polarizability of the material atoms (usually a gas). That is, $\kappa \sim 35.9/(\alpha m)^{1/2}$ cm^2/V/sec, if α is given in units of $(a_0)^3$ and the reduced mass m is in amu. A slow variation in κ with temperature is observed in some systems, as is seen in Figure 5.10, indicating that corrections to the long-range force and/or the repulsive barriers modify the diffusion cross section even at low temperatures.

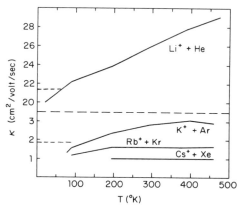

Figure 5.10. Ion mobilities versus temperature: Dashed lines are values of κ for He and Ar using the polarizabilities in Table 4.1. [From A. Dalgarno, in *Atomic and Molecular Processes*, ed. D. R. Bates, Academic Press, New York (1962).]

If the ion moves in its parent gas, the movement of the charge in the field is not just the motion of the heavy particle to which the charge is initially attached. At each encounter, electron exchange with a target molecule may occur, which slows the effective motion of the charge in the field and leads to lower mobilities. The diffusion cross section for such systems is found to be twice the charge-transfer cross section due to both elastic collisions and charge exchange (Problem 5.9). The charge exchange process is the topic of the next section.

Charge Exchange

The charge exchange process is one of the most extensively studied inelastic collision processes. This is due not only to its importance in plasma phenomena, but also because of the simplicity involved in identifying and separating the scattered charge-exchange species from the beam particles. Over the full energy range studied (MeV → thermal), measured charge-exchange cross sections provide a test of our ability to describe transition probabilities and interaction potentials at all levels of inelasticity in exothermic, endothermic, and symmetric-resonant collisions. At very high energies when the momentum transfer of the electron becomes important, this process becomes characteristic of fast reaction processes, as pointed out in Chapter 4. We therefore discuss charge exchange in this section as an example of inelastic collisions, considering curve crossings, weak coupling, interaction potentials, etc. In the following section we look at ionization cross sections and quantities that sum over all inelastic processes.

Integrated cross sections for charge transfer with incident singly and doubly charged ions have well-known general characteristics. For strong coupling, the cross terms in Eq. (4.71), which depend on the difference in action, generally go through many oscillations as a function of b (or τ) at intermediate and low velocities. These terms are often assumed to be self-cancelling (the random-phase approximation) out to some impact parameter b^* where the phase differences and/or the transition probabilities become very small. Using this, we write the integrated cross section in a form, like that in Eq. (2.26), which lends itself to direct comparison with experiment (see also Appendix I):

$$\sigma_{0 \to f} \sim \bar{P}_{0 \to f}\, \pi b^{*2} \qquad (5.11)$$

where $\bar{P}_{0 \to f}$ is the averaged transition probability.

For symmetric resonant charge transfer [e.g., Eq. (4.62)] the phase difference between the g and u states is determined by the exchange energy, which falls off exponentially at large R. Therefore, $b^* \sim \beta^{-1}[B - \ln v]$, where β is the screening constant indicating the overlap of the atomic wave functions [viz. Eq. (4.42a)] and B, which is independent of velocity, is related to the potential strength. As $\bar{P}_{0 \to f} \sim \frac{1}{2}$, this cross section increases monotonically with decreasing velocity at low and intermediate velocities (Figure 5.11). The experimental cross section can, therefore, be used to extract the parameters B and β. This procedure fails at high energies since the cosine of phase differences in Eq. (4.72) does not go through a number of oscillations, and, also, momentum transfer and transitions to excited states become important. At very low velocities b^* may be determined by the orbiting radius of the attractive potential, which will be considered in the final section of this chapter.

For multiply-charged ions colliding with atoms (e.g., $A^{+2} + B \to A^{+} + B^{+}$) the exothermic final-state repulsive potential will cross the more slowly varying initial state characterized by a polarization interaction, as indicated in Figure 5.12. For mutual neutralization $(A^{-} + B^{+} \to A + B)$ and the reverse process atom–atom charge exchange $(A + B \to A^{-} + B^{+})$ the coulomb attractive states often cross the rather slowly varying neutral–neutral interaction potential at long range. R. E. Olson, F. T. Smith, and co-workers found most of the cross section data for these reactions can be calculated with a simple potential form like that described in Eq. (4.43c). At the crossing point, the nonresonant exchange interaction to be used in the

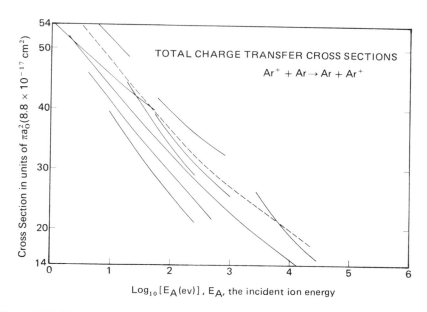

Figure 5.11. Charge-exchange cross section versus incident-ion energy for $Ar^{+} + Ar$. Solid lines, experiment; dashed lines, impact parameter calculation. [From R. E. Johnson, *J. Phys. B.* **3**, 539 (1970).]

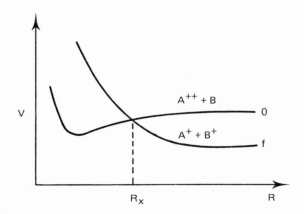

Figure 5.12. Diagram of the incoming and outgoing states for the $A^{+2} + B \rightarrow A^+ + B^+$.

discussion following Eq. (4.57) is

$$V_{0f}(R_x) \simeq V_0(\beta R_x) \exp(-\beta R_x) \tag{5.12}$$

where $V_0 = I_A^{1/2} I_B^{1/2}/0.86$, $\beta = 0.86(I_A^{1/2} + I_B^{1/2})/(2)^{1/2}$, and all quantities are in a.u. The screening constant β is similar to the expression used in Eq. (4.42a). In charge transfer due to a curve crossing, a reasonable approximation is obtained by setting $b^* \simeq R_x$. Using the LZS transition probabilities of Eqs. (4.58) and (4.59) we obtain the averaged transition probability of Eq. (5.11) as

$$\bar{P}_{0 \to f} \simeq \int_0^1 2p_{0f}(1 - p_{0f})\left(\frac{b}{R_x}\right) d\left(\frac{b}{R_x}\right) \tag{5.13}$$

which is velocity dependent. This expression has been applied to a large variety of charge-exchange collisions involving curve crossings. The general velocity dependence given by this prescription is characteristic of the observed velocity dependence for nonresonant collisions, (Figure 5.13). The cross sections usually go through a single broad maximum, falling off rapidly at low energies, and becoming orders of magnitude smaller than the maximum value of the cross section at energies well above the threshold energy (Appendix I).

Two-state, nonresonant charge-transfer reactions, which do not involve a curve crossing, also go through a single broad maximum. At low velocities, $b^* \simeq R_x$, the effective transition point; at high velocities, $\bar{P}_{0 \to f} \to \frac{1}{2}$ and b^* has a velocity dependence similar to the symmetric resonant case. The maximum occurs near that velocity predicted by the so-called Massey criteria discussed in Chapter 4. The transition probabilities are well ap-

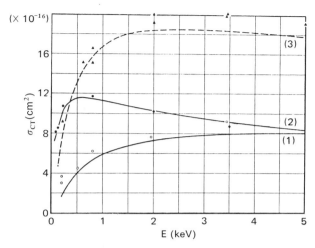

Figure 5.13. Single-electron capture by doubly charged ions, indicative of curve-crossing collisions. Curves, calculations using LZS: (1) $Ar^{+2}(^3P) + Ne(^1S) \rightarrow Ar^{+}(^2P) + Ne^{+}(^2p)$; (2) $N^{+2}(^2P) + He(^1S) \rightarrow N^{+}(^3P) + He^{+}(^2S)$; (3) $Ne^{+2}(^2P) + Ne(^1S) \rightarrow Ne^{+}(^1S) + Ne^{+}(^2P)$. Experiments: $\bigcirc$, $Ar^{+2} + Ne$; $\bullet$ $N^{+2} + He$; $\blacktriangle$, $N^{+2} + Ne$. [From R. A. Mapleton, *Theory of Charge Exchange*, Wiley, New York (1972), p. 212.]

proximated by the expressions of Demkov [Eq. (4.61c)], or Rosen and Zener [Eq. (4.61b)] (Appendix I).

For resonant charge transfer between unlike species or near-resonant charge transfer (a common occurrence when the target is a molecule or when a number of close-lying spin states are involved), similar concepts can be applied. However, now the behavior of the long-range forces critically affects the position of the maximum by producing splitting of the degenerate or nearly degenerate initial and final states of the collision system. For example, if there is significant spin–orbit splitting in the final-state multiplet, the exchange interaction, which increases exponentially with decreasing R, eventually dominates the spin–orbit interaction at some value of R. If b^* is less than that internuclear separation at which the spin–orbit and exchange interactions are approximately equal, the fraction of products in a given atomic J state of the multiplet can be determined from a statistical weighting, that is, all transitions among the spin states are equally likely. If b^* is much greater than that internuclear separation, the different J states should be treated as distinct final states, and transitions do not occur for the most part. This is indicated in the calculation for the $Xe^{+}(^2P_J) + Xe$ charge-exchange collision in Figure 5.14 in which the ground-state splitting of the ion is considered. At high velocities the splitting is unimportant and the cross sections are symmetric-resonant cross sections. At low velocities

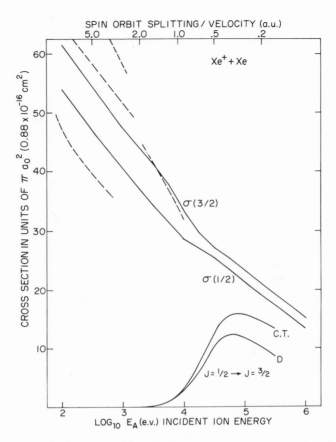

Figure 5.14. $Xe^{+} + Xe \rightarrow Xe + Xe^{+}$. Solid lines, impact parameter calculations: $\sigma(1/2)$, $J = 1/2 \rightarrow J = 1/2$, $3/2$; $\sigma(3/2)$, $J = 3/2 \rightarrow J = 3/2$, $1/2$; C.T., inelastic charge transfer; D, inelastic, no charge transfer. Dashed lines, experiments. [From R. E. Johnson, *J. Phys Soc. Jpn.* **32**, 1612 (1972).]

the cross sections are also symmetric-resonant for each of the spin states separately.

Detecting transitions between states that differ only slightly in energy is often quite difficult, particularly in fast collisions. The effect of fine structure or vibrational levels in fast charge-exchange collisions can be measured by crossing an ion beam with a near-thermal neutral beam. The slow ion produced from a charge-exchange collision will experience a significant deflection for even small inelastic energy differences (Problem 5.4). Velocity analysis of these ions allows one to measure small inelastic energy losses which occur during the collision. At a given scattered-ion energy the angular deflection is directly related to the inelastic energy loss. Therefore, the observed distribution in Figure 5.15 clearly indicates the vibrational energy

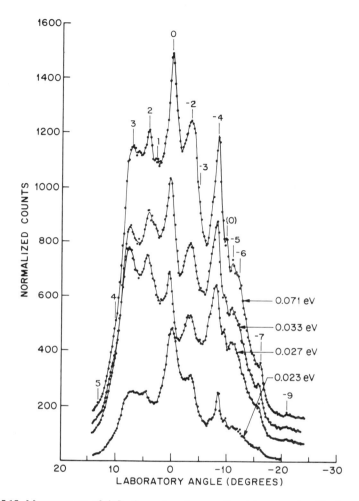

Figure 5.15. Measurement of deflections at various scattered ion energies, E'_B, in a crossed-beam experiment: $N_2^+ + N_2 \rightarrow N_2^* + N_2^*$. Ion beam 111 eV, neutral beam 0.071 eV, E'_B indicated on right. Deflections at these E'_B correspond to vibrational energy transfer as indicated by change number of vibrational quanta; minus sign indicates exothermic. (O) implies electronic transition in N'_2 with no vibrational exchange. [From K. B. McAfee, C. Szmanda, and R. Hozack, *J. Phys. B* **14**, L243 (1981).]

change in $N_2^+ + N_2 \rightarrow N_2 + N_2^+$ collisions. The change in vibrational state is a result of the electronic process, here charge transfer, and such effects will be considered further where molecular collisions are discussed.

Deviations from the general behavior of inelastic processes described above occur for a variety of reasons. For instance, if the difference in the potential functions involved, and, hence, the corresponding actions for two trajectories, goes through a maximum, the random-phase approximation would break down in the vicinity of this maximum as in the glory oscillations considered in the previous section. For the symmetric-resonant, two-state example, if a stationary point exists in Eq. (4.72) then

$$\sigma_{ct}(v) \approx \frac{\pi}{2} b^{*2} - \left[\frac{\pi^{3/2} b}{|(\partial^2/\partial b^2)[(A^{(g)} - A^{(u)})/2\hbar]|^{1/2}} \right.$$

$$\left. \times \cos\left(\frac{A^{(g)} - A^{(u)}}{\hbar} + \frac{\pi}{4} \right) \right]_{b=b_s} \tag{5.14}$$

where b_s is the position of the stationary point and $A^{(g)}$ and $A^{(u)}$ are the semiclassical phases. Similar expression can be obtained for inelastic collisions. The spacing and magnitude of the oscillatory structure Figure 5.16

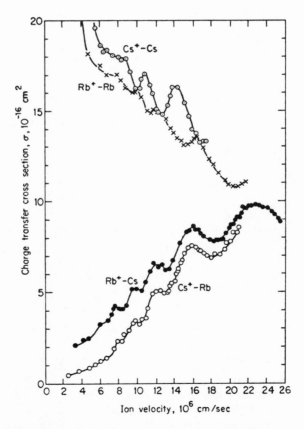

Figure 5.16. Charge-exchange cross sections showing oscillatory structure superimposed on the usual energy dependence for symmetric and nonresonant collisions. [From J. Perch, R. H. Vernon, and H. L. Daley, *Phys. Rev. A* **131**, 937 (1965).]

can be used to determine the behavior of the potential differences in the vicinity of the extremum using Eq. (5.14). Charge-exchange cross sections also differ from the simple forms discussed for very fast collisions. In this limit the momentum transfer required to exchange an electron between the colliding species decreases the average charge-transfer probability $\bar{P}_{0 \to f}$ used in Eq. (5.11), as described in Eq. (4.66a). Consequently, the charge exchange process, which is a dominant effect at intermediate velocities for collisions involving ions, becomes a very small effect compared to, for example, ionization at high energies.

In the charge-exchange collision of doubly charged ions with atoms or molecules the exiting fast ions can be energy-analyzed as described in Chapter 2. The resulting energy-loss spectra at each scattering angle (for example, Figure 2.12) is a measure of the transition probabilities as a function of impact parameter using $E_A \theta_A = \tau(b)$. For the $He^{+2} + Ne \to Ne^{+*} + He^{+*}$ collision there are three groups of exothermic states indicated in Figure 2.12. A transition to the ground state of $NeHe^{+2}$ is highly exothermic and is seen not to be populated significantly after the collision. Using the long-range $1/R$ exit channel potential and the polarizability of Ne (Table 4.1), we can estimate the crossing of the initial state and find states (Figure 5.12). The crossing of the initial potential with that of the $He^{+} + Ne^{+}[2s2p^6, {}^2S]$ potential, $Q = -5.94$ eV, occurs at $R_x = 4.8a_0$, and for the $He^{+} + Ne^{+}[2p^4 2s, {}^2D]$ potential, $Q = -2.2$ eV at $R_x = 12.3a_0$. As the coupling potential in Eq. (5.12) depends exponentially on R_x, at small scattering angles only transitions for which $Q = -5.94$ eV are observed. Transition probabilities to the channel with small exothermicity vary in angle more like the noncrossing, endothermic transitions. This should not be too surprising. At the crossing point $R_x = 12.3a_0$ almost no transitions occur at this collision energy because of the small coupling potential. At smaller R, the initial state lies below the repulsive final state; hence, any subsequent transitions would be a noncrossing type similar to the high-lying endothermic channels. Such transitions would occur, based on the discussion preceding Eq. (4.61), when $V_{0f} \approx \frac{1}{2}$(the level splitting). That is, as observed, the onset to this and higher-lying states will occur at larger τ.

The incident channel $He^{+2} + Ne[2p^6, {}^1S]$ leads to a single molecular state of symmetry ${}^1\Sigma^+$. According to the coupling mechanisms discussed in Chapter 4, ${}^1\Sigma^+ \to {}^1\Sigma^+$ or ${}^1\Pi$ transitions only are allowed in the absence of spin–orbit coupling. Two exit channels have exothermicity between 5 and 6 eV: $He^{+}[{}^2S] + Ne^{+}[2s(2p)^6, {}^2S]$ leads to ${}^{1,3}\Sigma^+$ molecular states, and $He^{+}[{}^2S] + Ne^{+}[({}^3P)3s, {}^2P]$ leads to ${}^{1,3}\Sigma^-$ and ${}^{1,3}\Pi$ states. At small angles the peak has been identified with the $Ne^{+}[2s(2p)^6, {}^2S]$ state, which is consistent with the fact that rotational coupling (between the Σ and Π states) is weak for slow collisions at large internuclear separations and ${}^1\Sigma^+ \to {}^{1,3}\Sigma^-$ is "forbidden".

The $He^{+2} + Ar$ system has very different energy characteristics, as is seen in Figure 5.17. The entrance channel lies in the continuum of the Ar^+ ion for charge transfer to the He^+ ground state. As in the previous case, the possible endothermic channels at large R diverge in energy from the incident $^1\Sigma^+$ state and, therefore, exothermic reactions are preferred. Three

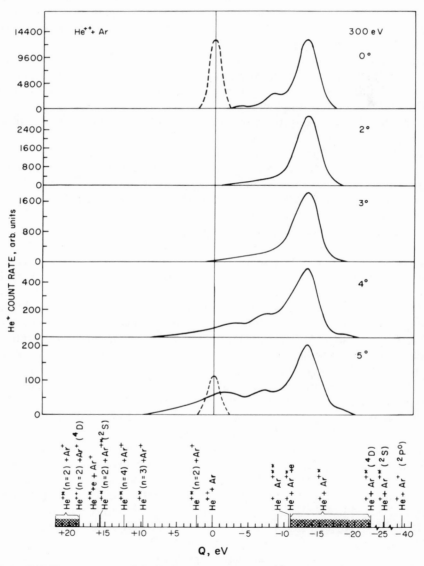

Figure 5.17. Energy spectra of single-electron capture in $He^{+2} + Ar$. Dashed peak indicates width of elastic peak. [From R. E. Johnson and J. W. Boring, in *Collision Spectroscopy*, ed. R. G. Cooks, Plenum Press (1979), p. 129.]

broad peaks, i.e., broader than the elastic peak, are distinguishable in Figure 5.17. The largest is associated with the closely spaced Ar^+ excited states. The next peak may be associated with doubly excited autoionizing states in the continuum of Ar, as these doubly excited states may lead to well-defined molecular potentials (e.g., Figure 4.7). The last peak is associated with single charge transfer into the continuum of states associated with $He^+(1s) + Ar^{+2} + e$ and those other bound states that lie in the continuum, e.g., $He^+(2s) + Ar^+$. This peak grows and broadens with increasing τ (decreasing impact parameter). Further, the maximum shifts toward endothermic energy transfers as τ increases, since the transitions are taking place at smaller values of R where the biasing effect of the $1/R$ long-range potential has diminished. This peak can be roughly described by the weak coupling probability, Eq. (4.54), multiplied by the density of states. In contrast, the maximum of the peak due to curve crossing at large R remains very nearly fixed in energy transfer over a broad range of τ.

Although the energy loss spectra above are useful, angular differential cross-section data will yield information not only on the relative likelihood of the possible transitions but also on the interaction potentials involved in the collision. The semiclassical procedure described in Chapter 4 has been employed with success by Smith and co-workers to a number of collisions involving charge transfer. They find that even though the procedure has severe limitations near the angular threshold region, it gives an excellent starting point for interpreting the angular dependence and interference oscillations in inelastic and, as we saw earlier, elastic cross sections. The quantities one wishes to find are the differences between potential energy curves, the position of transition regions, and the coupling strength.

The semiclassical relationship between the action and the angular momentum for a trajectory (q) (Problem 5.10)

$$\frac{\partial A^{(q)}}{\partial \chi} = -L^{(q)} \tag{5.15}$$

where $L^{(q)} = p_0 b^{(q)}$, relates the oscillations in the differential cross section [e.g., Eq. (4.71)], when they are resolved, to the spacing between the potential curves for the different trajectories. If the energy change, Q, in the transition is small compared with the CM energy E, then the spacing between the maxima in the measured cross section can be related to the difference in impact parameter for the two interfering trajectories associated with the scattering into angle χ. Using Eqs. (5.15) and (4.71), we see that this difference is

$$|\Delta b(\chi, E)| \simeq 2\pi\hbar\left(\frac{E}{2m}\right)^{1/2} \frac{\partial n}{\partial \tau} \tag{5.16}$$

where n is an indexing number for the maxima. Therefore, a first step in analyzing the collision is to choose potentials which yield deflection functions producing the correct spacing. This in itself does not uniquely determine the potential curves. One needs to know the value of the potentials at some R from an accurate calculation or adjust the trial potentials to yield accurate values for the cross section also. Since absolute differential cross-section measurements often have large errors associated with them, the latter check is sometimes difficult. The position of the maxima in the charge-transfer cross section as a function of τ at incident ion energies 45–300 eV are plotted in Figure 5.18 for the unsymmetric charge-transfer collision $He^+ + Ar \rightarrow He + Ar^+$ ($3s3p^6$, ^2S). It is seen that the slope of this curve is very nearly a constant, implying that the deflection functions are separated in b by a constant, $\Delta b \simeq 0.51 a_0$. This suggests that the interaction potentials, at those internuclear separations corresponding to the distances of closest approach involved in the collision, are nearly exponential functions of R with approximately the same screening constant but different multiplicative constants (Problem 5.11). The other reaction shown involves a nonexponential component in the potentials for close collisions.

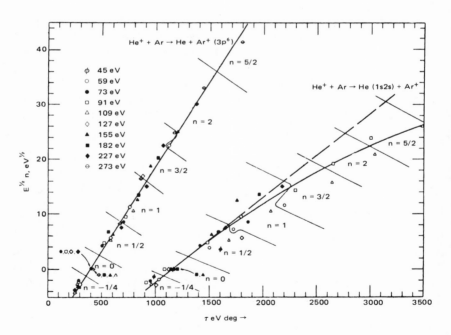

Figure 5.18. Position of maxima as a function of τ in the differential cross section for $He^+ + Ar \rightarrow He + Ar^+$ ($3s$, $3p^6$) and He ($1s$, $2s$) + Ar^+. Maxima labeled by index number n. [From F. T. Smith, H. H. Fleischman, and R. A. Young, *Phys. Rev. Sect. A* **2**, 379 (1970).]

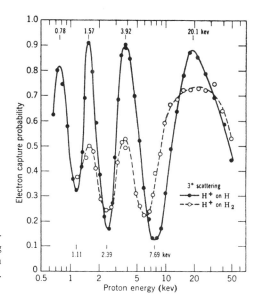

Figure 5.19. Electron capture probability versus energy at fixed scattering angle. Peaks nearly equally spaced in $1/v$. [From G. J. Lockwood and E. Everhart, *Phys. Rev.* **125**, 567 (1962).]

The above procedure has been applied most successfully for high-energy, small-angle collisions where the trajectories are simply approximated by straight lines. In this case the phase differences are described by the impulse approximation:

$$\frac{1}{\hbar} \left| A^{(1)} - A^{(2)} \right| \simeq \frac{1}{\hbar v} \left| \int_{-\infty}^{\infty} [V_1(R) - V_2(R)]\, dZ \right| \tag{5.17}$$

Further, τ is determined from some average interaction potential. In this limit, the oscillations have a simple v^{-1} dependence, as seen for the symmetric resonant, single-charge-transfer collision $H^+ + H$ in Figure 5.19, suggestive of the electron "oscillating" between the protons during the collision. Values for the exchange energy, $V_g - V_u$, as a function of internuclear separation extracted from these data agree closely with calculated potentials for this simple system. Nonresonant capture cross sections at intermediate and high energies also show a v^{-1} dependence in the oscillation of the charge-transfer probability. However, these oscillations are heavily damped at low energy, as can be seen for the $H^+ + H_2$ collision (Figure 5.19) and the $H^+ + He$ collision (Figure 5.20). These are examples of non-curve-crossing, strong-coupling collisions in the energy range shown. The damping, which is related to the collision time, has been shown to proceed roughly as $\exp[-v_0/v]$, where v_0 is a constant, for both crossing and noncrossing cases. Such a dependence on energy is indicated by Eqs. (4.58) and (4.61c) for the transition probabilities. Neglecting this damping,

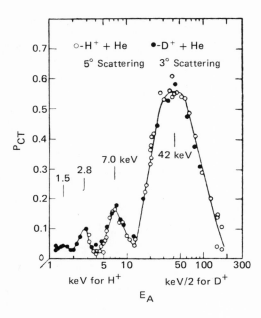

Figure 5.20. Electron capture probability versus energy for H^+ + He and D^+ + He showing approximate $1/v$ dependence and damping for nonresonant collisions. [From: F. P. Ziemba, G. J. Lockwood, G. H. Morgan, and E. Everhart, *Phys. Rev.* **118**, 1552 (1960).]

we find that the spacings between the oscillations are still a measure of the separation of the two lowest $^1\Sigma^+$ molecule–ion interaction potentials of HeH$^+$. It is clear from the spacing of the oscillations in the $H^+ + H_2$ collision that the exchange at these energies is between the incident H^+ and *one* of the target H atoms, as in an $H^+ + H$ collision.

As a final topic in this section we consider determining the location of transition regions from experiment. Such information will clearly help describe the interaction potentials between atoms. The onset in χ or τ of an inelastic process is related to the location of the transition region. However, as the transition region has a width, even for curve crossings, the onset for the charge transfer process is found to be somewhat energy dependent. This is indicated by the impact-parameter calculation of the transition probability for $Be^{+2} + H \rightarrow Be^+ + H^+$ illustrated in Figure 5.21, which is an example of the crossing described by Figure 5.12. Such an energy dependence means the extraction of R_x is not simple and calculations of the differential cross sections with trial potentials are needed.

In the measured elastic scattering cross section for the $He^{+2} + He$ collision shown in Figure 5.22, the competing *double* charge-transfer reaction is evident from the simple oscillating structure observed which resembles that of the $H^+ + H$ collision in Figure 5.19 and $He^+ + He$ in Figure 4.17. At τ equal to 3400 and 5000 eV-deg two perturbations appear in these observed oscillations. These perturbations can be associated with the crossing of the repulsive $^1\Sigma_g^+$ state of He_2^{+2} with higher states of the same

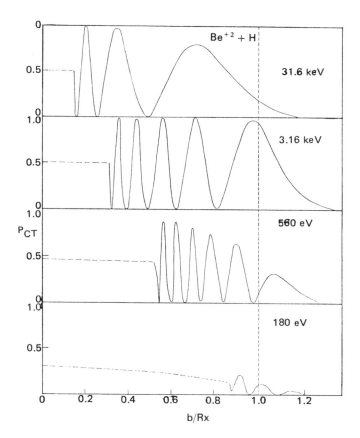

Figure 5.21. Calculated impact parameter of the charge transfer probability for $Be^{+2} + H$. Curve crossing indicated by dashed line. [From D. R. Bates, H. C. Johnson, and I. Stewart, *Proc. Phys. Soc.* **84**, 517 (1964)]

symmetry which correspond, at large R, to $He^{+}(1s) + He^{+}(n = 2)$(see Figure 4.7). At a crossing the elastic collisions can proceed via two separate trajectories (e.g., Figure 4.15) having somewhat different classical deflection functions. Olson and co-workers have shown that one of the deflection functions goes through two extrema which will produce perturbations on the elastic cross section if the transition probabilities are still small. Using exponential potential functions to estimate the actions in Eq. (4.72) which reproduce the oscillating structure (dashed curve), we find the crossing occurs at $R_x \sim 0.97a_0$ which helps locate the excited state curve in Figure 4.7.

At larger τ values the elastic cross section will begin to decrease as transitions to the excited states become very likely (Figure 5.23). At these

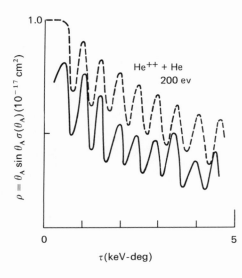

$\rho = \theta_A \sin \theta_A \sigma(\theta_A) (10^{-17}\ cm^2)$

He^{++} + He
200 ev

0   5

τ(keV-deg)

Figure 5.22. Elastic scattering of He^{+2} + He (solid curve). Calculation based on Eq. (4.72) using potentials to fit oscillatory structure (dashed curve). Two perturbations observed on peaks at high τ. [From Y. H. Chen, R. E. Johnson, R. V. Humphris, M. W. Siegel, and J. W. Boring, *J. Phys. B* **8**, 1527 (1979).]

energies the threshold in τ on the elastic channel is seen to be shifted considerably from the onset in the inelastic channel because the potentials of the final states are quite different. Since the thresholds for the elastic and inelastic cross sections are shifted, the total differential cross sections discussed earlier (e.g., Figure 5.1) are generally not smooth except at large τ. The outgoing potential for the *endothermic* reaction He^{+2} + He → He^{+}(1s) + He^{+}($n = 2$) is much less repulsive than the elastic channel for R near the transition region. The threshold seen in Figure 5.23 is consistent with the R_x determined from the elastic perturbations discussed above. Analysis of the cross section p_2 for the $n = 2$ state of He^{+} using the LZS approximation, Eqs. (4.58) and (4.71), yields a coupling strength of 0.17 a.u. and a value of 3.7 a.u. for the quantity $|(\partial/\partial R)(V_{ff} - V_{00})|_{R=R_x}$. The two peaks in the cross section are associated with the separate crossings of the incoming channel $^{1}\Sigma_g^{+}$ state with the two $^{1}\Sigma_g^{+}$ states (Figure 4.7) for the He^{+}(1s) + He^{+}($n = 2$) final states. Using the half-height of the peaks and the elastic potentials discussed earlier, we find that the second crossings occur at about $R_x \approx 0.85a_0$. The cross section labeled ρ_3 in Figure 5.23 is the sum of all unresolved higher single-charge-transfer processes. The onset of transitions to these higher states is $R_x \gtrsim 0.8a_0$.

The above discussion was intended to give the reader a sense of what can be learned about inelastic processes from experiments without making very detailed calculations. The charge exchange process was used as an example of such processes. In the following section we will discuss inelastic processes again, but will concentrate on measurements and calculations yielding quantities which are summed over inelastic events.

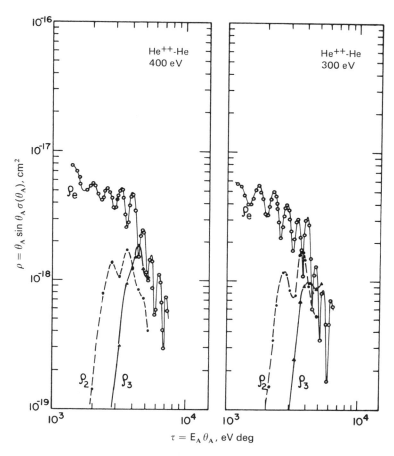

Figure 5.23. ρ vs τ plot of differential single charge transfer cross sections for $He^{++} + He$ collision: ρ_e, elastic collisions; ρ_2 and ρ_3, excited states. [From Y. H. Chen, R. E. Johnson, R. V. Humphris, H. W. Siegel, and J. W. Boring, *J. Phys. B.* **8**, 1527 (1979).]

Inelastic Energy-Loss Cross Sections

In calculating energy-loss cross sections that are summed over all inelastic events, the results are much less sensitive to the model. Bohr, therefore, used classical models with remarkable success in describing such quantities. When one is interested in the overall inelastic energy disposal in fast collisions, the principal contribution comes from collisions of low probability, but involving significant energy transfer. In this case, first-order (weak-coupling) approximations, either in the impact parameter or Born approximation, are useful, and their relationship to classical calculations will be considered.

An alternative, and more direct, approach to this problem involves calculating the kinetic energy *lost* by the incident particle due to the distortion of the target system, rather than calculating the inelastic energy transfer *to* the target system. For completely accurate calculations, these, of course, must be equal by energy conservation. But Lindhard has pointed out that in many instances the direct approach yields more accurate results when using only first-order perturbation theory. In this section we discuss total-cross-section quantities that involve sums over all inelastic processes and use the weak-coupling methods and the BEA to describe the measured results.

In many applications one would like to know the inelastic cross section $\sigma_{AB \to j}$, which we write here as σ_{Q_j}, Q_j, the inelastic energy transfer. Often it is easier to measure summed quantities: the total ionization cross section of the target, $\sigma_I = \sum_{Q_j > I_B} \sigma_{Q_j}$; the inelastic part of the stopping cross section from Eq. (2.18), $S_e = \sum_{Q_j} Q_j \sigma_{Q_j}$; or a quantity called the straggling cross section, $S_e^{(2)} = \sum_{Q_j} Q_j^2 \sigma_{Q_j}$. The quantity $S_e^{(2)}$ is related to the spread in inelastic energy loss during the passage of A through a target [i.e., $S_e^{(2)} = \Omega^2/(n_B \Delta x)$, where $\Omega^2 = \overline{(\Delta E - \overline{\Delta E})^2}$, the mean-squared deviation in the energy loss while traversing the target]. As the inelastic energy loss is weighted differently in each of these three quantities, they provide a good test of our knowledge of σ_{Q_j}.

In the simple BEA method of Chapter 2, Q_j is treated as a continuous variable, which, of course, it is for events leading to an ionization. At high collision energies the electrons can be considered very nearly stationary during the collision. The Thomson ionization cross section of Eq. (2.69) and corresponding stopping and straggling cross sections for an incident bare nucleus A are

$$\sigma_I \simeq 4\pi \frac{(Z_A e^2)^2}{(2m_e v^2)} \sum_{i,\,shell} N_{B_i} \left[\frac{1}{I_{B_i}} - \frac{1}{2m_e v^2} \right] \tag{2.69}$$

$$S_e \simeq 4\pi \frac{(Z_A e^2)^2}{(2m_e v^2)} \sum_{i,\,shell} N_{B_i} \ln \frac{2m_e v^2}{(Q_{min})_i} \tag{5.18}$$

and

$$S_e^{(2)} \simeq 4\pi (Z_A e^2)^2 \left[Z_B - \sum_{i,\,shell} N_{B_i} \frac{(Q_{min})_i}{(2m_e v^2)} \right] \tag{5.19}$$

where the target atoms are assumed to be neutral, $\sum N_{B_i} = Z_B$.

In these expressions the maximum energy transfer to any one electron was assumed to be the classical limit, $Q_{max} = 2m_e v^2$, and the electrons are in shells (i) with a removal energy I_{B_i} and minimum excitation energy $(Q_{min})_i$. For v large the ionization cross section decreases as $1/E$, the stopping cross section decreases more slowly, $(\ln E)/E$, and the straggling is a constant, the Bohr straggling cross section.

For the purpose of comparison with the above results, the Born approximation is used to calculate these sums employing the method of Bethe. In Chapter 4 we separated the Born cross section into close and distant collision contributions. The close-encounter part (large momentum transfers) had the same form as the Thomson cross section, but the distant, inelastic collisions resembled the light absorption process. That is, the transition probabilities were proportional to the dipole moment induced in the target atom by the time-varying field of the passing particle. Using Eqs. (4.79) and (4.80), we write the stopping cross section as $S_e = S_e^{close} + S_e^{dist}$. Defining $Q = \Delta p^2/2m_e$, we can then write the close-collision part in Eq. (5.18) in terms of the momentum transfer as

$$S_e^{close} \simeq 8\pi \frac{(Z_A e^2)^2}{(2m_e v^2)} Z_B \ln\left(\Delta p_{max}/\overline{\Delta p}\right) \qquad (5.20)$$

where $\overline{\Delta p}$ is the momentum-transfer cutoff between close and distant (hard and soft) encounters. Applying the sum rule for the oscillator strengths of Eq. (4.40b) (i.e., for a neutral target, $Z_B = \sum_j f_{0j}$), to the evaluation of S_e^{dist} from Eq. (4.80), we have

$$S_e^{dist} \simeq 8\pi \frac{(Z_A e^2)^2}{(2m_e v^2)} Z_B \ln\left(\overline{\Delta p}/\Delta p_{min}\right) \qquad (5.21)$$

where the lower bound, Δp_{min}, is an average value for the shells involved. The reader will notice immediately that Eqs. (5.20) and (5.21) have the same form, a rather important property of the coulombic interaction only. Therefore, one has the useful result that the sum of these two contributions is independent of $\overline{\Delta p}$. Bear in mind, however, that the physics of the close and distant collisions is very different.

The limits on the Bethe cross section are determined from the definition of the momentum transfer, $\Delta p = \hbar|\mathbf{K}_f - \mathbf{K}_0|$. The reader can verify that $(\Delta p_{max})_f = \hbar(\mathbf{K}_f + \mathbf{K}_0) \cong 2p_0 = 2m_e v$ is the classical limit, which is independent of state, and $(\Delta p_{min})_f = \hbar|\mathbf{K}_f - \mathbf{K}_0| \simeq Q_f/v$, which is Bohr's adiabatic cutoff described in Chapter 2 and discussed further in Chapter 4. A careful derivation of Eq. (5.21) results in a weighted average for the minimum Δp: $\Delta p_{min} = I/v$, with I defined by $Z_B \ln I = \sum_j f_{0j} \ln Q_j$, where f_{0j} are the oscillator strengths of Eq. (4.40b). Combining the close and distant collisions, we see that the inelastic stopping cross section becomes

$$S_e \simeq 8\pi \frac{(Z_A e^2)^2}{(2m_e v^2)} Z_B \ln\left(2m_e v^2/I\right) \qquad (5.22)$$

which is *twice* the Thomson (BEA) approximation in Eq. (5.18). Values for I

I (often referred to as the mean ionization energy) obtained by matching Eq. (5.22) with experiment are given in Figure 5.24.

The reader should note that if the lower bound on the momentum transfer, Q_f/v, is used to estimate a lower bound on the classical energy transfer to an electron, then the energy so calculated would, at large v, be lower than the first excited state of the atom! Such a limit does not suggest that the atom is able to absorb small energies. Rather, it emphasizes that energy and momentum are conserved for the system as a whole (all electrons and nuclei) and not separately for collisions of A with each electron on B, as in the BEA. Presumably a classical calculation can be made in which the target is an aggregate (electrons and nuclei), e.g., the collision of an ion with an electron oscillator shown in Appendix C. However, all such calculations must eventually be parameterized to correspond to the wave-mechanical target. In the BEA this only involves choosing a lower limit on energy transfer, the simplicity of which makes it an attractive model. If the cutoff between distant and close collisions in the Bethe approximation is chosen to be $\overline{\Delta p^2}/2m_e = I$, which is not unreasonable, then the close-collision contribution itself is equivalent to the BEA. This is consistent with the assumptions made for the BEA in Chapter 2.

Because a classical model allows any energy transfer, alternative parameterizations of the classical calculation can be used to "fix" the stopping cross section. For example, writing the classical Q in terms of the impact parameter,

$$S_e = 2\pi \int_{b_{min}}^{b_{max}} Q(b)b\,db$$

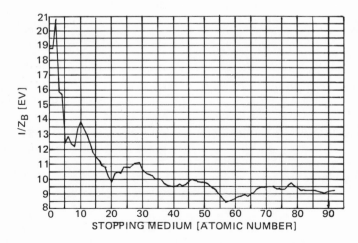

Figure 5.24. Experimental values for the mean ionization energy, *I*, obtained using Eq. (5.22) for the stopping of fast protons and hydrogen in various materials. [From H. H. Anderson and J. F. Ziegler, *Hydrogen: Stopping Powers and Ranges*, Pergamon Press, New York (1977).]

and letting $b_{max} \sim v/\omega$ (the adiabatic cutoff), and $b_{min} \sim \hbar/m_e v$ (the wavelength of the collision), a procedure discussed in Chapter 2, we see that the S_e obtained is equivalent to the Bethe expression. However, such a procedure relies on the fact that S_e^{dist} and S_e^{close} have the same form, and, therefore, cannot be extended to the calculation of σ_I or $S_e^{(2)}$ (as we will see below), nor for interactions other than the coulomb interaction.

Calculating the distant collision contribution to σ_I and $S_e^{(2)}$ for fast collisions, we have (here σ_I is the total inelastic cross section, ionization plus excitation)

$$\sigma_I^{dist} \simeq 8\pi \frac{(Z_A e^2)^2}{(2m_e v^2)} \left(\frac{2m_e \overline{z^2}}{\hbar^2} \right) \ln(2m_e v^2/I) \tag{5.23}$$

and

$$(S_e^{(2)})^{dist} \simeq 4\pi \frac{(Z_A e^2)^2}{(2m_e v^2)} \left(\frac{2\overline{p_z^2}}{m_e} \right) \ln(2m_e v^2/I) \tag{5.24}$$

where we have assumed the same lower limit, Δp_{min}, as in the S_e calculation. The quantities $\overline{z^2}$ and $\overline{p_z^2}$ are root-mean-square distortions of the ground state (e.g., $\overline{z^2} = \langle z - \bar{z} \rangle_{00}^2$, with $\bar{z} = \langle \Sigma_i z_{Bi} \rangle_{00}$) due to the passing particle. Comparing Eq. (2.69) with Eq. (5.23), we can see that the ionization cross section is dominated by dipole transitions (distant collisions) for incidentions at high v; hence the dominant energy dependence, $(\ln E)/E$, is close to that of S_e and not $1/E$ as in the classical calculation or for a screened coulomb (i.e., neutral–neutral) interaction [Eq. (4.55b)]. Conversely the straggling cross section is dominated by the close-collision, classical contribution in Eq. (5.19) which approaches a constant at high energies. This is basically what is observed, as indicated by the results in Figure 5.25. The similarity in energy dependence of σ_I and S_e for the incident particle–target pair permits one to define a quantity, the average loss per ionization produced, which is very nearly constant over a broad range of energies, although the values of I are slightly different in Eqs. (5.22), (5.23), and (5.24).

At intermediate and low velocities the above expressions fail. As v becomes comparable to v_e, the electrons in the target cannot be considered stationary. Therefore, a velocity distribution should be included in the BEA result [Figure (5.25)], or the Born approximation should be evaluated without the simplifications made in the above discussion. However, in this region two other effects also become quite important. First, the electrons do not simply behave as independent particles. The electron cloud distorts significantly in the direction of the incoming particle. This tends to screen the interaction of that particle with any one electron. Secondly, if the incoming particle in a stopping experiment is an ion, it may be neutralized rather quickly. Therefore, in passing through even a thin material the charge state of the particle will be changed, which is merely another manifestation of the distortion of the charge cloud.

In calculating single-collision values for σ_I, S_e, and $S_e^{(2)}$ the distorted-

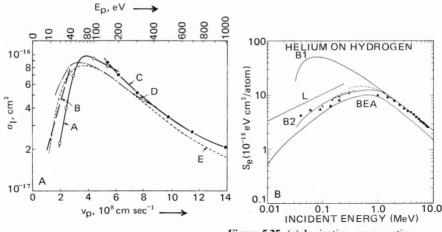

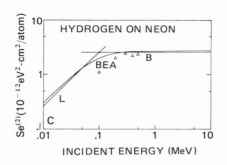

Figure 5.25. (a) Ionization cross section, σ_I, for p + He. Curves A, B, C, data; D, Born approximation; E, simple BEA (no velocity distribution). [From J. B. Hasted, *Physics of Atomic Collisions*, American Elsevier, New York (1972), p. 585.]; (b) Electronic stopping cross section for incident helium on hydrogen. B1, Bethe–Born formula, incident He^{+2}, Eq. (5.22); B2; Bethe–Born formula modified in hydrogen for equilibrium charge state; BEA, binary-encounter approximation for equilibrium charge state, including electronic velocity distributions; L, Lindhard low-energy expression [Eq. (5.31)]; dashed lines and other points, experiments. [From J. H. Harberger, R. E. Johnson, and J. W. Boring, *Phys. Rev. Sect. A* **9**, 1161 (1974).]; (c) Straggling cross section, $S_e^{(2)}$, for hydrogen on neon. Δ, experimental data; B, Born approximation; BEA, binary encounter approximation with electronic velocity distributions, equilibrium charge state, and contribution from incident particle excitation; L, Lindhard low-velocity expression [Eq. (5.32)]. [From R. E. Johnson, and A. M. Gooray, *Phys. Rev. Sect. A.***16**, 1432 (1977).]

wave corrections to the cross section have to be included, in which case the classical calculations become intractable. Further, the competing charge exchange processes have to be allowed for. That is, the quantities S_e and Ω^2 are usually measured as multiple collision quantities, with the expectation that, as the deflections are generally small, the energy losses are roughly additive. Therefore, measured cross sections are averages over the charge state of A *in* the material. If, for an incident A^+, a charge transfer occurs, the subsequent collision will be between a neutral A and a target atom. This neutralized particle may be re-ionized in further collisions, $A \rightarrow A^+$. The relative size of the neutralization cross section, σ_{ct}, and stripping cross sections, σ_s (ionization of A), will determine the fraction of collisions occurring in any particular charge state. Rate equations of the type discussed in Chapter 1 with rate constants for a beam of particles having a single velocity [e.g., Eq. (2.13a)] can be used to determine the charge-state fraction, where n_{A+} and n_A are the number densities of the beam

$$\frac{dn_{A+}}{dt} = -\sigma_{ct}vn_B n_{A+} + \sigma_s vn_B n_A \qquad (5.25)$$

with a corresponding equation for n_A. The reader can verify that the fraction of the beam ionized at anytime (or distance into the material) is

$$f^c_{A+} = [\sigma_s + \sigma_{ct}\exp(-\sigma' n_B v_A t)]/\sigma' \qquad (5.26)$$

with $\sigma' = \sigma_s + \sigma_{ct}$, if the atoms and ions are assumed to have lost relatively little energy. For a material of thickness $\Delta z \gg (\sigma' n_B)^{-1}$ (~ 10 Å in a solid at $V \sim V_e$) equilibrium fractions can be used: $f^c_{A+} = \sigma_s/\sigma'$ and $f^c_A = \sigma_{ct}/\sigma'$. Therefore, measurements of $(dE/dx)/n_B$ in a solid, for which $\Delta z \gg (\sigma' n_B)^{-1}$ but small enough that v changes little, should be compared to

$$\bar{S}_e \sim \sum_m f^c_{A+m}(S_e)_{A+m} \qquad (5.27)$$

where m sums over the possible charge states. In addition to this correction which allows for mixed charge states, nonadditive, multiple collision effects may influence the comparison with experiment, particularly for $S_e^{(2)}$. These have been treated by P. Sigmund and others.

The BEA calculations of $\bar{S}_e$ in Eq. (5.27), using a screened coulomb interaction for the neutral-particle collisions, including the velocity distribution of the target-particle electrons, and energy loss due to excitation of A is shown in Figures 5.26 and 5.25b. It is seen to give reasonable results at intermediate and low velocities for collisions of light atoms and ions and closely follows the Born calculation using Eq. (5.27) (Figure 5.25b). The contribution of the long-range dipole excitations of B by an ion becomes smaller as v decreases. In fact, at low velocities, where neutral collisions dominate, significant energy transfers *only* occur in close encounters of the target electrons with the incident particle. However, for large Z_B atoms, in which the mutual screening effect of the target electrons is important, the BEA approximation and, hence, the Born approximation fail hopelessly, unless a cutoff on Q [as in Eq. (3.66)] due to this electron screening effect is imposed. In this regard the reader should note that the BEA is too large for large Z_B targets at these velocities, unlike at high energies where it underestimates S_e.

The form found for the Bethe–Born approximation can be used to describe the measured stopping cross section at intermediate and low energies by replacing Z_A with an effective charge Z_A^*. Although this may seem more like curve-fitting than physics, the value of Z_A^* found as a function of

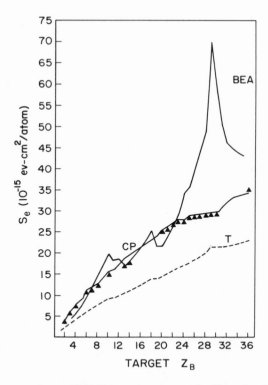

Figure 5.26. Helium stopping at 10 MeV in various targets, S_e. Triangles, data; CP, Chu and Powers [*Phys. Rev.* **187**, 478 (1969)] integration over Hartree–Fock velocities of the Lindhard–Winther plasma model for S_e; BEA, binary-encounter approximation with electron velocity distributions and equilibrium charge state, no plasma screening; T, simple BEA (Thomson) approximation [Eq. (5.18)]. [From R. E. Johnson and A. M. Gooray, *Phys. Rev. Sect. A.* **16**, 1432 (1977).]

energy has a rather sensible behavior, as seen in Figure 5.27, and it is in reasonable agreement with measured charge-state fractions of particles exiting from materials. Using Z_A^* is based on the fact that incident ion values of S_e are much larger than those for neutrals, allowing a rough scaling to proton values.

It is, of course, not clear that Z_A^* in the stopping expression should be anything like the average charge exiting the material. The latter may be a surface effect, with the state of the charge in the material being unknown. The justification for the procedure described above comes from a rather different description of inelastic energy loss. One can think of a solid as an electron gas with positive charges distributed throughout. As the average momentum transfer to the nuclei in fast collisions is small compared to Q/v the positive charges can be ignored when determining first-order density variations in the electron gas. The drag force on a positive ion traversing an

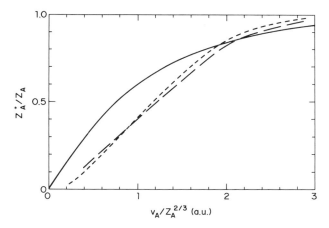

Figure 5.27. Effective equilibrium charge versus a scaled velocity. Solid line, $Z_A^* = Z_A[1 - \exp(-0.9v_A/Z_A^{2/3})]$, Z_A in atomic units; this expression gives a reasonable scaling for measured S_e to proton values for S_e [see, for example, J. Zeigler, *App. Phys. Lett* **31**, 544 (1977)]. Measured equilibrium charge-state fractions: short-dashed line, helium in nitrogen; long-dashed line, hydrogen in nitrogen. [Data from S. K. Allison, *Rev. Mod. Phys.* **30**, 1137 (1958).]

electron gas can be calculated classically, or quantum mechanically, by including the Pauli principle, i.e., a Fermi gas. Lindhard has obtained simple expressions for S_e and Ω^2 at high and low energies which yield rather good results for target atoms having more than a few electrons.

In an electron gas one can again consider close or distant collisions between the electrons and A. The separation in impact parameter between distant and close collisions is related to the electron density via $b \gtrsim k_0^{-1}$ (or k_F^{-1}), where k_0 is the Debye wave number for the classical electron gas and k_F is the corresponding value for a Fermi gas ($k = \omega_p/\langle u^2 \rangle^{1/2}$, where $\omega_p^2 = 4\pi n_e e^2/m_e$, n_e is the electron density, and $\langle u^2 \rangle$ is the mean-squared speed of the electrons in the gas). For distances greater than k_0^{-1} (k_F^{-1}) the electrons act as an aggregate. The corresponding cutoff for the Bethe–Born calculation is that point beyond which the electron and the nucleus appear to be an aggregate (i.e., the passing particle A "sees" the electron as being attached to the nucleus at $b > \langle r \rangle$, the mean radius of the electron cloud). The high-energy expression obtained from the electron gas model is

$$\left(\frac{dE}{dx}\right)_e = \frac{(Z_A e)^2}{v^2} \, \omega_p^2 \ln(2m_e v^2/\hbar\omega_p) \tag{5.28}$$

Using the definition of ω_p and writing $n_e = Z_B n_B$, we see that the stopping cross section is then

$$S_e = 8\pi \frac{(Z_A e^2)^2}{(2m_e v^2)} Z_B \ln(2m_e v^2/\hbar\omega_p) \tag{5.29}$$

Although this expression has the same form as the Bethe–Born approximation [Eq. (5.22)], the cut-off $\hbar \omega p$ different. The low-momentum cutoff is determined by the *local electron density* through the quantity ω_p and not by the binding energy of the electrons to the nucleus. Lindhard and Winther derived the corrections to this expression for a Fermi gas, and Chu and Powers averaged that expression over the electron densities of the target atom. Their results are also shown in Figure 5.26 and are in excellent agreement with experiment at the energy indicated.

At low velocities, $v \ll v_e$, the stopping power becomes proportional to v. This can be seen by determining the drag force on a slowly moving particle A which, at low speeds, is a result of the bombardment of A by the target electrons. As this drag force is just $(dE/dx)_e$ the electronic stopping power can be determined from

$$\left(\frac{dE}{dx}\right)_e = \frac{d(\mathbf{p_A})_{\|}}{dt} = \iint \rho(\mathbf{p_e})[vm_e(\mathbf{v_e} - \mathbf{v_A})]_{\|} \, d^3p_e \qquad (5.30a)$$

where $\mathbf{p_e} = m_e \mathbf{v_e}$, ρ is the momentum distribution of Chapter 3 [e.g., Eq. (3.72)], and v is the collision frequency for momentum transfer to A. For $v_e \gg v$, the velocity dependence is linear,

$$\left(\frac{dE}{dx}\right)_e \sim (v \overline{v_e \sigma_D} m_e n_e) \propto n_B v(Z_B m_e a_0^2 \bar{v}_e) \qquad (5.30b)$$

where σ_d is a diffusion cross section, Eq. (5.8) for electrons colliding with A, and the bar implies averaging over the electron velocities. Treating the electrons as a Fermi gas, Lindhard and Scharff obtained the useful expression

$$S_e \simeq \xi_e \, 8\pi e^2 \, a_0 \, \frac{Z_A Z_B}{Z} \frac{v_A}{v_0} \qquad (5.31)$$

where v_0 is the mean velocity of a ground-state hydrogen atom $(e^2/\hbar)$, a_0 is the Bohr radius, and $Z^{1/3} = (Z_A^{2/3} + Z_B^{2/3})^{1/2}$ is the effective atomic number discussed in Chapter 4. The quantity ξ_e is a slowly varying number ($\xi_e \approx Z_A^{1/6}$). The corresponding result for straggling is found to be

$$S_e^{(2)} \simeq S_e (5m_e v_A^2 \, \hbar \omega_B)^{1/2} \qquad (5.32)$$

Here $\hbar \omega_B$ is the lowest energy transfer possible to the gas; a reasonable approximation is to use the binding energy of the electrons. Results obtained using Eqs. (5.31) and (5.32) are shown in Figure 5.25.

Born and BEA methods apply to molecular-dissociation cross sections and nuclear stopping powers. As the reduced mass is thousands of times greater than m_e, these cross sections are small compared to the above except at low energies where the inelastic electronic effects diminish in importance. The cross-section approximations that are useful in this regime are just those used to describe integrated total cross sections and the diffusion cross section in an earlier section of this chapter. In the following section we will consider such low-energy molecular processes.

Molecular Processes

The simplest molecular event one generally thinks of is dissociation, a process comparable to ionization in that the target aggregate is split. However, even a peripheral discussion of this process indicates that the description can be very complex and species dependent, although in the limit that the molecular target is large, statistical theories become useful, as was the case for many-electron atoms. In the following, processes occurring in small molecules are emphasized.

From the potential curves for a simple diatomic molecule, shown schematically in Figure 5.28, dissociation is seen to be induced in two very different ways. Exciting the electrons on the target to a state which either dissociates directly (a repulsive state) or which decays after a period of time (predissociation) can produce a dissociation. Alternately, momentum transfer directly to the target nuclei, exciting the molecule into the vibrational-rotational continuum of the ground-state potential, will result in a dissociation event. Although the latter process generally requires a smaller energy transfer, the study of stopping powers in the previous section shows that fast particles transfer energy more efficiently to the electrons. If the states of the diatomic molecule are known, the probability of inelastic transitions leading to dissociation can be calculated using the techniques discussed in Chapter 4 and in the previous two sections. For example, in the charge-exchange dissociation of He_2^+ by Ne (i.e., $Ne + He_2^+ \rightarrow Ne^+ + He + He$), the capture of the electron from the Ne induces a transition from the bound $He_2^+(^2\Sigma_u)$ ground state to the primarily repulsive $He_2(^1\Sigma_g)$ ground state

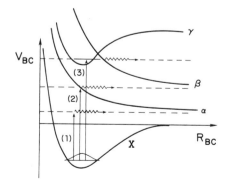

Figure 5.28. Three dissociation processes for molecule BC after a collision with another particle: (1) direct vibrational excitation into the continuum; (2) an electronic excitation to a repulsive state α; (3) predissociation, which is an excitation to a bound state γ and a subsequent radiationless transition (here tunneling) to a state β in which dissociation occurs.

shown in Figure 5.29. (Note that the g→u transitions in the *target* are allowed because the three-body collision is not axially symmetric.)

When applying the methods for calculating inelastic transition probabilities to molecules, the vibrational degrees of freedom provide an added complication. As vibrational periods are long compared with the times for electronic transitions (of the order of an electron orbital period), it is generally sufficient to assume that the nuclear coordinates are fixed during the *transition* (not necessarily during the collision). From the discussion in Chapter 3, the vibrational wave functions (or the classical vibrational speeds) indicate the probability of finding the oscillator at any possible separation; hence the transition probability will be modified by the position distribution of the initial vibrational state. In the example shown in Figure 5.29 the energy gap between the states is considerably reduced and, therefore, the transition probability is more likely if the transition occurs when the He_2^+ is at the maximum separation. By contrast, for the inelastic-dissociative collisions shown in Figure 5.30, although a significant transition energy is required, the relative change in energy between the upper and lower state is not very different at those internuclear separations classically allowed in the ground states.

The latter case is especially simple to treat as the transition probability can be separated into electronic and nuclear parts in the Born–Oppenheimer sense,

$$P_{0\to f} \simeq P_{0\to f}^{elec} W_{0\to f} \tag{5.33}$$

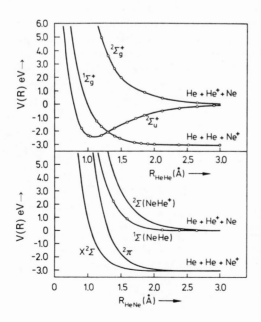

Figure 5.29. Two-body potentials for the $He_2^+ + Ne \to He + He + Ne^+$ reaction. Upper figure, $He^+ + He$ and $He + He$ ground-state potentials. Lower figure, $Ne + He$ and $Ne + He^+$ ground-state potentials. [From P. J. Kuntz and W. H. Whitton, *Chem. Phys.* **16**, 301 (1976).]

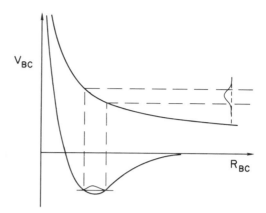

Figure 5.30. Dissociative transition between two molecular electronic levels produced by the collision of the molecule BC with an electron, atom, or molecule. The energy spectrum of the dissociating atoms, identified on the right of the figure, is seen to be a reflection of the probability distribution of the ground-state vibrational level.

Here $P^{elec}_{0 \to f}$ is the probability of the electronic transition, determined only by the electronic configuration, and $W_{0 \to f}$ is a weight factor determined by the vibrational states involved. From the form for the transition probabilities in Chapter 4, if the reader imagines the total wave function to be a product of electronic and nuclear wave functions, then the weight factor required in Eq. (5.33) is

$$W_{0 \to f} = |\langle \phi_0(R) | \phi_f(R)\rangle|^2 \qquad (5.34)$$

This factor is just the overlap between the initial vibrational wave function and the wave function describing the motion of the nuclei in the final state. Such a factor is referred to as the Franck–Condon factor for the transition. Classically this factor would be determined using the times of the nuclei in the initial and final states spent at the same R. The Franck–Condon factor in Eq. (5.34) indicates, for example, that the favored transitions to a bound state (which may subsequently dissociate) will not necessarily be to the lowest vibrational level of that state. The preferred state is that vibrational level having the greatest overlap with the initial, ground vibrational level. Transitions to a repulsive state are favored near the point where the relative nuclear velocity of the dissociating particles is zero and the oscillating continuum wave function amplitude is largest (see Figure 3.4). That is, the molecule tends to make an electronic transition to a classical turning point on the repulsive curve. Therefore, a measurement of the kinetic energy of the dissociating nuclei for such transitions reveals a distribution of energies reflecting the distribution of internuclear separations in the initial state of the vibrating molecule, as seen in Figure 5.30.

A vibrational-type dissociation, unlike the electronic dissociation, requires a direct transfer of momentum to one or both of the nuclei by the incident particle. When considering such energy transfers to heavy particles, a Pandora's box of molecular processes is often opened. Vibrational and rotational transitions are induced in this manner, as well as dissociation, and, depending on the complexity of the colliding particles, a variety of molecular reactions may occur. At the lowest collision energies, exothermic reaction processes generally dominate, as indicated in Figure 5.31 for the simple K + Br$_2$ collision. For an incident ion these are often followed in order of importance by near-resonant charge transfer processes with other inelastic processes, including vibrational dissociation, forming a small fraction of the total. As the energy increases, the relative importance of these competing processes changes rapidly.

To describe the direct momentum transfer to the target nuclei, the forces between all the heavy particles must be considered simultaneously. For the A + BC collision, knowing the two-body interaction potentials V_{AB}, V_{AC}, and V_{BC}, a simple, three-body diabatic potential can be produced which depends on the separations R_{AB}, R_{AC}, and R_{BC}. For very slow collisions, the outer-shell electrons will have time to make a number of orbits and adjust to the position of all three nuclei. This modifies the interaction between the three atoms, resulting in a fully adiabatic potential between the particles. A plot of either the diabatic or fully adiabatic three-body potential requires four dimensions. Figure 5.32 shows a cut of that plot for F + H$_2$ in which one parameter is fixed. In the cut chosen, the projectile F is allowed

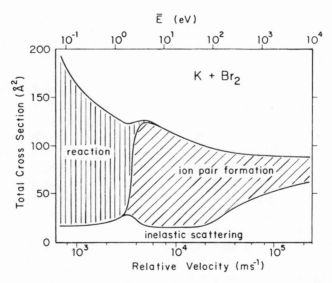

Figure 5.31. Summary plot of experimental total nonelastic scattering cross section for K + Br$_2$. Inelastic: vibrational, rotational, and electronic excitation; reaction: KBr + Br; ion pair: K$^+$ + Br$_2^-$. [From R. B. Bernstein, *Atom–Molecule Collision Theory*, Plenum Press, New York (1979), p. 30; J. Los and A. W. Kleyn, in *Alkali-Halide Vapors*, eds. P. Davidovits and D. L. McFadden, Academic Press, New York (1979), p. 275.]

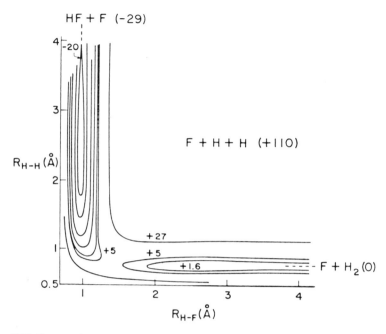

Figure 5.32. Computed potential contours for collinear FH_2 in kcal/mole. [From C. F. Bender, P. K. Pearson, S. V. O'Neil, and H. F. Scharfer, III, *J. Chem. Phys.* **56**, 4626 (1972).]

to approach along the axis of H_2, and hence the three internuclear separations are not independent. As the channel $F + H_2 \rightarrow H + HF$ is very exothermic ($Q = -29$ kcal/mole $= -1.26$ eV), reactions are likely, although a small energy barrier exists at $R_{H-F} \sim 1.4$ Å. The reactive channel is located on the potential surface where R_{H-F} is small and R_{H-H} is large. To achieve dissociation (i.e., where R_{H-H} and R_{H-F} are both large) requires a significant energy transfer ($Q = +110$ kcal/mole $= 4.8$ eV). The size of this barrier implies that the H_2 does *not* have to be dissociated first in order to form HF. The particles can approach slowly along the valley on the potential surface, and the strong HF bond will dominate on exiting. This is, in fact, the way many molecular reactions occur.

Using such potentials, one can carry out classical trajectory calculations of vibrational excitation, dissociation, and reaction cross sections. Making a Monte Carlo selection of initial conditions [(a) separation of B and C, (b) orientation of B–C, and (c) impact parameter], one can determine a classical cross section as described in Chapter 2. After the collision is completed, the relative separations and velocities of the colliding particles are noted, and, as in the BEA, these classical parameters are identified with a particular final state of the system. If, in addition, the corresponding classical actions are calculated, then semiclassical amplitudes can be constructed and, hence, cross sections including interference (viz. Chapter 3).

Although the above procedure is easily described, the calculations are not simple, particularly as a large number of initial conditions have to be considered. Therefore, the study of molecular collision phenomena is dominated by the search for simplifying approximations even when classical methods are employed. We consider below the results of a few such approximate methods. These methods can be divided into categories: (a) distant and/or fast collisions involving small deflections, with the coordinates of the bound nuclei being treated as internal variables which are perturbed by the passing particle, e.g., the dipole collision with a classical oscillator (Appendix C); (b) an inelastic effect requiring a close collision with one of the target nuclei, in which case BEA methods apply involving only two-body interactions; or (c) the projectile and target particles interact strongly and form highly excited quasi-bound particles, a collision complex, which eventually decays into one of a number of possible final states the relative probabilities of which are calculated statistically. The first two categories were considered when describing electronic effects in atoms. The third process can often be represented by a penetration of a potential barrier, as in Figures 3.2 and 3.4. While 'bound' a reaction may occur, as is clearly indicated by the classical-trajectory calculation of the $KCl + NaBr \rightarrow KBr + NaCl$ reaction in Figure 5.33. The four particles are seen to spend a long time in close proximity before separating to the final state. Such processes also occur when electrons collide with atoms, in which case the incident and target electrons can form a temporary, quasi-bound state on the atom (e.g., $e + A \rightarrow [A^-]^* \rightarrow A^* + e$).

That the above approximations often do correspond to actual phenomena is evidenced by observing the velocity distribution of the scattered

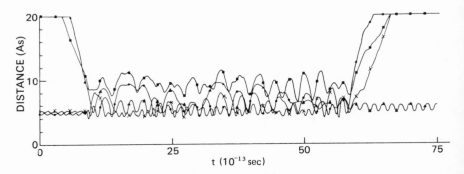

Figure 5.33. Calculated KCl + NaBr collision trajectory: separations vs time, ×, K–Cl; ■, Br–Cl; ▲, K–Br; ●, Na–Br. Quasi-bound complex indicated by intermediate region in which average separation stays about constant. [From P. Brumer, Ph.D. Dissertation, Harvard University (1972); P. Brumer and M. Karplus, *Discuss. Faraday Soc.*, **55** 80-92 (1973).]

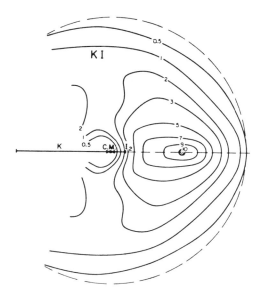

Figure 5.34. KI velocity distribution contours for the $K + I_2 \rightarrow KI + I$ reaction. Arrows indicate average CM velocities. [From K. T. Gillen, A. M. Rulis, and R. B. Bernstein, *J. Chem. Phys.* **54**, 2831 (1971).]

products from a reactive collision. The CM velocity distribution of KI, from the reaction $K + I_2 \rightarrow KI + I$, shown in Figure 5.34 is strongly peaked in the forward direction, indicating a capture of one of the I atoms by the passing K atom. The second I atom plays the role of a spectator, only involved in the sense that it carries off the excess momentum. This process is described as "harpooning" as the heavy-particle capture is preceded by an electron transfer, increasing the likelihood of a reaction. In contrast, the velocity distribution of the CsCl molecule, from the collision reaction $RbCl + Cs \rightarrow [RbClCs] \rightarrow CsCl + Rb$ shown in Figure 5.35, is seen to have a forward-backward symmetry indicative of a collision complex being formed. As a first guess one might expect the velocity distribution to be spherically symmetric after the complex "explodes" or releases its excess energy. However, total angular momentum conservation requires that the complex retain some information about the initial direction of approach. Our emphasis on fast collisions throughout the text should not prejudice the reader as to the form of the differential cross section. For near-thermal-energy collisions, even backward peaked CM cross sections are possible, indicating collisions in which the reaction occurs as the incident particle rebounds off the repulsive barrier. In the following, we consider collisions in which a collision complex is formed.

As long-range forces between ground-state molecules are generally attractive, collision complexes are often formed for those impact parameters less than b_0, the orbiting impact parameter discussed in Eq. (2.64) of Chapter 2. When this is the case, detailed knowledge of the interaction potential

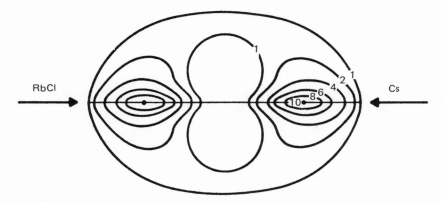

Figure 5.35. RbCl + Cs → [RbClCs] → CsCl + Rb. Contour map of CsCl velocity distribution, with forward–backward symmetry. [From R. D. Levine and R. B. Bernstein, *Molecular Reaction Dynamics*, Oxford University Press, Oxford (1974), p. 216; see also S. A. Safron, Ph. D. Thesis, Harvard University, Cambridge, Mass. (1969).]

at small internuclear separations is generally not required to estimate the cross section. From Eq. (2.26) the reaction cross section for collision complex formation is

$$\sigma_{0 \to f} \simeq \bar{P}_{0 \to f} \pi b_0^2 \tag{2.26}$$

where $\bar{P}_{0 \to f}$ is determined from a knowledge of the decay of the energetic complex. This expression is particularly simple to apply to ion–molecule collisions in which the long-range force is, to lowest order, also independent of the nature of the incident ion. Using Eq. (2.64), we see that the Langevin cross section for singly ionized, ion–molecule reactions is

$$\sigma_{0 \to f} \simeq \bar{P}_{0 \to f} \left[\frac{2\pi}{v} \left(\frac{\alpha e^2}{m} \right)^{1/2} \right] \tag{5.35}$$

where α is the polarizability of the target.

In our earlier discussion, long collision times were associated with small uncertainties in the energy and, hence, small transition probabilities. However, if a collision complex is formed, even significantly exothermic reactions may occur efficiently. Although the transition probability to a highly exothermic state is small on any one pass of a transition region, when the particles orbit for long times that region may be traversed frequently, producing a significant accumulative effect. The longer the excited collision complex lives, the less detailed is the information retained about the initial state, and the complex may decay into a number of lower-lying (exothermic) states. In general, the larger the number of states available, including the vibrational modes, the longer the complex lives. If a number of exothermic states exist, then setting $\bar{P}_{0\to f} \simeq 1$ in Eq. (5.35) gives an upper limit to the sum of the possible reaction cross sections.

The cross section for the $He^+ + CO \to C^+ + O + He$ reaction in Figure 5.36 is in reasonable agreement with the Langevin cross section with $\bar{P}_{0\to f} = 1$, as is the case for a host of other ion–molecule reactions. Averaging the Langevin cross section over a Maxwell–Boltzmann velocity distribution (cf. Problem 2.3), we obtain a temperature–independent rate constant, $k_{0\to f} \sim 2\pi(\alpha e^2/m)^{1/2}$. Using the polarizabilities in Table 4.1, we compare this rate constant to measured rate coefficients in Table 5.2 for the reactions discussed and those listed in Eq. (1.13).

Fair agreement between measurement and the Langevin result is indicated in Table 5.2 except for $O^+ + N_2$ and $O^+ + O_2$ reactions. In such systems a significant internal barrier to the reaction exists. This is indicative of the existence of an energy maximum in the reaction cross section like that for the inelastic processes discussed earlier. A plot of the reaction rate as a function of vibrational temperature is shown in Figure 5.37 along with the reaction cross section for $O^+ + N_2 \to NO^+ + N$. Here, increasing vibrational temperature implies a more likely population of higher vibrational states of N_2. The results show that the Langevin reaction rate is approached only in the limit of very high vibrational temperatures. That is, the N_2 must be very nearly dissociated *prior* to reacting with O^+ in order for the reaction to proceed readily. This is quite different from our earlier discussion of the $F + H_2$ reaction. A general feature of a process like

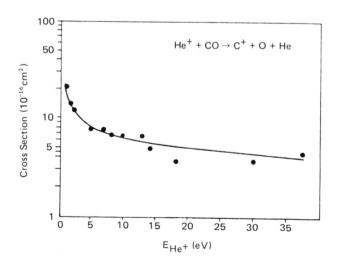

Figure 5.36. Measured reaction cross section versus laboratory energy. Line indicates Langevin orbiting cross section [Eq. (5.35)]. [From C. F. Giese and W. B. Maier, *J. Chem. Phys.* **39**, 197 (1963).]

Table 5.2. Comparison of Rate Constants

Reaction	k, measured (cm^3/sec)^a	$k = 2\pi(\alpha e^2/m)^{1/2}$
$He_2^+ + Ne \rightarrow Ne^+ + 2He$	6×10^{-10} (200°K)	6.0×10^{-10}
$He^+ + CO \rightarrow C^+ + O + He$	1.7×10^{-9}	1.8×10^{-9}
$O^+ + N_2 \rightarrow NO^+ + N$	1.3×10^{-12}	1.2×10^{-9}
$N_2^+ + O \rightarrow NO^+ + N$	2.5×10^{-10}	
$\quad\quad\quad \rightarrow O^+ + N_2$	$< 10^{-11}$	5.6×10^{-10}
$O^+ + CO_2 \rightarrow O_2^+ + CO$	1.2×10^{-9}	1.1×10^{-9}
$CO_2^+ + O \rightarrow CO_2 + O^+$	1.0×10^{-10}	5.2×10^{-10}
$\quad\quad\quad \rightarrow O_2^+ + CO$	1.6×10^{-10}	
$O^+ + O_2 \rightarrow O_2^+ + O$	2.0×10^{-11}	9.0×10^{-10}
$O_2^+ + NO \rightarrow O_2 + NO^+$	8.0×10^{-10}	7.6×10^{-10}

^{a} From E. W. McDaniel, V. Cermák, A. Dalgarno, E. E. Ferguson, and L. Fried–man, *Ion–Molecule Reactions*, Wiley, New York (1970), Chapter 6.

$O^+ + N_2 \rightarrow NO^+ + N$, which has a significant activation energy or an energy threshold, is the strong energy dependence of the reaction cross. Inserting a cross section with a threshold ΔE_0 into the rate-constant expression [Eq. (2.16)], we obtain the Arrhenius reaction rate,

$$k = A(T) \exp\left[-\Delta E_0/kT\right] \qquad (5.36)$$

In the above, $A(T)$ is slowly varying in T with the primary temperature dependence of k being exponential, in marked contrast to the Langevin result.

The Langevin procedure for evaluating the reaction cross section [Eq. (5.35)] provides no means for determining the relative population of the available exothermic levels after the decay of the collision complex. Often, the complex can relax into a number of possible final-state electronic configurations each involving a spectrum of vibrational and rotational energies. For the $O + CO_2^+$ reaction, two exothermic reaction channels are indicated in Table 5.2 in which the vibrational and rotational states of the CO_2, CO, and O_2^+ and the electronic spin state of O^+ are not specified. Determination of the final-state population [i.e., $\bar{P}_{0 \rightarrow f}$ in Eq. (5.35)] via the procedures for calculating transition probabilities in Chapter 4 is not very useful when a collision complex is formed. As a first guess at such transition probabilities, and in the absence of any selection rule or dynamical constraint, the disposal of energy is often estimated statistically.

In a statistical model, the probability of relaxing into a particular state is the ratio of the density of states of the same or nearly the same energy to the total number of states energetically available, with all states being equally likely. The densities in energy and momentum of translation "states" (i.e., continuum states) are simply related by $N(E)\,dE = N(\mathbf{p})4\pi p^2\,dp$, where E is the total translation energy in the CM system [$E = (1/2m)p^2$] and

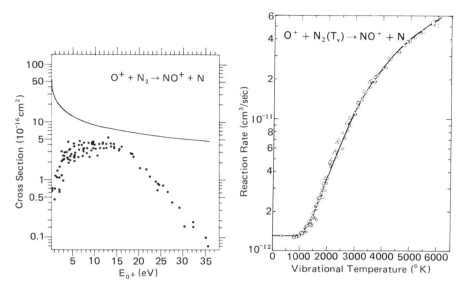

Figure 5.37. (left) Measured reaction cross section. Line indicates Langevin orbiting cross section. [From C. G. Giese, *Ion–Molecule Reactions in the Gas Phase*, Advances in Chemistry Series, No. 58, American Chemical Society, Washington, D. C. (1966), p. 20.]; (right) reaction rate versus vibrational temperature, T_v, of N_2. [From A. L. Schmeltekopf, E. E. Ferguson, and F. C. Fehsenfeld, *J. Chem. Phys.* **48**, 2966 (1968).]

we have assumed $N(\mathbf{p})$ is independent of angle. For continuum-like states, the uncertainty principle can be used to estimate the density of momentum states. For particles confined to some arbitrarily large volume V, $N(\mathbf{p}) \sim V/h^3$; hence, the number of translation states in the energy interval E to $E + dE$ has the form $N(E) \propto E^{1/2}$, which we use below.

When a collision complex is formed, the final CM translation energy E' is determined (cf. Table 2.4) from the initial CM energy E and the inelasticity Q ($Q < 0$, exothermic: $E = E' + Q$). Therefore, the statistical population of a particular bound state depends not only on the total number of bound states available but also on the density of the corresponding translational state of energy E'. Assuming, for simplicity, uniformly and closely spaced bound levels with the lowest level labeled Q_{min}, we see that the total density of states has the form $N(E', Q) \propto E'^{1/2}\delta(E - E' - Q)$, $Q \geq Q_{min}$. Defining the statistical probability density in the manner discussed above, we obtain the probability density for populating a state with energy between Q and $Q + dQ$:

$$\bar{P}(Q) = \tfrac{3}{2}(E - Q)^{1/2}/(E - Q_{min})^{3/2} \tag{5.37}$$

This expression is reasonable in that the probability density goes to zero for threshold reactions $(Q = E)$ and increases with increasing exothermicity. Such a result is of use for atomic systems, in which the complex may be formed with another heavy particle or an electron, if the final atomic levels are closely and uniformly spaced. For a collision like $A + BC \rightarrow (ABC) \rightarrow AB + C$, the product diatomic AB can rotate and vibrate. Assuming that a single, exothermic, electronic configuration is populated, the total CM energy available, $\varepsilon' = E - Q$, can be partitioned among the vibrational and rotational levels. The vibrational levels are uniformly spaced and each rotational energy has $2J + 1$ degenerate levels, where J is the rotational energy index discussed in Chapter 3. It is left as a problem for the reader (Problem 5.15) to show that the statistical vibrational probability distribution is

$$\bar{P}(E_v) = \tfrac{5}{2}(\varepsilon' - E_v)^{3/2}/\varepsilon'^{5/2} \tag{5.38}$$

where E_v is the vibrational energy. For large, very complex molecules the vibrational distribution becomes thermodynamic-like, $P(E_v) \propto \exp(-E_v/kT)$. Here, T is a "temperature" for the decaying collision complex determined from the excess energy available (ε') and the number of internal degrees of freedom. The deviation from such probability densities, often referred to euphemistically as the "surprise," is a measure of the role that dynamical constraints and/or selection rules play in determining the disposal of the energy of the excited collision complex.

As a final subject in this section, we consider processes that are the reverse of dissociation and ionization, that is, recombination and attachment processes. The simplest such process is dissociative recombination, $e + AB^+ \rightarrow A + B$, in which the excess energy of the unbound electron is disposed of as kinetic energy of the separating nuclei. In Figure 5.29, we see that an incident electron can induce transitions between the He_2^+, $^2\Sigma_u^+$, and He_2, $^1\Sigma_g^+$, curves of that figure by recombination. By comparison with the Ne collision, in which the Ne atom donates an electron via charge exchange, one would expect a rate constant for recombination of the order of 10^{-8} cm^3/sec. This is based on the size of the $He_2^+ + Ne$ rate constant and the fact that the average speeds of thermal electrons are much greater than those of atoms or molecules. In fact, even larger reaction rates are often found for such processes as the long-range force is coulombic.

In order to form a collision complex, a passing (incident) electron must penetrate toa distance of closest approach of the order of the radius of a highly excited state of the molecule AB. In this situation, the electron and molecular-ion can form a quasi-bound collision complex because of the attractive coulomb force between them. For this force there is no critical orbiting impact parameter equivalent to the Langevin case [cf. Eq. (2.64)]. In the present problem, the complex corresponds to an autoionizing state of AB. Such a state is temporarily bound, but lies higher in energy than the minimum ionization level by an amount equal to the initial energy of the electron. In an atom, an autoionizing state eventually decays by re-emitting the electron. However, if the target is a molecular ion, other transitions are possible. For instance, for $e + He_2^+ \rightarrow (He_2)^*$, the complex can make a radiationless transition to the ground, repulsive potential of He_2 shown in Figure 5.29, producing a dissociation. If the excited complex is long-lived with respect to autoionization, then the occurrence of dissociative recombination becomes likely.

The rate constant for dissociative recombination can be constructed as a product of the probability for formation of the excited molecular complex times the transition probability to the dissociating molecular state. The formation of an excited molecule is quite sensitive to the electron energy. However, for a thermal distribution of electrons, described by a temperature, T_e, the electron–ion collision frequency, ν_{ei}, can be used to characterize the *formation* probability for the complex. It is left as a problem for the reader (Problem 5.16) to show that $\nu_{ei} \propto T_e^{-3/2}$. The radiationless transition rate is inversely proportional to the lifetime of the autoionizing state and depends on the Franck–Condon factor between the bound and repulsive states. The transition occurs favorably near that value of R on the repulsive state, having an energy equal to the bound-state energy plus the electron energy E_e. Classically, this factor has the form $\int \rho[R(E_e)] f(E_e) dE_e$, where $f(E_e)$ is the electron energy distribution, characterized by T_e, and $\rho(R)$ is the probability distribution for the oscillator having a separation R. Such rate constants have a temperature dependence varying from roughly T_e^{-1} to T_e^0. For example, for the reaction of Eq. (1.13), $e + O_2^+ \rightarrow O^* + O$, $k_{e, O_2^+} \simeq 2.2 \times 10^{-7}(300/T_e)$ over the range $200 < T_e < 700°K$.

The dissociative recombination process is a two-body reaction, as is dissociative attachment (e.g., $e + O_3 \rightarrow O^- + O_2$). The attachment process, however, often involves three bodies (e.g., $e + O_2 + X \rightarrow O_2^- + X$ or $O + O_2 + X \rightarrow O_3 + X$). In the first process, the electron is attracted to the O_2 by the polarization force and forms a quasi-bound negative ion in a vibrationally excited state. During the lifetime of that complex, a collision with an ion may produce neutralization via charge exchange. However, in a weak collision with a third body X (generally a neutral), the excess energy of the complex can be carried off, producing a stable bound O_2^- molecular ion. The three-body rate coefficient for the process can be estimated [cf. Eq. (2.17)] in terms of the two-body rate constant for the collision of the complex with X

$$(k_{e, O_2, X}) n_e n_{O_2} n_X \approx (k_{O_2^-*, X}) n_{O_2^-*} n_X \qquad (5.39)$$

The number density of excited O_2^- needed in Eq. (5.39) depends on the e, O_2 reaction rate and the lifetime of the O_2^- state. For relatively short lifetimes τ^*, $n_{O_2^-*} \approx (k_{e, O_2}) n_e n_{O_2} \tau^*$ as in Eq. (2.15). Substituting this into

Eq. (5.39), we find

$$k_{e,O_2,X} \approx k_{O_2^-*,X} k_{e,O_2} \tau^* \qquad (5.40)$$

Assuming $k_{O_2^-*,X} \approx 10^{-10} \, \text{cm}^3/\text{sec}$, $k_{e,O_2} \sim 10^{-8} \, \text{cm}^3/\text{sec}$, and $\tau^* \sim 10^{-12}$ sec, the vibrational period, we have $k_{e,O_2,X} \sim 10^{-30} \, \text{cm}^6/\text{sec}$. Experimentally, an Arrhenius-type reaction rate is found; for X an O_2 molecule, we have $k_{e,O_2,O_2} \sim 1.5 \times 10^{-29}(300/T_e)\exp(-600/T_e) \, \text{cm}^6/\text{sec}$ with T_e in degrees Kelvin. Such a result implies that a small activation energy exists for forming O_2^-*. It is also found that the reaction rate for attachment to O_2 is smaller when the third body is not O_2 (e.g., N_2 and He). Because the vibrational levels of the O_2^-* are close in energy to those of O_2, the complex is most readily de-excited by near-resonant, vibrational energy exchange to another O_2 molecule.

A similar two-step, three-body reaction, involving O rather than e as the attached species, will form ozone. To estimate the reaction rate for this process, we assume the k_{O,O_2} step is more than a factor of 10^{-2} smaller than k_{e,O_2} because of the slower thermal speeds (O vs. e) and shorter-range force (exchange versus polarization). This suggests a rate constant, $k_{O,O_2,X}$, of the order of 10^{-32}–$10^{-33} \, \text{cm}^6/\text{sec}$. For X an N_2 molecule, measurement yields $k_{O,O_2,N_2} \sim (5 \pm 2.5) \times 10^{-34}(300/T)^3 \, \text{cm}^6/\text{sec}$, with T in degrees Kelvin, which is consistent with our estimate.

The processes discussed in this chapter were chosen to exemplify the use of simple cross-section models for interpreting measurements of collision quantities. In the following chapter, we use the methods and results discussed here to interpret and understand various macroscopic processes.

Exercises

5.1. From the ρ vs τ plot for $\text{He}^+ + \text{Ne}$ in Figure 5.1, obtain an interaction potential. Use the power-law inversion procedure in Chapter 2 over successive steps in log τ of 0.5. Compare the result to the screened potential of Smith *et al.* discussed in the text.

5.2. Show that the extremum in the radial phase shift for a 6-12 potential has the form given in Eq. (5.4).

5.3. Verify the optical theorem that relates the total cross section and the differential cross section at zero degrees using Eqs. (3.35) and 3.38): $\sigma_{AB} = (4\pi/K) \times \text{Im} \, f_{AB}(0)$ [see Eq. (3.41)].

5.4. Consider a crossed-beam experiment in which a fast ion beam $(E_A \cong 100 \, \text{eV})$ is incident perpendicular to a slow, superthermal beam $(E_B \sim 0.1 \, \text{eV})$. For the fine-structure splitting in Ar^+ (^2P_j) $(Q = 0.068 \, \text{eV})$ estimate the laboratory deflection angle for an inelastic charge-exchange collision $\text{Ar}^+ + \text{Ar}$ as a function of χ for χ small. Construct the Newton diagram [the diagram of laboratory + CM vectors] see K. B. McAfee, Jr., W. E. Falconer, R. S. Hozacks, and D. J. McClure, *Phys. Rev. A* **21**, 827 (1980).

5.5. Derive the stationary phase result for the symmetric-resonant charge-exchange cross section in Eq. (5.14) and the glory oscillations [Eq. (5.3)].

5.6. Use the BEA method for stationary target electrons to obtain the results in Eqs. (5.18) and (5.19).

5.7. Calculate the ion mobility for an interaction potential of the form $V = -\alpha/2R^4$.

5.8. Use the results in Appendixes B and D to determine Born and classical impulse cross sections for σ_d of Eqs. (5.6) and (5.9) for various potentials.

5.9. Show that the diffusion cross section for an ion in its parent gas is twice the charge-exchange cross section.

5.10. Derive the semiclassical expression in Eq. (5.15) using the expressions in Chapter 3 for the action.

5.11. For two exponential potentials with the same screening constant show that $\partial n/\partial \tau$ in Eq. (5.16) is very nearly a constant at a given E.

5.12. Use the Born expressions in Eqs. (4.79) and (4.80) to obtain the close and distant contributions to the Born approximation for the ionization cross section, stopping cross section, and straggling cross section.

5.13. Solve the rate equations for A and A^+ in Eq. (5.25) to obtain the charge state fraction of a beam in a material. Find and plot the fraction as a function of depth for protons in some realistic material.

5.14. The electron gas model for stopping is based on the premise that the average momentum transfer to the electrons is larger than to the target nuclei. Over what velocity range is this valid?

5.15. For a collision $A + BC \rightarrow (ABC)^* \rightarrow AB + C$ obtain the statistical vibrational distribution of final states in Eq. (5.38).

5.16. For the electron–ion collision show that the collision frequency for momentum transfer, v_{ei}, goes as $v_{ei} \propto T_e^{-3/2}$.

Suggested Reading

For many references to the literature see the figure captions for various experiments and tables. Other references of interest which either summarize data in certain areas or discuss the new phenomena considered are listed below (see also Chapter 4 references).

Elastic Scattering of Ions at High Energies

I. AMDUR and J. E. JORDAN, in *Advances in Chemical Physics*, Vol. 10, ed. J. Ross, Wiley, New York (1966), Chapter 2.

J. B. HASTED, *Physics of Atomic Collisions*, 2nd edn., American Elsevier, New York, (1972), Chapter 10.

K. L. BELL and A. E. KINGSTON in *Atomic Processes and Applications*, ed. P. G. Burke and B. L. Moiseiwitsch, North-Holland, Amsterdam (1976), Chapter 14.

Low-Energy Elastic Scattering of Atoms and Ions

R. B. BERNSTEIN and J. J. MUCKERMAN, in *Advances in Chemical Physics*, Vol. 12, ed. J. O. Hirschfelder, Wiley, New York (1967), Chapter 8.

U. BUCK, in *Advances in Chemical Physics*, Vol. 30, ed. K. P. Lawley, Wiley, New York (1975), p. 313.

H. S. W. MASSEY, *Electronic and Ionic Impact Phenomena*, Vol. 3, 2nd edn., Oxford University Press, London (1974).

H. PAULY and J. P. TOENNIES, in *Advances in Atomic and Molecular Physics*, Vol. 1, ed. D. R. Bates, Academic Press, New York (1965), Chapter 5.

K. SMITH, *Calculation of Atomic Collision Processes*, Wiley-Interscience, New York (1972).

Diffusion and Mobilities

A. DALGARNO, in *Atomic and Molecular Processes*, ed. D. R. Bates, Academic Press, New York (1962), Chapter 16.

J. O. HIRSCHFELDER, F. CURTISS, and R. B. BIRD, *Molecular Theory of Gases and Liquids*, Wiley, New York (1964), Part II.

E. A. MASON, R. J. MUNN, and F. J. SMITH, in *Advances in Atomic and Molecular Physics*, Vol. 2, ed. D. R. Bates, Academic Press, New York (1966), Chapter 2.

Inelastic Energy Loss: Born, BEA, Electron Gas

U. FANO, *Ann. Rev. Nuc. Sci.* **13**, 1 (1963).

H. A. BETHE and R. JACKIW, *Intermediate Quantum Mechanics*, 2nd edn., W. A. Benjamin, New York (1968), Part III.

L. VRIENS, in *Case Studies in Atomic Collisions*, Vol. 1, ed. E. W. McDaniel and M. R. C. McDowell, North-Holland, Amsterdam (1969), p. 337.

D. R. BATES, in *Atomic and Molecular Processes*, ed. D. R. Bates, Academic Press, New York (1962), Chapter 14.

A. BURGESS and I. C. PERCIVAL, in *Advances in Atomic and Molecular Physics*, Vol. 4, ed. D. R. Bates, Academic Press, New York (1968), p. 109.

D. R. BATES and A. E. KINGSTON, in *Advances in Atomic and Molecular Physics*, Vol. 6, ed. D. R. Bates, Academic Press, New York (1970) Chapter 6.

J. LINDHARD and A. WINTHER, *K. Dan. Vidensk. Selsk. Mat. Fys. Medd.*, **34**, No. 4 (1964).

Potential-Energy Curves and Surfaces

K. P. LAWLEY, ed., *Potential Energy Surfaces, Advances in Chemical Physics*, Vol. 42, Wiley, New York (1980).

W. A. LESTER, JR., ed. *Potential Energy Surfaces in Chemistry*, IBM, San Jose, California (1970).

Excitation by Incident Ions

E. W. THOMAS, *Excitation in Heavy-Particle Collisions*, Wiley, New York (1972).

Q. C. KESSEL, E. POLLACK, and W. W. SMITH, in *Collision Spectroscopy*, ed. R. G. Cooks, Plenum Press, New York (1978), Chapter 3; J. T. Parks, *ibid.*, Chapter 1.

Charge-Exchange Collisions

R. A. MAPLETON, *Theory of Charge Exchange*, Wiley, New York (1972).

R. E. JOHNSON and J. W. BORING, in *Collision Spectroscopy*, ed. R. G. Cooks, Plenum Press, New York (1978), Chapter 2; J. Appell, *ibid.*, Chapter 4.

Dissociative Collisions, Incident Ions

G. W. McCLURE and J. M. PEEK, *Dissociation in Heavy-Particle Collisions*, Wiley, New York (1972).

J. LOS and T. R. GROVERS, in *Collision Spectroscopy*, ed. R. G. Cooks, Plenum Press, New York (1978), Chapter 6; R. G. Cooks, *ibid.*, Chapter 7.

Atom–Molecule Collisions and Reactions

R. B. Bernstein, ed. *Atom–Molecule Collision Theory*, Plenum Press, New York (1979).

R. D. LEVINE and R. B. BERNSTEIN, *Molecular Reaction Dynamics*, Oxford University Press, London (1974).

E. E. Nikitin, *Theory of Elementary Atomic and Molecular Processes in Gases* (translated by M. J. Kearsley from Russian), Oxford University Press, London (1974).

H. K. SHIN, in *Dynamics of Molecular Collisions*, Vol. 2, ed. W. H. Miller, Plenum Press (1976), Chapter 4.

Ion–Molecule Reactions

E. W. McDANIEL, V. CERMÁK, A. DALGARNO, E. E. FERGUSON, and L. FRIEDMAN, *Ion–Molecule Reactions*, Wiley, New York, (1970).

S. G. LIAS and P. AUSLOOS, *Ion–Molecule Reactions, Their Role in Radiation Chemistry*, American Chemical Society, Washington, D. C (1975).

R. B. BERNSTEIN and R. D. LEVINE, *Ad. At. Mol. Phy.*, **11** 215 (1975).

Incident-Electron Collisions

J. B. HASTED, *Physics of Atomic Collisions*, 2nd edn., American Elsevier, New York (1972), Chapters 4, 5, 6, and 8.

N. F. MOTT and H. MASSEY, *The Theory of Atomic Collisions*, 3rd edn., Oxford University Press, London (1965), Chapters XVI, XVII, XVIII.

H. S. W. MASSEY and E. H. S. BURHOP, *Electronic and Ionic Impact Phenomena*, Vol. 1, 2nd edn., Oxford University Press, London (1969).

S. GELTMAN, *Topics in Atomic Collision Theory*, Academic Press, New York, (1969), Part II.

B. L. MOISEIWITSCH, in *Lectures in Theoretical Physics*, Vol. 3, eds. W. E. Brittin, B. W. Downs, and J. E. Downs, Interscience, New York (1961), p. 142.

Recombination

E. W. McDaniel, *Collision Phenomena in Ionized Gases*, Wiley, New York (1964), Chapter 12.

H. S. W. Massey and H. B. Gilbody, *Electronic and Ionic Impact Phenomena*, Vol. 4, 2nd edn., Oxford University Press, London (1974).

J. N. Bardsley and M. A. Biondi, in *Advances in Atomic and Molecular Physics*, Vol. 6, ed. D. R. Bates, Academic Press, New York (1970), Chapter 1.

M. R. Flannery, in *Atomic Processes and Applications*, ed. G. Burke and B. L. Moiseiwitsch, North-Holland, Amsterdam (1976), Chapter 12.

6

Application of Results

Introduction

In this concluding chapter the concepts developed for describing cross sections and rate constants are applied to a few of the problem areas discussed in Chapter 1. This discussion, being consistent with the rest of the text, is not intended to be a state-of-the art presentation on each topic and the ideas presented are, for the most part, well established. Further, the cross-section models used are quite simple. In the previous chapters the relationship between the simple and more accurate models for cross sections have been considered. Therefore, the reader should have a feel for the kinds of errors introduced by the choice of cross-section model and is encouraged to include improved estimates. Care should be taken that the increased complexity caused by using a more accurate approximation for cross section is warranted by the correctness of the description of the physical problem. In the following no statement is implied about this. I have only chosen models which lend themselves to either analytic solutions for the problem or to simplifying the presentation. This merely follows the lead of a number of authors who have made wide use of simple cross-sectional forms in discussing, for example, complex radiation transport problems. In many other cases I have simply taken results for certain rate constants generally used in the literature without further discussion, particularly for processes considered in Chapter 5.

The discussion in the following sections begins with a consideration of the penetration of fast particles into a material and the subsequent diffusion of these implanted particles. In addition, entering fast particles cause displacements in the material, which is discussed in the context of a problem from radiation biology, and also the very interesting phenomenon, sputtering. Sputtering involves the etching of the surface layer generally by fast-ion ejection of the target material. Up to this point it was assumed that the material penetrated had a surface, and inside the surface a nearly constant density exists. Although this provides a certain simplicity, it is not a necessary restriction. Therefore, it is pointed out that similar ejection occurs in a planetary atmosphere in which the material is contained by gravity rather than the short-range atomic forces. The radiation primarily responsible for ejection now is "sunlight," although in some situations charged particles play a role. This leads quite naturally to the discussion of other processes induced by solar radiation, e.g., those chemical processes determining the density and constituents in the earth and Martian ionospheres.

Energetic Particle Transport

Two cases of energetic particles traversing a material are shown in Figure 6.1. The first indicates the occurrence of a hot particle created in a material. This could be the result of the absorption of radiation (ultraviolet, X-rays, neutrons, etc.), a spontaneous fission, or because the atom of interest is set in motion by a collision with a fast particle. The second situation is the radiation example we considered earlier in which, typically, ions are incident on a material as in ion implantation experiments. These situations are not that different. In the second case the net effect will be a sum of the effects of the moving secondary particles created in the material. Differences arise, however, when one considers phenomena near the surface. In either case the first question one asks is how much of the material is affected by the particle or, on the average, how far does it travel.

To answer the above question a simulation procedure could be constructed in which the material atoms are assigned places, randomly for amorphous materials, or in an ordered manner for crystalline materials. The motion of the atom is then followed using the collision cross sections at each encounter of the moving atom with a material atom. The initial impact parameter is chosen randomly for the first collision. In Chapter 1 we took an alternate approach to case b in Figure 6.1, determining the partical

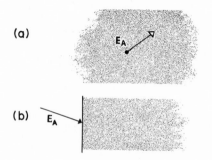

Figure 6.1. Events initiating a particle cascade. (a) Radiation created in a material; (b) radiation incident on a surface.

flux in the cascade, as a function of depth, from an integral equation. This procedure applies to amorphous materials or to crystalline materials for particles not moving along crystal planes. In many instances both of these procedures provide more detail than may be warrented. For instance, the mean penetration depth of an incident particle of energy E_A is

$$\bar{R}_p(E_A) = \iint I(\mathbf{p}, z)z \, dz \, d^3p \tag{6.1}$$

using the solution to Eq. (1.2). However, solving for $I(\mathbf{p}, z)$ to obtain $\bar{R}_p$ would be superfluous. Therefore, we begin the discussion by considering a simpler, historically useful approach, and build upon it when more detail is required. We will ignore the simulation approach which, though straightforward, requires a large programming effort to calculate even rather simple quantities such as R_p.

The Continuous Slowing-Down Approximation

Earlier in Eqs. (2.18) and (2.17) the energy loss per unit pathlength, the stopping power of the material, was defined in terms of a quantity called the stopping cross section, S_{AB},

$$-\frac{dE}{ds} \equiv n_B S_{AB} \tag{6.2}$$

where s is the path length and n_B the target number density. The energy loss processes are, as usual, separated into elastic collisions involving energy transfer to the target center of mass (the nuclei), and inelastic processes involving energy transfer to the electron, i.e.,

$$S_{AB} \simeq S_n + S_e$$

Equation (6.2) can be integrated directly to yield an estimate of the average path length traveled by the incident particle before stopping,

$$\bar{R}(E_A) = \int_0^{E_A} \frac{dE'}{n_B S_{AB}(E')} \tag{6.3}$$

This is the continuous slowing-down approximation (CSDA) to $\bar{R}$. That is, it is a calculation in which the energy is assumed to be dissipated continuously as in a drag force. For a particular choice of incident particle and target material, Eq. (6.3) can be integrated numerically using measured or calculated stopping cross sections. For our purposes the behavior of $\bar{R}$ with energy can be investigated using the power law potentials of Chapter 2 either classically or in the Born approximation. The resulting electron or nuclear stopping cross section can be written as $S_{AB} = \xi_x E_A^{1-x}$, with x

determined from the potential (e.g., $x = 2/n$, classically). From the discussion and results in Chapter 5 for S_{AB}, we see that x is slowly varying in energy over broad regions in energy. Treating x as a constant over the predominant stopping region for each E_A, we have

$$\bar{R}(E_A) \sim \frac{1}{x n_B \zeta_x} E_A^x = \frac{E_A}{x n_B S_{AB}} \tag{6.4}$$

Therefore $\bar{R}$ can be estimated from the size of S_{AB} and the slope of S_{AB} (e.g., x) at each energy. By fitting S_{AB} to an accurate result in the vicinity of E_A in order to obtain x, the approximation in Eq. (6.4) is compared in Figure 6.2 to a numerical integration over the stopping cross section shown. The mean range is seen to be monotonically increasing in energy, a not very surprising result, and the simple approximation is in reasonable agreement.

There are two errors in the above description. The first is the use of an oversimplified approximation to S_{AB}, which is easily remedied. The second arises from the fact that the target medium is not continuous, the targets are discrete, and the collisions occur statistically. This implies that there is a variation of the particle ranges from the mean range as each particle has a different history, that is, a different energy loss in each collision. The root-mean-squared deviation in the range, $\overline{\Delta R^2}$, in the CSDA approximation is determined from the mean-squared deviation in the energy transfer in an individual collision, $S_{AB}^{(2)}$. This quantity, referred to as the energy straggling cross section, has the form

$$S_{AB}^{(2)} \simeq \int \Delta E^2 (d\sigma/d\Delta E)\, d\Delta E \tag{6.5}$$

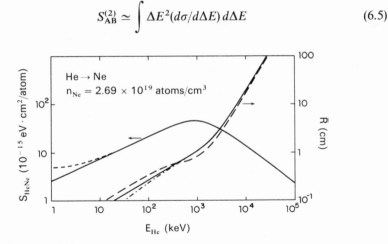

Figure 6.2. He → Ne stopping for equilibrium charge state of He, (see Chapter 5; from J. F. Ziegler, *Helium: Stopping Powers and Ranges in All Elements*, Pergamon Press, New York (1978). S_{HeNe}: (left-hand scale) solid line, electronic; short-dashed line, electronic plus nuclear. R: (right-hand scale) solid line, accurate calculation using Eq. (6.13a) (with $\cos \Theta_A = 1$); long-dashed line, using approximation to CSDA in Eq. (6.4); $\bar{R}_p$: (right-hand scale) dot-dashed line, using Eq. (6.13b).

For power laws, we have $S_{AB}^{(2)} = \xi_x' E_A^{2-x}$, where ξ_x' can be determined from the potential. The effect on the range, due to the uncertainty in energy, ΔE, at each collision, j, is

$$(\overline{\Delta R})_j \approx \left(\overline{\Delta E}\,\frac{ds}{dE}\right)_j$$

where s is measured along the path of the particle. The accumulated range straggling, $\overline{\Delta R^2}$, is

$$\overline{\Delta R^2} = \sum_j \overline{(\Delta R)_j^2} = \sum \left(\overline{\Delta E}\,\frac{ds}{dE}\right)_j^2$$

where the sum is over all collisions that occur while the particle is stopping. In the CSDA, we have $\overline{\Delta E_j^2} \approx S_{AB}^{(2)} n_B \Delta s_j$, where Δs_j is the distance between encounters. Hence, replacing the sum by an integral, we have

$$\overline{\Delta R^2} = n_B \int_0^{E_A} S_{AB}^{(2)} \left(\frac{ds}{dE}\right)^3 dE \qquad (6.6)$$

which for power laws has the form

$$\overline{\Delta R^2} \approx \frac{\xi_x'}{2x\xi_x^3 n_B^2}\, E_A^{2x} = \frac{E_A\, S_{AB}^{(2)}}{2xn_B^2\, S_{AB}^3} \qquad (6.7)$$

The range straggling, like the mean range, is seen to be a monotonically increasing function of energy in each region of x. To estimate the ion implantation distribution inside of the material, it is often sufficient to write the distribution of stopped particles as a gaussian using $\bar{R}$ and $\overline{\Delta R^2}$, as indicated in Figure 6.3, *if* the incident particles are not deflected significantly. This is the case when $M_A \gg M_B$ or when the stopping is primarily due to the electrons.

The mean range, $\bar{R}$, is not an easily measurable quantity if the deflections are large. The range provides, however, a measure of the total damage produced in the material. That is, for a low flux of incident particles, $n_B \bar{\sigma}_D \bar{R}$ is the number of damage sites produced directly by the incident ion, where $\bar{\sigma}_D$ is an average damage cross section (e.g., breaking a bond, causing an

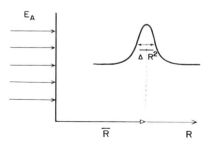

Figure 6.3. Schematic diagram of stopped-particle distribution when the deflections are small: $\bar{R}$ is the mean range and $\overline{\Delta R^2}$ is the range straggling or root-mean-squared deviation in the range.

ionization, or displacing a target atom). This will be treated in more detail shortly.

For the penetration problem, a more useful quantity is the mean penetration depth, $\bar{R}_p(E_A)$, also called the mean projected range and defined in Eq. (6.1). The relationship between $\bar{R}$ and $\bar{R}_p$ is indicated schematically in Figure 6.4. Extending the CSDA one would write

$$\bar{R}_p \sim \int_0^{E_A} \frac{\overline{\cos\theta}}{n_B\, S_{AB}}\, dE \tag{6.8}$$

where $\overline{\cos\theta}$ is defined to be an average angle between the incident direction and the direction of motion in the material when the particle has slowed to energy E. The quantity $\overline{\cos\theta}$ therefore depends both on E_A and E, with $\overline{\cos\theta}$ decreasing with depth of penetration, i.e. as E decreases. A useful approximation is to write $\overline{\cos\theta} \simeq (E/E_A)^{\mu/2}$, with $\mu = M_B/M_A$, where nuclear collisions dominate, and $\mu \to m_e/M_A \simeq 0$, where the energy loss to the electrons dominates. For the power-law cross section Eqs. (6.3) and (6.8) yields the ratio

$$\frac{\bar{R}}{\bar{R}_p} \sim 1 + \frac{\mu}{2x} \tag{6.9}$$

This ratio, the mean range vs the mean projected range, provides a useful estimate of the increase in damage density when the deflections are significant. Remembering that in the approximations used here, μ and x vary slowly in energy, we see that for incident ions the ratio approaches unity at high energies when energy loss to the electrons dominates.

The approximation in Eq. (6.8) for $\bar{R}_p$, although useful for indicating the general behavior of this quantity, has limited validity. In the following section the more accurate integral equation method for determining $\bar{R}$ and $\bar{R}_p$ is presented. This method incorporates the statistical nature of the stopping process, which is especially important in determining $\bar{R}_p$.

Integral Equations for Range and Projected Range

The mean range and mean projected range of a particle undergoing a series of collisions is obtained by summing the average distances between

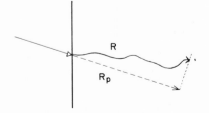

Figure 6.4. R, the range, is total distance traveled by an incident particle, and R_p, the projected range, is the distance penetrated along the initial direction of motion.

collisions. If the mean free path of the projectile at energy E_A is

$$\lambda_{AB} = [n_B \sigma_{AB}(E_A)]^{-1}$$

(viz. Chapter 2), the mean range can be written as the mean free path for the first collision plus the mean range of the particle after this collision. As the distance traveled after the first collision depends on the energy loss in that collision, the range is written

$$\bar{R}(E_A) = 1/n_B \sigma_{AB} + \int_0^{\Delta E_{max}} \bar{R}(E_A - \Delta E) \, d\sigma_{AB}/\sigma_{AB} \qquad (6.10)$$

where ΔE accounts for elastic and/or inelastic energy loss. In Eq. (6.10) $d\sigma_{AB}/\sigma_{AB}$ is the probability of an energy loss ΔE occurring in the first collision, where $d\sigma_{AB} \equiv (d\sigma_{AB}/d\Delta E) \, d\Delta E$. After the first collision, the particle has a mean range $\bar{R}(E_A - \Delta E)$ and the integral implies an average over all possible first collisions. As the particle is moving in a new direction after the first collision, the corresponding equation for $\bar{R}_p$ is

$$\bar{R}_p(E_A) = 1/(n_B \sigma_{AB}) + \int_0^{\Delta E_{max}} \bar{R}_p(E_A - \Delta E) \cos \theta_A \, d\sigma_{AB}/\sigma_{AB} \qquad (6.11)$$

where $\cos \theta_A$ is the deflection angle in Table 2.4 for a given energy loss ΔE.

When the cross section $\sigma_{AB}(E_A)$ is finite, and the collision energy low, Eqs. (6.10) and (6.11) can be solved by iteration. That is,

$$\bar{R}(E_A) = \lambda_{AB}(E_A) + \int_0^{\Delta E_{max}} \lambda_{AB}(E_A - \Delta E) \, d\sigma_{AB}/\sigma_{AB}$$

$$+ \int_0^{\Delta E_{max}} d\sigma_{AB}/\sigma_{AB}(E_A)$$

$$\times \int_0^{\Delta E'_{max}} \lambda_{AB}(E_A - \Delta E - \Delta E') \, d\sigma_{AB}/\sigma_{AB}(E_A - \Delta E) + \cdots$$

$$(6.12)$$

Noting that $d\sigma_{AB}/\sigma_{AB}$ is the probability of a collision with energy loss between ΔE and $\Delta E + d\Delta E$, we see that the sum above implicitly represents the total path as a series of collisions separated by mean free paths, λ_{AB}. Although this iterative method is instructive, it is not particularly useful for determining $\bar{R}$ or $\bar{R}_p$, particularly for singular cross sections.

An alternate procedure, which yields analytic solutions for the power-law cross sections, begins with a rearrangement of Eqs. (6.10) and (6.11) in the form

$$1 = n_B \int_0^{\Delta E_{max}} d\sigma_{AB}[\bar{R}_p(E_A) - \cos \theta_A \bar{R}_p(E_A - \Delta E)] \qquad (6.13a)$$

where the equation for $\bar{R}$ is obtained by replacing $\cos \theta_A$ by one. In this form, as $\Delta E \rightarrow 0$ (hence $\cos \theta_A \rightarrow 1$) and $d\sigma/d\Delta E \rightarrow \infty$, the two terms in brackets cancel. When small energy transfers dominate, expanding $\cos \theta_A$ and $\bar{R}_p(E_A - \Delta E)$ about $\Delta E = 0$, we can recover the CSDA expressions in Eqs. (6.3) and (6.8). This emphasizes that for small energy transfers, which is the case when the targets are the atomic electrons, the stopping *is* like a drag force, and this is the justification for the separation of the electronic energy loss term in the integral equation [Eq. (1.2)], in Chapter 1. It is also common to separate these processes when determining $\bar{R}_p$; that is, one writes

$$
1 = n_B \int_0^{\gamma E_A} (d\sigma/dT)\, dT [\bar{R}_p(E_A) - \cos \theta_A\, \bar{R}_p(E_A - T)]
$$

$$
+ n_B S_e [d\bar{R}_p/dE_A] \tag{6.13b}
$$

where, as usual, T is the elastic nuclear energy transfer which produces the deflections.

The general form for the power-law cross section is $d\sigma/d\Delta E = C_{AB}/(E_A^a \Delta E^{1+b})$, which yields the expression used earlier, $S_{AB} = \xi_x E_A^{1-x}$, when $x = a + b$ and $\xi_x = \gamma/(1 - b)$. (In the Born approximation, $a = 1$, $b = 2 - n$; in the classical impulse approximation, $a = b = 1/n$, both for $V \propto R^{-n}$.) Substituting this expression into Eq. (6.13a) and writing $y = \Delta E/E_A$, we solve this equation in the form $\bar{R}_p(E_A) = \bar{R}_p^0 E_A^x$. With $\Delta E_{max} = \bar{\gamma} E_A$, where $\bar{\gamma}$ is determined by the dominant energy transfer mechanism (i.e., to nuclei or electrons), the constant $\bar{R}_p^0$ is seen to be

$$
(\bar{R}_p^0)^{-1} = n_B C_{AB} \int_0^{\bar{\gamma}} \frac{dy}{y^{1+b}} [1 - (1 - y)^x \cos \theta_A] \tag{6.14}
$$

When applying Eq. (6.14) to elastic nuclear-energy transfer, $\Delta E \rightarrow T$, $\bar{\gamma} = \gamma$, and, from Eq. (2.41), we have $\cos \theta_A = \{1 - [(1 + \mu)/2](T/E_A)\}/(1 - T/E_A)^{1/2} = \{1 - [(1 + \mu)/2]y\}/[1 - y]^{1/2}$.

For simple isotropic scattering of equal mass particles, the cross section can be described using $b = -1$, $a = 1$, and $C_{AB} = \sigma_{AB}$, in which case $\bar{R}_p = 3/n_B \sigma_{AB}$. That is, a superthermal molecule of a gas will move about three mean free paths in the initial direction of motion while slowing down, which the reader should verify, is $3/4$ the result obtained using Eq. (6.8). Of course, the total path length, $\bar{R}$, is infinite with such a cross section. That is, the particle rattles around a lot before thermalizing, but does not progress very far. The energy dependence obtained from the integral equations does not differ from that obtained in the CSDA; however, the multiplicative constant is different (we leave as an exercise the problem of comparing the magnitude of the result here to the CSDA for power-law potentials: Problem 6.2.

The general procedure for solving Eq. (6.13), when a more complicated cross section is used or both electronic and nuclear energy loss contribute, is to assume an analytic form for $\bar{R}_p$. That is, we write $\bar{R}_p$ as a sum of terms each having a different energy dependence with variable coefficients. The "best" coefficients are then obtained by minimizing the difference between the left- and right-hand side of the integral equation. The result of such a calculation is given in Figure 6.2. For implantation of ions we would also like to know the range straggling. Rather than calculate this quantity directly, we proceed in the following section to consider equations describing the stopped particle distribution and obtain the straggling as a product.

Stopped-Particle Distribution

In the last two sections we constructed methods for obtaining two of the parameters which characterize the stopped-particle distribution, that is, $\bar{R}$ and $\overline{\Delta R^2}$. It would, of course, be preferable to obtain the distribution itself, or at least have a scheme for determining it as accurately as possible.

To describe the distribution of stopped particles another integral equation is constructed. We define $F(E_A, \hat{p}_A, z)$ to be the number density of stopped particles at depth z that had initial energy E_A and direction $\hat{p}_A$. This distribution is made up of contributions from particles that had different collision histories as they come to a stop. Following Chartres, Lindhard, Sigmund, and others we consider the collisions occurring in the first layer of material. From Figure 6.5 it is clear that the distribution, $F(E_A, \hat{p}_A, z)$, at a depth $z > \Delta z$ determined by the flux of particles starting at $z = 0$ is equivalent to the distribution determined by the flux of particles emerging at Δz. Some of these particles have a collision in the layer and emerge with a new energy and direction. Recalling that $(\Delta z\, n_B / \hat{p}_A \cdot \hat{z})\, d\sigma_{AB}$ is the probability of a collision occurring in the layer resulting in an energy loss between ΔE and

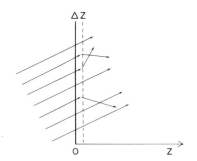

Figure 6.5. Change in the incident-particle flux after passing through a layer Δz thick.

$\Delta E + d\Delta E$, we obtain the distribution at $z > \Delta z$:

$$F(E_A, \hat{p}_A, z) = (\Delta z \, n_B/\hat{p}_A \cdot \hat{z}) \int d\sigma_{AB} F(E_A - \Delta E, \hat{p}'_A, z - \Delta z)$$

$$+ [(1 - \Delta z n_B \sigma_{AB})/\hat{p}_A \hat{z}] F(E_A, \hat{p}_A, z - \Delta z)$$

The first term on the right accounts for those particles scattered in the layer, and the second for those emerging without a scattering. The quantity $F(E_A - \Delta E, \hat{p}'_A, z - \Delta z)$ is the stopped-particle distribution at z for particles starting at Δz with energy $E_A - \Delta E$ and momentum $\hat{p}'_A$. Rearranging terms and letting $\Delta z \to 0$, we obtain the integral equation

$$- \cos\theta \frac{\partial}{\partial z} F(E_A, \hat{p}_A, z) = n_B \int d\sigma_{AB} [F(E_A, \hat{p}_A, z) - F(E_A - \Delta E, \hat{p}'_A, z)$$

$$\tag{6.15}$$

where $\cos\theta \equiv \hat{p}_A \cdot \hat{z}$.

It is customary to separate electronic and nuclear energy transfers as in Eq. (1.2), giving the following integral-differential equation for F:

$$- \cos\theta \frac{\partial}{\partial z} F(E_A, \hat{p}_A, z) = n_B \int_0^{E_A} \frac{d\sigma}{dT} [F(E_A, \hat{p}_A, z) - F(E_A - T, \hat{p}'_A, z)] \, dT$$

$$+ n_B S_e \frac{\partial}{\partial E_A} F(E_A, \hat{p}_A, z) \tag{6.16}$$

Although the equation is similar in form to Eq. (1.2), the quantities to be determined, F and I, have different boundary conditions. On comparing these equations, we see that the probability per unit pathlength of a momentum change, postulated in Eq. (1.2), has the form (see also Appendix E)

$$\omega(\mathbf{p}, \mathbf{p}') \, d^3p' = n_B \left(\frac{d\sigma}{dT}\right) dT \, \delta[\hat{p}' - \hat{p}(T)] \, d\hat{p}' \tag{6.17}$$

and the integral over $d\hat{p}'$ has been carried out in Eq. (6.16). The delta function arises from the classical collision kinematics of Chapter 2 as each T corresponds to a particular direction change.

Before examining the distribution in detail it is worth noting that if $M_A \gg M_B$, so that T and the scattering angle $\hat{p}_A \cdot \hat{p}'$ are small, then Eq. (6.16) reduces to

$$- \cos\theta \frac{\partial}{\partial z} F \sim n_B (S_e + S_n) \frac{\partial}{\partial E_A} F \tag{6.18a}$$

Now depth and energy are simply related and all the particles stop at the same point, producing a delta-function distribution. This is simply the continuous slowing-down approximation. In the opposite extreme, if each collision leads to an absorption of an incident particle (stopping), then from

Eq. (6.15) we have

$$F \sim F_0 \exp[-n_B \sigma_{AB} z/\cos\theta] \tag{6.18b}$$

where F_0 is a normalization constant, n_B is assumed constant, and the $\cos\theta$ allows for other than perpendicular incidence. This is the situation when incident light or slow neutrons are absorbed by a material. In fact this distribution is not unlike that for slow, heavy atoms which lose their forward motion in a couple of collisions. In the following the more general case is considered.

As Eq. (6.16) is very difficult to solve, the distribution is often expressed by its moments

$$F^n(E_A, \hat{p}_A) = \int_{-\infty}^{\infty} z^n F(E_A, \hat{p}_A, z)\, dz \tag{6.19}$$

The zeroth moment, which is the total number of stopped particles, provides one condition on the solution. For sufficiently thick targets, all particles are stopped unless they are backscattered from the material. Normalizing F to be the probability distribution per incident particle, we have

$$\langle z^0 \rangle \equiv F^0 = 1 - \text{probability of backscattering} \tag{6.20}$$

In many examples the backscattering probability is small, in which case $F^0 = 1$. The mean projected range is simply related to F^1, i.e., $\langle z \rangle \equiv \bar{R}_p(E_A)\cos\theta = F^1/F^0$, and the projected range straggling is $\langle \Delta z^2 \rangle \equiv \langle (z - \langle z \rangle)^2 \rangle = (F^2 - \langle z \rangle^2)/F^0$.

The simplest approximation to $F(E_A, \hat{p}_A, z)$ is a gaussian, as in Figure 6.3, using $\langle z^0 \rangle$, $\langle z \rangle$, and $\langle \Delta z^2 \rangle$,

$$F(E_A, \hat{p}_A, z) \approx \frac{\langle z^0 \rangle}{[2\pi < \Delta z^2 >]^{1/2}} \exp\left[-\frac{(z - \langle z \rangle)^2}{2\langle \Delta z \rangle^2}\right] \tag{6.21}$$

which involves determining F^0, F^1, and F^2. Integral equations for these quantities are obtained by multiplying Eq. (6.15) by z^n, integrating and using Eq. (6.19). Applying this procedure, we find that the equation for F^1 is equivalent to that for $\bar{R}_p$, Eq. (6.13a), when the simple angular dependence associated with the incident particle direction is removed. The angular dependence in the higher terms, F^n, is generally expressed using the Legendre polynomials $P_l(\cos\theta)$. Noting that z very nearly scales as $\cos\theta$, one can write $F^n(E_A, \hat{p}_A) = \sum_l P_l(\cos\theta) F_l^n(E_A)$, where $l \leq n$ and l is odd for n odd, even for n even. Winterbon and others have constructed codes for determining the moments, F_l^n, for simple cross sections. These are then used to reconstruct the distribution, with the gaussian distribution in Eq. (6.21) being the simplest form. When moments higher than F^2 are used, a number of reconstruction methods may be employed, which lends a certain arbitrariness to the result. However, comparisons of reconstructed (from moments) distributions and simulation calculations generally show reasonable agreement, except near $z = 0$.

The stopped-particle or projected range distribution for normal incidence is shown in Figure 6.6 for 1-keV protons entering a biological-like target. The moments were obtained using the Lindhard–Thomas–Fermi approximation to $d\sigma/dT$ discussed in Appendix H and the Lindhard-Scharff low-energy stopping formula [Eq. (5.31)]. The range distribution is seen to be skewed only slightly for this case, although this is clearly not typical of all distributions or combinations of projectile and target, e.g., the extreme case is the solution (6.18b). When M_A is large compared to the target mass the width of the distribution narrows, approaching a delta function in the limit where the CSDA is valid, the solution to (6.18a). In these examples the material is treated as infinite and $F^0 = 1$.

Diffusion of Implanted Particles, Backscattering, and Transmissions

The gaussian form for F in Eq. (6.21), although approximate, provides a useful initial condition for describing the diffusion of the stopped particles. These particles are said to be stopped in the sense that they have attained temperature equilibration with the target material. However, as the distribution of these particles is nonuniform, thermal diffusion will occur. The "stopped", and by now neutralized, particle will migrate until it leaves the material, reacts with a material component, or is trapped (at a lattice or interstitial sight, or in a physical gap or hole in the material).

The diffusion-reaction process is described by the rate equations developed in Chapter 1. That is, in a uniform material, the particle distribution $F(\mathbf{r}, t)$ is determined from

$$\frac{\partial}{\partial t} F(\mathbf{r}, t) = + D(T)\nabla^2 F(\mathbf{r}, t) - \nu_R(T)F(\mathbf{r}, t), \qquad (6.22)$$

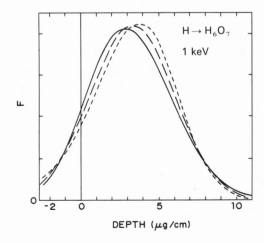

$$H \rightarrow H_6O_7$$

1 keV

F

DEPTH (μg/cm)

-2   0   5   10

Figure 6.6. Stopped-particle distribution, F, for $H \rightarrow H_6O_7$ (target: RNase-like weighting of light and heavy atoms): short-dashed line, using three moments; long-dashed line, using four moments, solid line, using five moments. [From R. Romanelli, Ph.D. thesis, University of Virginia (1980).]

where the diffusion coefficient, $D(T)$, and the reaction frequency, $v_R(T)$, both depend on the temperature of the material and the species involved. In Chapter 5 the diffusion coefficient for a gas was found to have a temperature dependence, $D \propto T^{x+1/2}$ for power laws. In a solid, the movement of atoms (or vacancies) between sites occurs in jumps. Because an energy barrier exists for jumping between sites, the diffusion coefficient for a particle in a solid has a form like the Arrhenius reaction rate in Eq. (5.36). That is, $D(T) = f(T) \exp[-\Delta E_a/kT]$, where ΔE_a is the activation for a jump between sites, and $f(T)$ is slowly varying with $D(T)$ approaching the gaseous diffusion limit at high temperatures or small ΔE_a.

As D and v_R are assumed to be independent of depth, Eq. (6.22) has a simple solution when the initial spatial distribution is gaussian. In the one-dimensional problem we have been solving, the distribution of unattached "stopped" particles is

$$F(z, t) = \frac{\langle z^0 \rangle}{[2\pi \langle \Delta z^2 \rangle_t]^{1/2}} \exp\left[-\frac{(z - \langle z \rangle_0)^2}{2\langle z^2 \rangle_t} \right] \exp(-v_R t) \qquad (6.23)$$

where $\langle z^0 \rangle$ is F^0 of Eq. (6.18) and the subscripts indicate the time dependence. In this expression $\langle z \rangle_0 = \langle z \rangle$, the stopped-particle mean penetration depth of the previous section, and $\langle \Delta z^2 \rangle_t = \langle \Delta z^2 \rangle + 2D(T)t$, so that Eq. (6.23) reduces to Eq. (6.20) as $t \to 0$. A quantity of greater interest in ion implantation is the distribution of bounded or trapped, implanted particles at $t \to \infty$, (Problem 6.12).

A fraction of an ion beam is backscattered from surface or interior atoms. This fraction increases with decreasing ion velocity and in Figure 6.6 is the part of the distribution to the left of the surface. Of interest for material analysis are those backscattered particles having experienced a single, hard collision. For fast, light ions ($\gtrsim 1$ meV/amu) the reflections are a result of close ($b \simeq 0$) Coulomb collisions with the nuclei of the target atoms, [Eq. (2.61b) with $n = 1$, $T = \gamma E_A$]. The total energy loss of such backscattered particles is roughly $\gamma E_A + n_B S_e(2\bar{z})$, where $\bar{z}$ is the depth at which the collision occurred. As γ depends on the mass of the target atom and S_e is tabulated, measuring the energies of the particles backscattered can provide a depth distribution for near surface impurity atoms or the thicknesses of thin films deposited on a known substrate.

The number of particles transmitted through a foil can be calculated by extensions of the techniques discussed above. In Figure 5.6 such particles were used to estimate the interaction potential with target atoms. Transmission experiments are also used to determine spectra of ions and structures of molecular ions. As fast ions are further ionized (stripped) and excited on transmission through thin carbon foils, the electronic relaxation processes are monitored after the ions exit the foil. Highly stripped molecules, for instance, will "explode" due to the Coulomb repulsion between the nuclei. Although, this repulsive energy, Q_r, is a small fraction of the incident kinetic energy, E_A, in the laboratory frame (Table 2.3), significant changes in kinetic energies of the fragments ($\sim (Q_r E_A)^{1/2}$) are observed if the repulsive force is nearly parallel to the incident velocity. Knowing the charge states produced by stripping and monitoring the exiting-ion energies and angles due to the repulsion, initial bond lengths and dissociation energies can be deduced for simple molecular ions which may be difficult to examine spectroscopically.

Energy Deposition Effects

The previous sections have dealt with the behavior of the incident particles and not the description of the alterations produced in the target material by incident radiation. Alterations, such as net ionization or number of displaced particles, are often characterized by the energy deposition per unit volume, or for uniformly irradiated materials, the energy deposited per unit pathlength.

Using the CSDA and power laws, the penetration distance and particle energy can be simply related. Neglecting deflections, the penetration depth, z, at which the incident particles have been slowed to an average energy $\bar{E}$ is

$$z \sim \int_{\bar{E}}^{E_A} (1/n_B S_{AB})\, dE \tag{6.24}$$

Employing the simple approximation for S_{AB} that we have been using and Eq. (6.4), we reduce Eq. (6.24) to the form $z \sim \bar{R}_p - \bar{E}^x/(x\xi_x n_B)$. We can now estimate the general form for the energy deposition rate at any depth z using $\bar{E}$ from Eq. (6.24) and Eq. (6.2) for the energy loss rate,

$$-dE/dz \sim (n_B \xi_x)^{1/x}/[x(\bar{R}_p - z)]^{1 - 1/x} \tag{6.25}$$

For slow collisions in the electronic stopping regime where $x \sim \frac{1}{2}$, the energy deposition decreases slowly with depth, whereas for fast collisions (i.e., $1 < x < 2$), it is strongly peaked toward the end of the path. This is seen in Figure 6.7 for fast heavy particles on tissue-equivalent material, which yields the typical Bragg curve for energy deposition. The localized, high-density radiation deposited at the end of the path is something one tries to exploit when using ion beams to selectively irradiate targets in a material. That is, one would guess that for fast particles the major damage to the material occurs at the end of the path, where, of course, the deposited particles are also found. In slow collisions for which elastic nuclear energy loss is important, the deposited energy distribution is obtained from an integral equation and resembles the range distribution in form (viz. Figure 6.6).

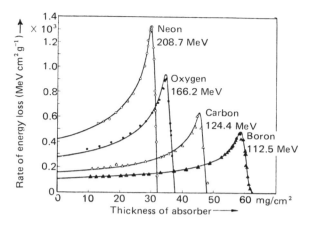

Figure 6.7. Bragg curves for heavy ions in tissue-equivalent material (use $\rho \simeq 1$ g/cm^3 for density conversion). [From T. Brustad, *Radiat. Res.* **15**, 139 (1961).]

Damage to a material may be associated with a number of changes induced by the incident particles. One measure of damage is the number of target atoms displaced from their original site. Such damage may be repaired by diffusion processes, but this discussion will concentrate on the initial damage produced. Dividing the amount of energy deposited in *elastic* collisions by twice the energy required to displace a target atom in a dynamic collision has been shown to give a reasonable estimate of the number of displaced atoms. This model, based on the Kitchin–Pease approximation for solving the integral equations, is widely used for estimating displacement damage. The corresponding concept in ionization production has been extensively verified. That is, over a broad range of energies, where $S_e \gg S_n$, the average energy deposited per ionization produced is a constant, which we write as W_e. This number, referred to as the W value, is roughly related to the ionization potential (i.e., the energy required to displace an electron), as indicated for a few targets in Table 6.1. If the total energy E is separated into that part deposited elastically, v, and that deposited inelastically, η (i.e., $E_A = v + \eta$), the number of ionizations is η/W_e and the number of displacements is v/W_n, where W_n is approximately twice the displacement energy. (It is traditional to use v for the elastic energy deposited and η for the electronic energy, and the reader should not confuse these with a collision frequency or a phase shift.)

Based on the above notions, a rough approximation to the number density of displaced particles at various depths is

$$n_D(z) \simeq \phi_A n_B S_n(\bar{E})/W_n \qquad (6.26)$$

with $\bar{E}(z)$ determined from Eq. (6.24). Equation (6.26) applies only for small fluences ϕ_A; ϕ_A is the particle flux (assumed perpendicular and uniform)

Table 6.1. W Values at High Energies: Average
Energy Required to Produce an Ion Pair for a
Particle Stopping in a Gas

Target	I^b	$W_e(H^+)$	$W_e(He^{++})$	$\sim W_e/I$
He	24.5	45.	44.	1.8
Ne	21.6	39.	36.8	1.8
Ar	15.8	26.6	26.3	1.7
Kr	14.0	23.0	24.1	1.7
Xe	12.1	21.	21.6	1.8
N_2	15.6	37.	36.4	2.4
Air^c	14.9	35.2	35.1	2.4
CO_2	13.8	34.4	34.2	2.5
TE^d	13.1	30.	31.	2.3
CH_4	12.6	31.	29.	2.4

[a] H. Bischel, D. H. Pearson, J. W. Boring, A. Green, M. Inokuti,
and G. Hurst, ICRU Report No. 31, Washington, D.C. (1979).
[b] I is the first ionization potential.
[c] I based on average mixture of N_2, O_2, and Ar.
[d] TE is tissue-equivalent gas: 64.4% CH_4, 32.4% CO_2, and 3.2%
N_2.

times the exposure time. For fast collisions, as the mean energy $\bar{E}$ is determined by the electronic stopping power and S_n increases with decreasing energy, $n_D(z)$ is also peaked near the end of the path. In this example it is assumed that all the energy deposited by the incident particle in elastic collisions goes into nuclear motion. However, the fast secondaries produced in these collisions may in fact lose a considerable fraction of their energy inelastically and will deposit their energy away from the initial collision site. The integral equation approach will correct both of these oversights. Here we use the fact that the production of slow secondaries dominates. As these particles lose *most* of their energy elastically (i.e., $v(E) \to E$ as $E \to 0$), and do not travel far from the initial production point, a first-order improvement to the result in Eq. (6.26) is

$$n_D(z) \sim \frac{\phi_A n_B}{W_n} \int_0^{\gamma\bar{E}(z)} v(T) \frac{d\sigma}{dT} \, dT \qquad (6.27)$$

with $v(T) \sim \int_0^T [S_n/(S_e + S_n)] \, dE$. The reader should note that these "simple" methods rapidly become complex and the approximations made rather restrictive. The integral equation method, which at first sight seems rather involved, begins to appear more attractive as the accuracy required increases. Equations (6.26) and (6.27) can both be integrated to yield the total number of displacements produced per incident particle v/W_n.

A useful quantity, related to the displacement problem, is the number of recoils with energy between T and $T + dT$ produced at any depth in a cascade initiated by a particle of energy E_A which we write $G(E_A, T)$. Solving the collision cascade equation, which includes all recoils, Sigmund has shown that

$$G(E_A, T) \sim \beta_n \frac{v(E_A)}{T^2} \quad \text{for } E_A \gg T \tag{6.28}$$

where n labels the power law used to describe collisions of the secondaries with atoms of the target material. Using an exponential potential ($n \to \infty$) to describe the slow elastic collisions, we have $\beta_0 = 6/\pi^2$ in Eq. (6.28). The number of displacements per incident ion is found by integrating Eq. (6.28) over all energies greater than the displacement energy, E_D. This procedure yields the result

$$\int_{E_D}^{E} G(E_A, T) \, dT \sim \frac{6}{\pi^2} \frac{v(E_A)}{E_D} \tag{6.29}$$

implying $W_n \sim 1.64 E_D$. Accounting for replacement collisions which occur when the incident-particle's energy is less than E_D after the collision, one finds that $W_n \sim 2.4 E_D$. These results are both close to the Kitchin–Pease result, $W_n \sim 2 E_D$, assumed earlier.

In the following section we extend these ideas to consider, first, damage to a biological sample and, then, the amount of target material removed by an incident ion, or the sputtering of the target.

Biological Damage

Simple enzymes, involving a single, long chain of molecules (strand), folded by hydrogen bonds, are known to be damaged by a single "hit" from incident, heavy-particle radiation. Defining exactly what constitutes a " hit" is rather difficult, however. It appears that radicals such as OH^-, H_3O^+ and electrons in an aqueous solution are very effective in damaging biological molecules. These radicals are products of the ionization and dissociation of H_2O by the incident radiation. The radicals apparently diffuse to a sensitive site of the enzyme and react destructively. This type of damage, referred to as indirect, is controlled to a large extent by diffusion and chemistry occurring after the initial energy deposition. For comparison, samples are often freeze-dried to remove the water content, so the direct effect of the radiation can be measured. Although a considerable body of knowledge has accumulated regarding the chemical kinetics that follow the initial energy deposition events, obtained using ingenious experimental procedures, this discussion will deal with initial events only.

If a material is irradiated for a time, t, it was shown in Chapter 2 that the probability of a given target, within the material, receiving a hit increases exponentially. Conversely the survival probability is, based on Eq. (2.12), the simple Poisson distribution function

$$P_S = \exp[-\sigma_h \phi_A] \qquad (6.30)$$

where Φ_A is the fluence ($\phi_A = It$) and σ_h is the hit cross section. We note in passing, that in double-stranded material (e.g., DNA) it is thought that two separate hits, presumably one on each strand, are required for damage, in which case

$$P_S = 1 - (1 - \exp[-\sigma_h \Phi_A])^2 \qquad (6.31)$$

which has a different dependence on fluence, as indicated schematically in Figure 6.8. In the following an expression for σ_h is constructed.

We define $F_D(z, r, E_A)$ to be the number density of damage sites produced in a biological material (consisting of a large number of enzymes, either freeze dried or in an aqueous solution) by an incident ion of energy E_A. These damage sites may be produced directly by ionization and dissociation, or indirectly by diffusing radicals resulting from ionization and dissociation events. In the quantity $F_D(z, r, E_A)$, r indicates the radial distance from the incident particle direction (presumed normal to the surface for simplicity), and z the depth into the bulk material. Such a quantity is calculated from the integral equation describing the cascade (or the CSDA) and subsequent diffusion kinetics, as in the previous examples. Since the fast secondary electrons produced, often called delta rays, can travel quite far before stopping and, in addition, diffusion acts to distribute the product radicals, damage may take place at significant distances, on a molecular scale, from the initial track. If the energy deposited per unit path length by the incident particle is large, then the damage has roughly a uniform, cylindrically symmetric distribution about the incident particle track. Using a power cross section in the Born approximation for the production of ioni-

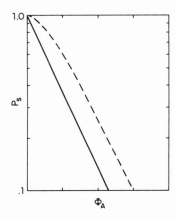

Figure 6.8. Survival probability, P_S, vs. fluence, Φ_A (flux × time): solid line, single-hit targets; dashed line, double-hit targets.

zation, we leave it to the reader (Problem 6.15) to show that the distribution of the inelastic energy deposited by a fast primary is

$$\left(\frac{dE}{dV}\right)_{e} \sim \frac{1}{2\pi R_{max}^{2}} \left|\frac{dE}{ds}\right|_{e} \left(\frac{n-1}{2-n}\right) \left(\frac{R_{max}}{r}\right)^{3-n} \left[1 - \left(\frac{r}{R_{max}}\right)^{2-n}\right] \quad (6.32)$$

for $1 < n < 2$. In Eq. (6.32) it was assumed, for simplicity, that the secondary electrons lose energy at a constant rate and move perpendicular to the track, with R_{max} the maximum range for the electrons. As $F_D(z, r, E_A)$, according to the previous discussion, is simply related to $(dE/dV)_e$, the damage distribution is seen to be peaked close to the track and decreases slowly with distance from the track. The experimental results for deposited dose in a gas shown in Figure 6.9 indicate the behavior expected from Eq.

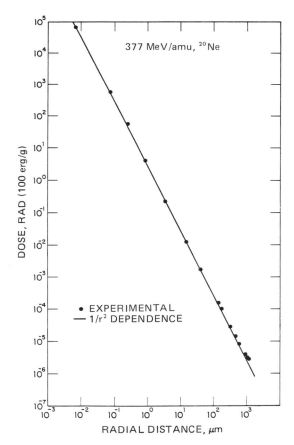

Figure 6.9. Dose deposited $(dE/dV)_e$ in a tissue-equivalent gas versus radial distance from incident Ne direction. Distance scale based on an assumed density of 1 g/cm³. [From: M. N. Varma and J. W. Baum, *Radiat. Res.*, **81** 335 (1980)].

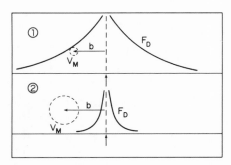

Figure 6.10. Examples of target, V_M, damaged by passing particle: case 1, dimensions of target small; case 2, target dimensions large compared to spread in energy density.

(6.32). For very low energy-density radiations, effects produced by the individual secondary electrons do not overlap, a tacit assumption in obtaining a continuous distribution like that in Eq. (6.32). If each fast secondary produces a separate damage cascade or spur off the primary particle track, then $(dE/dV)_e$ or F_D only defines the average distribution of events. A more careful treatment would consider the size and distribution of energy deposited within the cascades, and the random distribution of the cascades. The two limits, separate and overlapping cascades, will be considered more carefully in the next section when discussing sputtering.

In constructing the damage cross section, attention has to be paid to the fact that even simple biological systems, like enzymes, have more than one sensitive site. That is, damage can be incurred by "hitting" any one of a number of "targets." Calling the volume of the biological molecule (here an enzyme) V_M, the *average* number of damages, $\eta_D(z, b, E)$, produced by *one* incident particle in an enzyme centered a distance b from the incident direction is obtained by integrating F_D over the volume of the enzyme, $\eta_D \sim \int_{V_M} F_D \, dV$. Consider the two cases shown in Figure 6.10. In case 1, the dimensions of the molecules are small compared to the extent of F_D and, therefore, $\eta_D \sim V_M F_D(z, r, E_A)$. In the opposite limit, V_M is large (or F_D narrowly distributed), in which case either the particle misses ($b > L_M/2$, $L_M \sim$ linear dimensions of a molecule) and $\eta_D \sim 0$, or it passes through the molecule ($b < L_M/2$) and $\eta_D \sim L_M F_D(z, E)$. Here, $F_D(z, E)$ is the damage deposited per unit path length [$F_D(z, r, E)$ integrated over the radial dimensions], a quantity we described in the previous section. As an enzyme has a number of sensitive sites, the *average* number of damages in V_M may be greater than one. Using this average, we see that the Poisson probability for the occurrences of *at least* one damage in V_M, initiated by a particle passing at a distance b away, is

$$P_D(z, b, E_A) = 1 - \exp\left[-\eta_D(z, b, E_A)\right] \qquad (6.33)$$

and P_D is seen to be less than one no matter how large η_D is. This fact—one molecule can only be counted as one damaged molecule—is often referred to as saturation when the damage density, hence η_D, is large. For low damage density when η_D is small, which is often the case at distances away from the incident particle path, $P_D \approx \eta_D$.

By what may appear to have been a sleight of hand, we have constructed an impact-parameter probability for producing an effect (here damage) in a multiple-target molecule. Now the cross section for the effect is constructed as always [viz. Eq. (2.24)]

$$\sigma_D = 2\pi \int_0^\infty b \, db \, P_D(z, b, E) \tag{6.34}$$

In the limit that the molecule has only one target, Eqs. (6.34) and (6.33) reduce to our previous definitions of cross section. The above discussion merely generalizes the concept of cross section, defining it for complex molecular systems. For instance, in calculating the ionization cross section for a large molecule or a molecular cluster such saturation effects have the be included. (The reader should note that binary encounter calculations, in which the probability of producing an effect in each component is summed, apply only if these probabilities are small and hence may grossly overestimate inelastic cross sections (viz. Figure 5.26). The quantity σ_D in Eq. (6.34) can now be used for the "hit" cross section in Eq. (6.30) when describing the survival probability of a colony of enzymes.

For case 2 in Figure 6.10, the damage cross section in Eq. (6.34) becomes

$$\sigma_D \sim \sigma_g \{1 - \exp[-L_M F_D(z, E)]\} \tag{6.35}$$

where σ_g is the average geometric cross-sectional area of the enzyme ($\sigma_g \approx L_M^2$). For high-damage-density radiation in a large target volume Eq. (6.35) yields $\sigma_D \sim \sigma_g$, i.e., either the particle passes through the target, damaging it with unit efficiency, or not. For low-damage-density radiation ($\eta_D \ll 1$) both cases 1 and case 2 yield

$$\sigma_D \sim V_M F_D(z, E) \tag{6.36}$$

where $V_M \sim L_M \sigma_g$. This is a frequently used approximation for σ_D. Combining Eq. (6.36) with the fact that the damage density F_D scales like the stopping power [Eq. (6.3.2)] or in the radiation damage literature as LET (a part of $|dE/ds|$), we have $\sigma_D \propto |dE/ds|$. Using plots of active fraction versus fluence (Figure 6.8) for thin targets, $z \sim 0$ in Eq. (6.34), we can obtain damage cross sections. In Figure 6.11 the measured damage cross section for various incident particle radiations on a number of enzymes is seen to depend nearly linearly on the stopping power, confirming this description.

The reader may be surprised by the fact that the damage cross section is many times the geometric cross section at high LET. The derivation of Eq. (6.36) (giving σ_D as proportional to F_D) was obtained assuming the

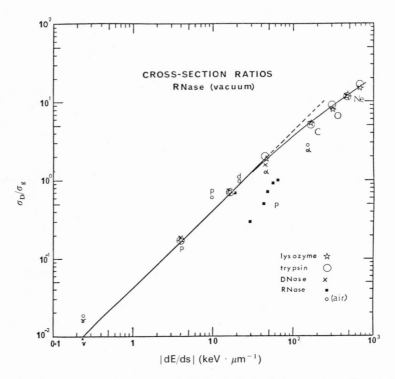

Figure 6.11. Damage cross section versus dE/dx for a variety of enzyme targets. Cross sections scaled to the geometric cross section. Incident particles and targets are labeled. Low-energy protons < 10 keV. [From D. E. Watt, A. T. Al-Kazwini, H. M. A. Al-Shaibani, and D. Twaij, in: 7th Symp. on Microdosimetry, Oxford, September, 1980.]

average number of damages occurring in any one enzyme was small. Although the probabilities may be small in case 1 of Figure 6.10, damage can be generated by particles passing at rather large impact parameters from the enzyme [e.g. Eq. (6.34)]. These impact parameters are weighted heavily in determining the cross section. Therefore, a large damage cross section, which is also proportional to LET, corresponds to the energy deposited per unit path being large but spread significantly in the radial direction by either the delta rays or diffusion. The falling-off of the cross section at high dE/dx values can be understood to be related to damage saturation in the molecules lying close to the track where the damage density is high, as in seen in Figure 6.10 and Eq. (6.32).

For the enzymes, one has the simple result that over a broad range of particle energy and type $\sigma_D \propto |dE/ds|$. In fact, converting the energy deposition section for incident γ-rays into a stopping power, we see that the damage cross section for γ-ray radiation falls on the curve in Figure 6.11.

Applying the results of the previous section, in the simple CSDA, we have

$$\sigma_D \sim V_M n_B[(S_e/w_E) + (S_n/w_n)] \tag{6.37}$$

where w_n and w_e are the energies deposited per displacement and ionization, respectively, which lead to a biological damage. From Figure 6.11 w_e is found to be about 60 eV in the energy region shown, i.e., $S_e \gg S_n$. For every 60 eV of inelastic energy deposited (roughly twice the W value for ionization), an average of one damage site is produced in an enzyme. The situation for displacement damage to enzymes is not clear at present; however, in the limit of small fluence the total number of damages produced has the same form as Eq. (6.26). The problem considered in the subsequent section is again related to energy deposition in a material, but the damage effect is a surface effect—the removal or sputtering of atoms from the surface of the target.

Sputtering

A result of considerable interest related to the displacement of target particles is the sputtering of the surface material by incident ions. Material is ejected by recoil cascades occuring close to the target surface. When the density of recoils is small enough that the secondary particle cascades produce in the target all decay in energy independently, then the flux of particles leaving the surface can be treated in a straightforward manner. The sputtering yield, Y, is defined to be the number of material particles exiting the target per incident particle. This yield consists of contributions from all cascades produced near the surface, each cascade associated with a secondary particle produced by the incident ion. For simplicity, we assume the cascades are spherically symmetric spurs about the ion's path located at random distances from the surface. From each cascade (hot spot) the collision energy moves outward, eventually transporting atoms across the target surface.

Using these ideas, we construct the sputtering yield as follows. Defining $S_p(z, \Delta E)$ to be the sputtering from a cascade at depth z, due to the deposition of energy ΔE, and $(n_B d\sigma_{AB})$ to be the number of cascades of energy ΔE produced per unit path length, we obtain the total yield

$$Y = n_B \int_0^{\bar{\gamma}E_A} d\sigma_{AB} \int_0^{\infty} Y(z, \Delta E)\, dz \tag{6.38}$$

where ΔE is T or Q depending on whether the source of sputtering is nuclear or electronic energy deposition. To construct the sputtering contribution from each cascade we need to know the energy distribution of particles in the cascade. Any recoil atom B will have an energy E' for a time

v_B^{-1}, where v_B is the collision frequency. Therefore, the secondary-particle energy spectrum at any time, in those cascades initiated by an event ΔE, is determined from the recoil spectrum, G, discussed earlier [Eq. (6.28)],

$$f(\Delta E, E') \simeq (\psi/v_B)G(\Delta E, E')$$

where ψ is the number of such cascades produced per unit time.

Generally one assumes a sputtering occurs if a recoil particle is produced at the surface or reaches the surface with its perpendicular component of velocity large enough to overcome a potential barrier, U_0; that is $E' \cos^2 \theta' > U_0$, where θ' is the direction of motion of the recoil measured from the normal to the surface. Noting that the cascades are randomly distributed and the velocities randomly oriented, we see that those particles with energy E' at a depth $v' \cos \theta'/v_B$ with $\cos \theta' > (U_0/E')^{1/2}$ will be able to exit the material. As $G(\Delta E, E')$ is the number of particles of energy E' in the cascade, the first integral in Eq. (6.38) becomes

$$\bar{z}(\Delta E) \equiv \int_0^\infty Y(z, \Delta E)\, dz$$

$$= \frac{1}{2} \int_0^{\Delta E} dE' \int_{(U_0/E')^{\frac{1}{2}}}^1 d\cos\theta' \, (v' \cos\theta'/v_B) G(\Delta E, E') \qquad (6.39)$$

which is a mean sputtering depth for the cascade, ΔE. It was assumed in Eq. (6.39) that the positive direction is out of the material when defining the angular limits. Writing the collision frequency as $v_B = n_B \bar{\sigma} v'$ and using Eq. (6.28b), one finds

$$\bar{z} \sim \frac{3}{4\pi^2} \frac{v(\Delta E)}{n_B \bar{\sigma} U_0} \qquad (6.40)$$

where it is assumed that σ is energy independent, a condition valid for low-energy recoils and from Eq. (5.5) $\sigma \simeq \bar{\sigma}_d/2$ as particles have the same mass. Substituting Eq. (6.40) into Eq. (6.38), we obtain the net sputtering yield

$$Y \sim \frac{3}{4\pi^2} \frac{1}{\bar{\sigma} U_0} \int_0^{\bar{\gamma} E_A} v(\Delta E)\, d\sigma_{AB} \qquad (6.41)$$

For $\Delta E \to T$, Eq. (6.41) is the general form obtained by Sigmund from the transport equations for sputtering, and $\bar{\sigma} \sim 1.8\ \text{Å}^2$ is used for many atomic materials. Approximating the integral in Eq. (6.41) by S_n, we have $Y \sim (0.042/U\ \text{Å}^2)S_n$. A frequently used result for estimating sputtering due to collision cascades based on the above is $Y \sim (0.042/U_0\ \text{Å}^2)\alpha S_n$, where α is a factor which accounts for the geometry of the cascade near the surface, the angle of incidence, and the effect of inelastic energy loss in Eq. (6.41). In the keV region α varies from about 0.2 for incident heavy atoms to about 1 for $M_B \gg M_A$ and at high energies it approaches 0.5. As α depends on the mass of the target atoms, in alloys preferential removal of one atomic species may occur, changing the character of the surface layer. The above expressions are valid when the cascades can be considered to be noninteracting and when $\Delta E \gg E'$, as required in Eq. (6.28). For molecular solids other expressions for $\bar{\sigma}$ are required.

Electronic processes may also initiate heavy-particle motion, hence sputtering or desorption of atoms from a material. One such process is dissociative recombination in gases, a process resulting in sputtering from small planetary bodies, which will be discussed in the following section (e.g., $e + O_2^+ = O + O + \Delta E$). Heavy-particle motion, hence sputtering, may be produced in certain nonconducting materials by similar electronic processes or by the coulomb repulsion between closely spaced ions resulting from the incident ionizing radiation, which is thought to produce tracks in insulators. If the cascade density is low, then the energy deposited in heavy-particle motion produced by such processes can be used to determine $\nu(\Delta E)$ in Eq. (6.41). For example, as the ionization density along a heavy-particle track is $W_e^{-1} | dE/dz |_e$, the energy per unit path length available in coulomb repulsion is $C \cdot (e/W_e)^2 | dE/dz |_e^2$, where C is a factor depending on the track width, which varies slowly with energy. The sputtering yield then would have the form $Y \propto | dE/dz |_e^2 \cdot C$ if charge separation (electrons and ions) is maintained for times of the order of 10^{-13} sec (i.e. the time for the ions to repel a lattice spacing).

Generally, electronic processes would be associated with low-energy cascades, in which case the approximation to $G(\Delta E, E')$ should be improved [e.g. for thermal cascades, Eq. (6.40) becomes $\bar{z} \sim .04 \, (\Delta E / U_0)^{5/3} \bar{\sigma} n_B^{1/3}$]. Similarly, if the cascade density produced either by elastic or inelastic processes is high, then sputtering also becomes a thermal process. That is, the cascades deposit energy in overlapping volumes leaving a heated region. The energy of the incident particle is deposited symmetrically in a narrow region about the path, with a hot spike formed at some depth in the material, as indicated by the Bragg curve in Figure 6.7. When the spike depth is large enough that its effect at the surface is negligible, the sputtering yield is determined by the energy density in the cylindrically symmetric part of the track near the surface. This is the case we treat here, with the energy density parametrized by a temperature.

For high energy densities, the heated material can be considered, to a reasonable approximation, to behave locally like a very hot gas. If the temperatures are not very high, little erosion occurs anyway. It is left as an exercise for the reader (Problem 6.6) to show that the flux of particles crossing the material surface from a hot region of temperature T and at a density equal to the material density, n_B, is

$$\phi_B \sim n_B kT \exp(-U_0/kT)/(2\pi M_B kT)^{1/2} \tag{6.42}$$

where U_0 is the surface barrier potential, or the sublimation energy of the material, and M_B is the mass of the material particles. The sputtering yield can now be written

$$Y = \int_0^\infty dt \int_0^\infty \pi \, dr^2 \, \phi_B[T(r, t)] \qquad (6.43)$$

where the problem remaining is to determine the temperature as a function of time and radial distance from the track, r.

Based on the notions developed in treating particle diffusion, the equation for thermal diffusion is

$$\nabla K(T) \nabla T = C(T) \frac{\partial}{\partial t} T \qquad (6.44)$$

where $K(T)$ is the thermal conductivity and $C(T)$ the heat capacity. For an atomic vapor $C(T) = \frac{3}{2} n_B k$ and $K(T)$, like the diffusion coefficient, depends on the collision frequency. Using mean free path arguments, we have $K(T) \propto \bar{v}/\bar{\sigma}_d$, which for hard-sphere collisions is proportional to $T^{1/2}$. The resulting nonlinear equation for a cylindrically symmetric temperature distribution has the solution

$$T(r, t) = \varepsilon[1 - r^2/3\Delta^2(t)]^{1/2}/\Delta^2(t), \quad \text{and} \quad T = 0 \text{ for } r^2 > 3\Delta^2(t) \qquad (6.45)$$

where

$$\Delta^2(t) = \left[\frac{4K_0}{C}(t + t_0)\varepsilon^{1/2}\right]^{2/3} \quad \text{and} \quad \varepsilon = \frac{1}{\pi C}\left|\frac{dE}{dz}\right|_{\text{eff}}$$

In the above we have written $K = K_0 T^{1/2}$, with $K_0 \approx (k/\bar{\sigma}_d)(2k/M_B)^{1/2}$, and assumed the distribution had an initial width determined by t_0. The part of the energy deposited per path length in the region that leads to the sputtering is indicated by $|dE/dz|_{\text{eff}}$. For example, for a high density of elastic scatterings, $|dE/dz|_{\text{eff}} \sim n_B S_n$. Lastly, in obtaining Eq. (6.45), we required that the total energy be conserved, i.e., for a temperature-independent specific heat

$$\int_0^\infty CT\pi \, dr^2 = |dE/dz|_{\text{eff}} \qquad (6.46)$$

On substituting the calculated temperature profile of Eq. (6.45) into the equation for sputtering, Eq. (6.43), one finds

$$Y = \left[\left(\frac{dE}{dz}\right)_{\text{eff}}^2 \bigg/ 8\pi(K_0 L_0^{3/2})(2\pi M_B k L_0)^{1/2}\right][1 - g(t_0)] \qquad (6.47)$$

where $L_0 = U_0/k$ and

$$g(t_0) = x_0^3 \int_{x_0}^\infty dx \, e^{-x}/x^2, \quad \text{with } x_0 = L_0/T(0, 0)$$

For a narrow spike (i.e. $T(0, 0) \gg L$ or $g(t_0) \approx 0$) the expression for sputtering in Eq. (6.47) is quite simple. This occurs if the energy deposited per unit path length is high, as determined by $|dE/dz|_{\text{eff}} \gg \pi(C_0 L)\Delta^2(0)$. The $(de/dz)^2_{\text{eff}}$ factor in Eq. (6.47) does *not* depend on the assumed forms for K and C; it is a result only of the cylindrical symmetry involved. Therefore, as the cascade density increases, collisional sputtering changes from being proportional to (dE/dz) to being proportional to a higher power of (dE/dz). This fact is again independent of the sputtering mechanism.

For frozen gases and alkali-halides it appears that the dominant sputtering mechanism is related to electronic energy deposition, S_e, rather than to the elastic collision cascade, i.e. S_n. (Recently, ion-induced desorption of large biological molecules has also been associated with electronic energy deposition in material.) Further, the sputtering yield for frozen gases depends nearly quadratically on S_e, as in Figure 6.12. Again, as in the biological damage, the initial events probably do not directly eject material. Rather, the electronic events initiate a "hot" spot, are a source of coulomb repulsion, or are responsible for the production of unattached species which diffuse to the surface.

In the following section we apply some of the ideas discussed here to a related problem, the escape of atoms and molecules from a planetary atmosphere. Although the problem is similar, there are some interesting differences, and it provides a transition into discussing some of the atomic and molecular processes occurring in a planetary atmosphere.

Planetary Escape

Hot atoms in the upper levels of a planetary atmosphere may have enough energy to overcome the gravitational barrier and escape the planet in much the same way atoms sputter or are evaporated from materials. In the latter case the potential barrier is a result of the atomic interactions between the material particles, whereas in the former it is simply the gravitational force of the planet. For the planetary case the conditions for escape, in the absence of other forces, is $\frac{1}{2}M_B v_B^2 > U$, where U is the gravitational potential energy of the particle, and the direction must be "upward," $\cos \theta > 0$. Noting that $U \sim GM_B M_p/\mathscr{R}_p$, we see that this condition is the same for rockets and atoms, as the mass, M_B, cancels. In the above relationship M_p, $\mathscr{R}_p$, and G are the planetary mass, planetary radius, and gravitational constant respectively.

Of course rockets and atoms behave rather differently if other effects are included, because of their huge mass differences. Whereas the atmospheric constituents only provide an overall drag force on the rocket, they easily deflect each other or any fast atom. Therefore, escape can only occur from the upper levels of the atmosphere where it is unlikely that an atom or

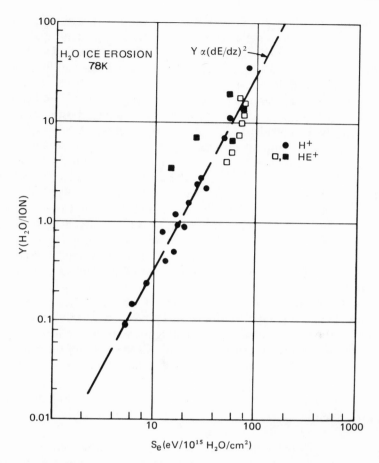

Figure 6.12. Sputtering yield versus electronic stopping cross section, S_e, plotted for incident H^+ and He^+. Dashed line indicates proportionality to the square of the stopping power. [From W. L. Brown, W. M. Augustyniak, L. J. Lanzerotti, R. E. Johnson, and R. Evatt, *Phys. Rev. Lett.* **45**, 1632 (1980).]

molecule will collide with another atom or molecule while exiting the planetary field. This is not unlike the solid-material example we just discussed. It was always assumed for a material that sputtering occurred from the "surface." For sputtering, the thickness of the "surface" is determined by the depth at which an atom can exit without making another collision. This is clearly of the order of a few mean free paths and therefore a few atomic layers. For metals, Sigmund finds the effective depth of origin to be $\Delta z = 3/(2n_B \bar{\sigma}_d)$, using an $n \to \infty$ power law for the low-energy collisions. Earlier we showed that the mean projected range for hard-sphere collisions was $\bar{R}_p = 3/(n_B \sigma_{AB})$, which, averaged over possible angular directions leading to escape, would give a similar result for the effective depth of origin of a sputtered particle.

For solid materials it was assumed that the density n_B was constant up to some plane, the "surface," where it vanished. However, for an atmosphere, n_B decreases roughly exponentially [viz. Eq. (1.10)]. Rather than discussing the depth of the material, one therefore considers the density of atoms "stacked" above any particular point. This is referred to as the column density

$$N_B(z) \equiv \int_z^\infty n_B(z')\,dz' \sim H_B n_B(z) \tag{6.48}$$

where the exponential approximation to the atmospheric density is used [Eq. (1.10)], with H_B the scale height. The effective column density from which particles will escape is, based on sputtering theory,

$$H_B n_B(z) \sim 3/(2\bar{\sigma}_d) \tag{6.49}$$

This atmospheric region is referred to as the exosphere, for obvious reasons, and this "depth" is called the critical level. Although the escape fluxes often are small, over a geological time period, it is felt that they are very important in determining the evolution of a planet's atmosphere. In the following we delineate these processes contributing to escape. The separation made is somewhat artificial in that the escaping "hot" atoms may have attained the required kinetic energy by a combination of processes.

Because an atmosphere has a background temperature, due predominantly to heating by the incident solar radiation, there is a probability that any particular atom may have an appropriate velocity to escape. This is roughly equivalent to evaporation from a warm surface and is referred to as Jean's escape. Using the Maxwell–Boltzmann velocity distribution, the reader can verify that the thermodynamic escape flux in Eq. (6.42), subject to the escape conditions given above, becomes

$$\phi_T \sim n_B(U_c + kT)\exp(-U_c/kT)/(2\pi M_B kT)^{1/2} \tag{6.50}$$

Here, $U_c = GM_B M_P/\mathscr{R}_c$, is the gravitational potential energy, $\mathscr{R}_c$ is the distance from the critical level to the planet center ($\mathscr{R}_c \sim \mathscr{R}_p$), n_B is the density, and T is the temperature—all evaluated at the critical level, defined by Eq. (6.49).

In Table 6.2 are listed the escape energies in eV/amu for a number of planetary bodies. Remembering that room temperature corresponds to a mean particle energy of the order of 1/40 eV, it is seen that escape of heavy elements is unlikely from any of the listed bodies. Further, although the upper atmospheres of some planets may attain temperatures of a couple of thousand degrees Kelvin, escape of even hydrogen from a heavy planetary body like Jupiter is unlikely. Bodies as small as the moon, of course, are known to have lost almost all outgassed material unless there is a continuous source of gas. Using Viking data for Mars, and number densities for O and H estimated by McElroy and others, we see that the thermal escape flux for O, a dominant constituent at high altitudes, is $\phi_T \sim 10^{-38}$ parts/cm^2/sec. That is, at the present temperatures, oxygen atom escape is negligible even over long evolutionary time periods. On the other hand, $\phi_T \sim 10^7$ parts/cm^2/sec for hydrogen atoms, a rather significant flux. Even though the hydrogen density is much smaller than that of oxygen, the exponential in Eq. (6.50) changes rapidly with particle mass.

The absorption of ultraviolet radiation and, to a much smaller extent, ionization by incident electrons and ions results in the formation of an ionized region, a weak plasma of electrons and ions. As the processes in the upper atmosphere can be considered to be roughly in equilibrium at any time, the plasma density is determined by the balance of the ionization processes with electron–ion recombination. Such processes as $A^+ + e \to A + h\nu$ (radiation recombination) are very slow compared to the molecular process $AB^+ + e \to A^* + B^*$ (dissociative recombination) because of the weak coupling between the charged particles and the radiation field. Therefore, a large fraction of the recombination occurs at those altitudes where the molecular ion density is significant. In dissociative recombination the separating atoms attain kinetic energies determined by the exothermicity, $|Q|$, and momentum conservation:

$$E_A = \frac{M_B}{M_A + M_B} |Q| \quad \text{and} \quad E_B = \frac{M_A}{M_A + M_B} |Q|$$

For small planets, where a significant amount of dissociative recombination occurs around the critical level, the dissociation energies may be sufficient for one of the particles to escape the gravitational field of the planet. As the dissociation energies are of the order of a few electron volts, it is clear from Table 6.2 that on the heavier planets not even hydrogen is likely to escape by this process. However, on a planet like Mars, where there is still enough

Table 6.2. Escape Energy for Planetary Bodies

Planet or satellite	Escape energy (eV/amu)	Planet or satellite	Escape energy (eV/amu)
Mercury	0.096	Neptune	2.9
Venus	0.56	Moon	0.029
Earth	0.65	Callisto	0.029
Mars	0.13	Ganymede	0.040
Jupiter	19.0	Europa	0.023
Saturn	6.8	Io	0.035
Uranus	2.5	Titan	0.040

gravitational attraction to hold a molecular atmosphere, this escape process probably plays an important role in the evolution of the atmosphere. The ionosphere is predominantly O_2^+ above the critical level, which means that for $|Q| > 4.2$ eV escape of O is likely. This is the case for the first two of the possible recombination processes listed below:

$$e + O_2^+ \rightarrow O(^3P) + O(^3P) + 6.96 \text{ eV}$$

$$\rightarrow O(^3P) + O(^1D) + 5.00 \text{ eV}$$

$$\rightarrow O(^1D) + O(^1D) + 3.04 \text{ eV} \qquad (6.51)$$

$$\rightarrow O(^3P) + O(^1S) + 2.80 \text{ eV}$$

$$\rightarrow O(^1D) + O(^1S) + 0.84 \text{ eV}$$

The escape flux for such processes is calculated from the contributing recombination rates as

$$\phi_e \sim \tfrac{1}{2} P^e \alpha_i n_i^2 H_i \qquad (6.52)$$

where n_i is the ion number density at the critical level, H_i the scale height, α_i the recombination coefficient, and P^e an escape probability, a number between 0 and 1. In Eq. (6.52) it was assumed, for simplicity, that $n_e \sim n_i$, n_e being the electron number density. Assuming all recombination occurs by the first two processes (which is not likely viz. Chapter 5), $P^e \sim 1$, and using α_i given in Chapter 5, we find that the average recombination escape flux for O from Mars is $\phi_e \sim 1 \times 10^7$ parts/cm^2/sec, obtained by extrapolating the Viking measurements of ion density. This flux, being much larger than that for thermal escape of O and comparable to the thermal escape of H, may control the loss of H_2O from the Martian atmosphere. Owing to the considerably larger gravitational fields, this process will be unimportant on the Earth and Venus.

Escape can also be driven by particle bombardment. That is, on nonmagnetic planets, the solar wind interacts strongly with the upper atmosphere of the planets, and some satellites of the outer planets lie within the planetary radiation belts and are subject to heavy-particle and electron impact. If these incident particles do penetrate into an atmospheric region, then escape due to sputtering is initiated by electronic energy deposition (ionization), as just discussed, and by collision cascades. However, it is known that the upper ionospheric region of a nonmagnetic plant interacts strongly with the solar wind, deflecting a large fraction of the incident particles around the planet via induced fields in much the same way the magnetosphere of the earth deflects the solar wind. Further, as electrons are easily deflected by these field and, in fact, ions do penetrate, a net charging may develop which repels subsequent ions and/or causes ejection of planetary ions. (Such charging, of course, is controlled in a laboratory sputtering experiment when one irradiates materials with ions.) Fast neutral particles which are impervious to these planetary fields are created if that charge exchange occurs at high altitudes, leaving a slow ion in the magnetic field and a fast neutral directed into the atmosphere. As the resonant and near resonant charge transfer cross sections (e.g., $H^+ + H \rightarrow H + H^+$ and $H^+ + O \rightarrow H + O^+$) can be quite large compared to other processes, neutralization may occur at sufficient heights to allow penetration of a significant fraction of the solar wind.

From the above discussion it is apparent that sputtering and/or field ejection due to particle bombardment is quite complex. Estimates for Mars put the escape flux of atoms due to such processes a couple of orders of magnitude below that for thermal ejection of H or recombination ejection of O. On the Galilean satellites of Jupiter, ejection due to particle bombardment either of tenuous "atmospheric" gases, e.g., on Io, or condensed gases on the surfaces of the satellites may be important sources of heavy atoms for the Jovian magnetospheric plasma. Similarly, charged particles or U.V. ejection of molecules from ice grains and comets may be a source of large molecules in space. In the following section we consider some of the chemical processes involved due to ultraviolet irradiation of planets. Whereas the process we have been discussing above affect the atmosphere only over long periods of time, the processes to be discussed determine the day-to-day character of the atmosphere.

The Ionosphere

Historically the earth's ionosphere has been divided into three major shells in which the ion type, density, and temperature are controlled by various molecular processes. Although the following discussion of these processes is well known and the presentation oversimplified, it still provides a good example of a macroscopic phenomenon controlled by molecular processes. We begin with a brief discussion of solar radiation absorption which initiates the molecular processes occurring in the ionosphere.

In the solar spectrum, radiation of wavelength, λ, less than about 1000 Å is responsible for ionization of atmospheric species. This comprises a very small fraction of the incident radiation. At longer wavelengths dissociation and other molecular processes dominate the absorption of solar energy. Associating a cross section $\sigma_{\lambda, AB}$ with the ionization process ($v = c/\lambda$, c being the velocity of light),

$$hv + AB \rightarrow AB^+ + e$$

where hv is the photon energy, we can express the ionization production as a function of altitude. Assuming a simple exponential, single-species atmosphere, we write the production rate as

$$\frac{dn_{AB^+}}{dt} = \sigma_{\lambda, AB}\, n_{AB}\, I_\lambda(z) \tag{6.53}$$

where I_λ is the radiation intensity at altitude z. Neglecting any sources of radiation of wavelength λ that may exist in the atmosphere, we see that the intensity decreases with depth due to absorption, as in Eq. (2.2). That is,

$$I_\lambda(z) = I_\lambda(\infty) \exp\left[-\sum_i n_i\, \sigma_{\lambda i}\, z_i / \cos\chi_s \right] \tag{6.54}$$

where the sum is over all species in the atmosphere, and χ_s is the solar zenith angle, the angle that the incident radiation forms with the vertical. In Figure 6.13 the functions dn_{AB^+}/dt, I_λ, and n_{AB} from Eq. (1.10) are illustrated and, as one would expect, the production rate of ions has a maximum. This production profile is often referred to as a Chapman profile. At high altitudes the production decreases like the density of absorbers, n_{AB}, and at low altitudes it goes to zero because the light intensity has diminished. For small wavelengths, the absorption cross section tends to decrease with decreasing wavelength. Therefore, the maximum ionization production for X-rays tends to occur at a lower altitude than that for ultraviolet light. This is only roughly true, however, as radiation absorption is quite specific, with the absorption cross section for atmospheric molecules being very high (or small) at certain frequencies.

If electron–ion recombination is fast compared to other molecular processes and to diffusion, then from the production profiles one would expect to find a layered ionosphere, with the ionospheric constituents determined by the absorption process. Although one does indeed find such layered

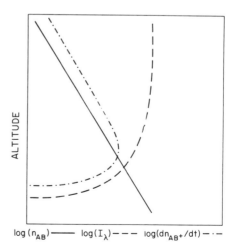

Figure 6.13. Schematic diagram of the molecular density, n_{AB}, incident radiation flux, I_λ, and the ion-production rate dn_{AB^+}/dt.

$\log(n_{AB})$ ——— $\log(I_\lambda)$ – – – $\log(dn_{AB^+}/dt)$ — · —

ionospheres, as indicated in Figure 6.14, the chemical and diffusion processes are not slow and, therefore, the dominant ionic species found is not the ion of the dominant absorbing species in some regions. In the simple photochemical-equilibrium model, a *single* atmospheric molecule, AB, is ionized and recombination, $e + AB^+ \rightarrow A^* + B^*$, is assumed to be fast compared to reactions of AB^+ with other species and to diffusion. Assuming charge equilibrium, i.e., $n_e = n_{AB^+}$, with production and recombination in balance, we obtain

$$n_{AB^+}^2 = J_{AB}\, n_{AB}/\alpha_{AB^+} \tag{6.55}$$

where α_{AB^+} is the recombination coefficient discussed in Chapter 5, and $J_{AB}(z) = \sum_{\lambda} \sigma_{\lambda AB}\, I_{\lambda}(z)$ is the sum over all absorption processes leading to ionization of AB. A situation close to this exists on Mars, but the dominant ion is not the ionic form of the dominant absorbing molecule.

The Martian atmosphere is predominantly CO_2 up to about 200 km where atomic oxygen begins to dominate. The most important ionization processes are $CO_2 + h\nu \rightarrow CO_2^+ + e$ and $CO_2 + h\nu \rightarrow CO + O^+ + e$ below 230 km and $O + h\nu \rightarrow O^+ + e$ above this. However, the molecular pro-

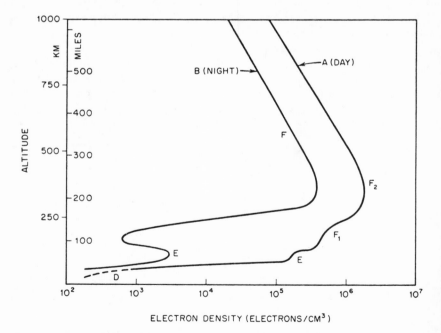

Figure 6.14. Electron-density profiles in the earth's atmosphere around the solar maximum. [From S. Glasstone, *Sourcebook on the Space Sciences*, Van Nostrand, New York (1965), p. 487.]

cesses discussed earlier [viz reactions (1), (2), and (8) of Eq. (1.13) and Table 5.2]

$$CO_2^+ + O \rightarrow CO + O_2^+, \qquad k_1 = 1.6 \times 10^{-10} \text{ cm}^3/\text{sec}$$

$$CO_2^+ + O \rightarrow CO_2 + O^+, \qquad k_2 = 1 \times 10^{-10}$$

$$O^+ + CO_2 \rightarrow O_2^+ + CO, \qquad k_8 = 1.2 \times 10^{-9}$$

are fast compared to recombination as the neutral densities are much larger than the electron and ion densities. It is seen from these equations that both the CO_2^+ and O^+ ions rapidly produce O_2^+. As the recombination processes

$$e + O_2^+ \rightarrow O + O, \qquad \alpha_1 = 2.2 \times 10^{-7}(300°\text{K}/T_e) \text{ cm}^3/\text{sec}$$

$$e + CO_2^+ \rightarrow CO + O, \qquad \alpha_2 = 3.8 \times 10^{-7} \text{ cm}^3/\text{sec}$$

where T_e is the mean electron temperature, are in turn fast compared to diffusion, the ionosphere is very nearly in photochemical equilibrium. The rate equations, like Eq. (1.14), neglecting diffusion and assuming equilibrium, become

$$\frac{\partial}{\partial t}[CO_2^+] = 0 = J_{CO_2}[CO_2] - (k_1 + k_2)[CO_2^+][O] - \alpha_2 n_e [CO_2^+]$$

$$\frac{\partial}{\partial t}[O^+] = 0 = J_O[O] + k_2 [CO_2^+][O] - k_8 [O^+][CO_2]$$

$$\frac{\partial}{\partial t}[O_2^+] = 0 = k_1[CO_2^+][O] + k_8 [O^+][CO_2] - \alpha_1 n_e [O_2^+]$$

In the above equations J_{CO_2} and J_O account for absorption by CO_2 and O for all wavelengths. Using the incident solar spectrum, measured photoionization cross sections, and densities for CO_2 and O with the above rate constants, one obtains the densities of O^+, CO_2^+, and the predominant O_2^+. As usual, charge equilibrium is also assumed, i.e., $n_e = [O_2^+] + [CO_2^+] + [O^+]$. Calculated ion densities are plotted in Figure 6.15 and compared to Viking I measurements of the ion densities on Mars. An approximate expression for the dominant O_2^+ density similar to Eq. (6.55) can be found. Noting that $n_e \approx [O_2^+] \gg [CO_2^+]$ and the production of CO_2^+ is much greater than O^+, we obtain

$$[O_2^+]^2 \sim \frac{J_{CO_2}[CO_2]}{\alpha_1} \qquad (6.56)$$

Therefore the O_2^+ density follows the production profile, having a scale height, above the maximum, roughly half the CO_2 scale height when diffusion is neglected. At higher altitudes than shown here, the ions presumably

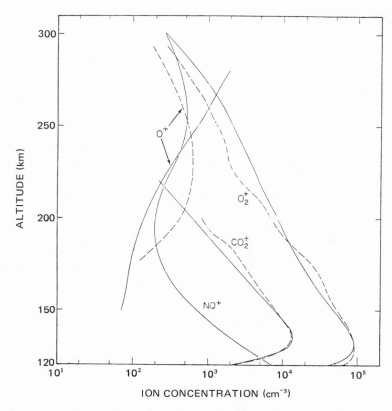

Figure 6.15. Ion concentration versus altitude in the Martian atmosphere: dashed lines, Viking I measurements; solid lines, calculation based on chemical rate equations including nitrogen reactions and a solar zenith angle of $45°$. [From W. B. Hanson, S. Sanatani, and D. R. Zuccaro, *J. Geophys. Res.* **82**, 4351 (1977).]

attain diffusive equilibrium separately, and at very high altitude the solar wind–ionosphere interaction, discussed earlier, dominates.

The earth's ionosphere differs from that of Mars in that different molecular processes dominate at different altitudes, forming a layered ionosphere as mentioned earlier. We begin with the lowest layer in which ionization is due to hard X-rays and Lyman α radiation and, hence, is influenced by solar flares. In the so-called D region (Figure 6.14), where the molecular densities are large, three-body processes can occur with significant probabilities. For example, O_2^- is formed by attachment, $O_2 + X + e \rightarrow O_2^- + X$. This in turn leads to a significant fraction of the recombination occurring via mutual neutralization, $A^+ + O_2^- \rightarrow A^* + O_2^*$. We saw in Chapter 5 that such a process has a large rate coefficient, but the electron is also easily detached by photons or collisions with neutrals. Therefore, although recombination depends on the negative-ion concentration (as well as free-electron concentration), the density of negative ions is controlled by the electron detachment cross sections. Simple theoretical models of the D region would predict the presence of varying amounts of O_2^+, NO^+, and O_2^-. However, measurements show the presence of significant fractions of H_3O^+, $H_5O_2^+$, and other water vapor ion clusters, as well as CO_3^-, HCO_3^-, NO_3^-, and CO_4^- due to the presence of small amounts of CO_2 and water vapor particularly at the lower levels. The chemistry in this region can, therefore, be quite complex.

In the E region ionization proceeds by the following processes: X-rays ($100 < \lambda < 31$ Å) on O_2, N_2, and O; ultraviolet in the Lyman continuum ($\lambda < 910$ Å) on O_2 and O; and Lyman β (at 1025.7 Å) on O_2. However, reactions (3)–(7) in Eq. (1.13) rapidly remove N_2^+ and O^+, leaving an ionosphere dominated by O_2^+ and NO^+. In the daytime O_2^+ is the dominant ion, and simple photochemical equilibrium predicts densities

$$[O_2^+]^2 \approx n_e^2 \approx J_{O_2}[O_2]/\alpha_1 \qquad (6.57)$$

where α_1 is the recombination coefficient considered earlier. As the effect of solar radiation diminishes in the evening and hence the density of electrons decreases, the recombination process eventually gives way to the charge-exchange process

$$O_2^+ + NO \rightarrow O_2 + NO^+ \qquad k_{10} = 8.0 \times 10^{-10} \text{ cm}^3/\text{sec}$$

which controls the removal of O_2^+. As the recombination process

$$e + NO^+ \rightarrow N^* + O^*, \qquad \alpha_3 = 3 \times 10^{-7} \text{ cm}^3/\text{sec}$$

is relatively slow, the E region is eventually dominated by NO^+ ions. Apparently the background radiation of the night sky produces a sufficient amount of O_2^+ to maintain an ionosphere consisting of O_2^+ and NO^+. It is important to note here that the time constant for decay or formation of the E region is still small, so that transport processes for the most part play a small role. From the rate equations, the time constant for loss of O_2^+ is $\tau \simeq (2\alpha_1 n_e)^{-1}$, which for values of n_e in the E region is of the order of a few seconds. Therefore, total ion densities closely follow changes in the ion production rate.

In the lower F region, the F_1 region, the dominant ions produced are N_2^+ and O^+ as the O_2 density is now diminished (i.e., above about 125 km $[O] > [O_2]$ due to gravitational separation). Invoking the same processes for O^+ and N_2^+ as before, it is clear that NO^+ will be the dominant ion. Allowing removal only via dissociative recombination, $e + NO^+$, and production of NO^+ via ionization of O and reaction with N_2, one finds

$$[O^+] = J_O[O]/k_4[N_2]$$
$$[NO^+] = \tfrac{1}{2}[O^+][(1 + 4J_O[O]/\alpha_3[O^+]^2)^{1/2} - 1] \qquad (6.58)$$

At higher altitudes the production-to-loss ratio for O^+ in Eq. (6.58) gradually increases due to the differing scale heights for N_2 and O and the increased intensity of ionizing radiation. The O^+ density eventually dominates NO^+ in the vicinity of 180 km. It is useful to evaluate the time constant for obtaining equilibrium. The reader can verify that, in the absence of diffusion, the equilibrium time constant for losing O^+ is $\tau = (k_4[N_2])^{-1}$, which increases with altitude. At 250 km, $\tau \sim 1/2$ hour, and, therefore, transport processes must begin to be important in determining the character of the ionosphere.

The upper F region, called the F_2 region, will be dominated by O^+ ions. From our previous discussion both the O^+ density and the time constant increase with increasing altitude. That is, as radiation recombination is very slow, recombination is still controlled by the formation of molecular ions, as it was in the F_1 region. Such an increasing density would not, however, be in gravitational equilibrium; eventually a downward, diffusive flow of ions will occur. As the time constant for recombination increases, this downward flow will become more important, enhancing recombination by bringing O^+ ions into a region of higher N_2 density. An ion density peak will occur when the times for chemical processes become comparable to the characteristic transport times.

To maintain charge equilibrium, small electric fields exist in the ionosphere. By adding such a term to the force equation [Eq. (1.7)] for both electrons and ions, and combining their equations of motions, the reader can verify (Problem 6.11) that the ion diffusive flux in a steady state is

$$n_i w_i = -\frac{1}{M_i \nu_i}\left(\frac{\partial P_i}{\partial z} + \frac{\partial P_e}{\partial z} + M_i g n_i\right) \tag{6.59}$$

where P_i and n_i are the ion pressure and density respectively, P_e the electron gas pressure, and ν_i the ion–neutral collision frequency; here we have used $M_i \gg m_e$. Equation (6.59) can be rewritten in the form

$$w_i = -D_p\left(\frac{1}{n_i}\frac{\partial n_i}{\partial z} + \frac{1}{T_p}\frac{\partial T_p}{\partial z} + \frac{1}{H_p}\right) \tag{6.60}$$

where $T_p = T_e + T_i$, the sum of the electron and ion temperatures, $D_p = kT_p/M_i \nu_i$, and $H_p = kT_p/M_i g$. In equilibrium, if T_p is nearly constant, then

$$n_i = n_i^0 \exp[-(z - z_0)/H_p] \tag{6.61}$$

which is the situation achieved at high altitudes up to the exosphere. The reader should note that since the electrons and ions reach equilibrium together, the scale height H_p is roughly twice the scale height of the parent atom (assuming $T_i \sim T_e$), and, similarly, the effective diffusion constant is increased. This should not be surprising, as an electric field, which was eliminated from the equations, is acting to affect the densities and motion of the ion.

From the discussion in Chapter 5, the ion–neutral collision frequency in Eq. (6.59), is determined by the polarization interaction, $V = -\alpha_B e^2/2R^4$,

$$v_i = 2.6 \times 10^{-9} n_B (\alpha'_B/m')^{1/2} \text{ sec}^{-1} \quad (6.62)$$

where α'_B is the polarizability, α_B, in units of 10^{-24} cm^3, m' is the reduced mass, m, in amu, and n_B in cm^{-3}. At altitudes where the dominant neutral gas is O the dominant momentum-change process is, however, not elastic scattering but resonant charge exchange, giving an expression for v_i which is temperature dependent. In resonant charge exchange, $O^+ + O \rightarrow O + O^+$, the ionic species emerges, moving at roughly the speed of the target neutral. Therefore, the momentum change for the ion is approximately $M_A v$, as charge transfer takes place predominantly at large impact parameters. Remembering from Chapter 5 that the resonant charge-exchange cross section had the form $\sigma_{ct}^{1/2} \sim (a - b \log v)$, the thermally averaged collision frequency integrates to

$$v_i \simeq \tfrac{8}{3} n_B (8k/\pi M_A)^{1/2} (T_i + T_n)^{1/2} [(a + 3.96b) - b \log_{10}(T_i + T_n)]^2 \quad (6.63a)$$

For $O^+ + O$ this is

$$v_i \sim 4 \times 10^{-12} n_B (T_i + T_n)^{1/2} [11 - 0.7 \log_{10}(T_i + T_n)]^2 \text{ sec}^{-1} \quad (6.63b)$$

with $T_i + T_n$ in degrees Kelvin and greater than about 500°K. At lower temperatures the polarization expression in Eq. (6.62) would dominate.

Rather than attempt to solve the diffusion–recombination problem, the above results can be used to estimate a characteristic diffusion time to be compared to the recombination rate. Using the neutral scale height as a characteristic distance over which significant atmospheric changes occur, we can define a characteristic diffusion time $\tau_d \sim H_n(v_i/g)$. This time *decreases* as the density of neutrals decreases, which is opposite in behavior to the recombination time discussed earlier. Assuming $T_i \sim T_n \sim 500$°K and O is the dominant molecule, we have $\tau_d \sim 6 \times 10^{-6} n_O$ cm^3 sec. This is equal to about a half hour at 250 km in the vicinity of the electron density maximum, or it is comparable to the O^+ removal time in that region. At higher altitudes the diffusive process becomes much faster than removal of O^+ via reaction, resulting in a downward transport of ions. At still higher altitudes (the "top-side" ionosphere), the ions seek diffusive equilibrium separately, with the light ions H^+ and He^+ gradually becoming more important. In this region the behavior of the ions is to a large extent controlled by the earth's magnetic field.

Final Summary

In this book I have presented a broad range of topics related to collision phenomena from processes involving fast ions at MeV energies, as in radiation effects in material, to the thermalized chemical reactions which determine the nature of planetary atmospheres. In all cases simple models for the cross sections and rate constants can be used to describe observations in an approximate, quantitative way. In the Chapters 2–4 the physical basis of many commonly used collision models was discussed in order to establish the relationship between simple collision models and accurate theories and descriptions of molecular collisions. It is hoped, therefore, that this book will have provided the foundation for reading more detailed materials both on collision processes and on macroscopic phenomena determined by molecular interactions. The reader should be assured that considerable progress, far beyond that indicated here, has been made in terms of understanding the phenomena described. On the other hand, the lack of good atomic and molecular cross-section measurements on many systems means there are a large number of very interesting problems to solve before any of these areas is understood in great detail. As every problem area creates its own body of literature, I also hope that, on having completed this text, readers working in rather disparate areas of applied research will have obtained a vantage point that will allow them to exploit ideas and methods applied in the various fields of research related to atomic and molecular collisions.

Exercises

6.1. Use the integral equations for $\bar{R}$ and $\bar{R}_p$ to obtain the CSDA expressions of Eqs. (6.3) and (6.8) in the limit of small energy transfers.

6.2. Compare the power-law-approximation solutions of the integral equation for $\bar{R}$ and $\bar{R}_p$ [Eq. (6.14)] to the CSDA solutions for $n = 2$ and $n \to \infty$.

6.3. Derive the integral equation for $\bar{R}_p$ [Eq. (6.13b)] from Eq. (6.16) using $\bar{R}_p \cos \theta = F^1/F^0$, where F^1 and F^0 are defined in Eq. (6.19).

6.4. Derive the Poisson distribution, $P(n)$, for n "hits" occurring on a target if the average number of "hits" is $\sigma_h \cdot \Phi$, where $\Phi = It$. You can construct rate equations for $P(n)$ like Eq. (2.7).

6.5. Using a power-law potential in the Born approximation to describe the energy transfer to the electrons, obtain the radial distribution of deposited energy by the secondary electrons given in Eq. (6.32). Assume the electrons lose their energy at a constant rate and travel perpendicular to the particle track. Assume the minimum energy for ionization is negligible. How is the distribution modified if this constraint is relaxed?

6.6. Derive the escape flux from a very hot material in Eq. (6.42) using a Maxwell–Boltzmann distribution (Problem 2.3).

6.7. Demonstrate that an initially narrow, cylindrically symmetric, heated region produced by an energy deposition $|dE/dz|$ has a sputtering yield proportional to $(dE/dz)^2$, independent of the forms for K and C. That is, use the definition of thermal sputtering [Eq. (6.43)], the form of the thermal diffusion equation [Eq.

(6.44)], and the integral boundary condition, as in Eq. (6.46), to obtain the $(dE/dz)^2$ dependence for any K and C. What is the energy-deposition dependence of sputtering from a spherical spike?

6.8. Derive the thermodynamic escape flux in Eq. (6.50) from a planet with an atmosphere at temperature T and a gravitational energy barrier to escape, U_c, at the critical level.

6.9. Estimate the sputtering yield for a 1-keV proton impinging on an atmosphere consisting primarily of O above the critical height.

6.10. Justify the form of the time constants used in the discussion of the earth's ionosphere.

6.11. Derive Eq. (6.59) using equations of motion like Eq. (1.7) for the ion and electrons that include a small, vertical electric field.

6.12. The sum of all reactions at depth z, using Eq. (6.22), is $\int_0^\infty F(z, t)v_R(T)dt$. Evaluate and discuss when $v_R \gg 2D/\langle \Delta z^2 \rangle$ and $v_R \ll 2D/\langle \Delta z^2 \rangle$. When v_R and D have the same temperature, show this quantity is temperature independent.

Suggested Reading

Many useful references are included in the figure captions and tables. Others are listed below.

Radiation Transport: Ranges and Energy Deposition

A. DALGARNO in *Atomic and Molecular Processes*, ed. D. R. Bates, Academic Press, New York (1962), Chapter 15.

J. LINDHARD, M. SCHARFF, and H. E. SCHIOTT, *K. Dan. Vidensk. Selsk. Mat. Fys. Medd.* **33**, No. 14, 1 (1963).

K. B. WINTERBON, *Ion Implantation Range and Energy Deposition Distributions*, Vol. 2, *Low Incident Ion Energies*, Plenum Press, New York (1975).

P. SIGMUND, *Rev. Roum. Phys.* **17**, No. 7, 823 (1972).

P. SIGMUND, *K. Dan. Vidensk. Selsk Mat. Fys. Medd.*, 40, No. 5, 1 (1978).

D. W. PALMER, M. W. THOMPSON, and P. D. TOWNSEND, eds., *Atomic Collision Phenomena in Solids*, American Elsevier, New York (1970).

M. T. ROBINSON and I. M. TORRENS, *Phys. Rev. B*, **9** 5008 (1974).

P. D. TOWNSEND, J. C. KELLY, and N. E. W. HARTLEY, *Ion Implantation, Sputtering and their Applications*, Academic Press, New York (1976).

J. JACKSON, J. ROBINSON, and D. THOMPSON, eds. *Atomic Collisions in Solids*, North-Holland, Amsterdam (1980).

Biological Damage

R. D. COOPER and R. W. WOOD, eds., *Physical Mechanisms in Radiation Biology*, Technical Information Office, USAEC, Oak Ridge, Tennessee (1974).

H. DERTINGER and H. JUNG, *Molecular Radiation Biology*, Springer-Verlag, New York (1970).

R. A. ROTH and R. KATZ, *Radiat. Res.* **83**, 499 (1980).

Beam-Foil Transmission Experiments

I. A. SELLIN and D. J. PEGG, eds. *Beam Foil Spectroscopy*, Plenum, New York (1975).

H. G. BERRY, R. DeSERIO, and A. E. LIVINGSTON, *Phys. Rev. Lett.*, **41** 1652 (1978).

D. S. GEMMEL, *Chem. Rev.*, **80** 301 (1980).

J. REMILLIEUX, *Nucl. Inst. and Meth.*, **170** 31 (1980).

R. LEVI-SETTI, K. LAM, and T. R. FOX, *Nucl. Inst. and Meth.*, **194**, 281 (1982)

Sputtering and Thermal Spikes

D. W. PALMER, M. W. THOMPSON, and P. D. TOWNSEND, eds. *Atomic Collision Phenomena in Solids*, American Elsevier, New York (1970), Session V.

P. SIGMUND, *Phys. Rev.* **184**, 383 (1969).

P. SIGMUND, in *Proc. Symp. on Sputtering*, Vienna, April 1980, ed. P. Varga et al, (in press).

R. D. MACFARLANE and F. TORGENSON, *Science*, **191** 920 (1976).

D. A. THOMPSON, R. S. WALKER and J. A. DAVIES, *Rad. Eff.*, **32** 135 (1977).

R. Kelley, *Radiat. Eff.*, **32**, 91 (1977).

R. BEHRISCH, Sputtering by Particle Bombardment, in *Topics in Applied Physics, Vol. 47*, Springer-Verlag (1981).

C. C. WATSON and P. K. HAFF, *J. Appl. Phys.*, **51** 691 (1980).

J. B. SANDERS, *Rad. Eff.*, **51** 43 (1980).

R. E. JOHNSON and R. EVATT, *Rad. Eff.*, **52** 187 (1980).

M. SZYMONSKI, *Rad. Eff.*, **52** 9 (1980).

P. SIGMUND and C. CLAUSSEN, *J. Appl. Phys.* **52**, 990 (1981).

Atmospheres and Escape

P. M. BANKS and G. KOCKARTS, *Aeronomy*, Parts A and B, Academic Press, New York (1973).

J. W. CHAMBERLAIN, *Theory of Planetary Atmospheres*, Academic Press, New York (1978).

A. E. S. GREEN and P. J. WYATT, *Atomic and Space Physics*, Addison-Wesley, Reading, Massachusetts (1965).

J. C. G. WALKER, in *Atomic Processes and Applications*, ed. G. Burke and B. L. Moiseiwitsch, North-Holland, Amsterdam (1976), Chapter 3.

P. K. HAFF and C. C. WATSON, *J. Geophys. Res.* **84**, 8436 (1979).

J. FOX and A. DALGARNO, *J. Geophys. Res.*, **84** 7315 (1979).

Appendixes

A. Delta Functions

To represent the deterministic processes discussed in Chapter 2 we introduce the concept of a delta function, $\delta(x - x_0)$. By definition,

$$\delta(x - x_0) = \begin{cases} 0, & x \neq x_0 \\ +\infty, & x = x_0 \end{cases}$$

and

$$\int_{-\infty}^{\infty} \delta(x - x_0)\, dx = 1$$

such that

$$\int_{-\infty}^{\infty} f(x)\delta(x - x_0)\, dx = f(x_0) \tag{A.1}$$

There are a number of mathematical representations of $\delta(x - x_0)$ using functions symmetric about x_0 which have finite widths and which are used by taking the limit as this width goes to zero, e.g.,

$$\begin{aligned} \delta(x - x_0) &= \frac{1}{\pi} \lim_{\varepsilon \to +0} \varepsilon/[(x - x_0)^2 + \varepsilon^2] \\ &= \lim_{\Delta \to +0} (2\pi\Delta)^{-1/2} \exp[-(x - x_0)^2/2\Delta] \\ &= \frac{1}{\pi} \lim_{L \to \infty} \frac{\sin L(x - x_0)}{(x - x_0)} \end{aligned} \tag{A.2}$$

An often used expression from Fourier analysis, which we employ in Chapters 3 and 4, is

$$\delta(x) = \frac{1}{2\pi} \int_{-\infty}^{\infty} \exp(ikx)\,dk \tag{A.3}$$

For further discussion [see A. Messiah, *Quantum Mechanics*, Vol. 1, North-Holland, Amsterdam (1961), p. 468].

The integral property of the one-dimensional delta function for a variable defined on $(-\infty, +\infty)$, as in Eq. (A.1), is often modified in practice to $\int_{W_-}^{W_+} f(w)\delta(w - W_0)\,dw = f(W_0)$, if $W_+ > W_0 > W_-$. For the classical collision probability of Eq. (2.31), $p_{AB}(b, \theta) = (1/2\pi)\delta[\cos\theta - \cos\Theta(b)]$ with $\cos\theta$ defined on $(-1, 1)$, the total collision probability is

$$P_{AB}(b) = 2\pi \int_{-1}^{1} p_{AB}(b, \theta)\,d\cos\theta = 1$$

B. CM Deflection Function and Semiclassical Phase Shift

In the classical impulse approximation of Chapter 2, Eq. (2.57a), we have

$$\bar{\chi}(b) \simeq -\frac{d}{dR_0}\left\{\frac{1}{2E}\int_{-\infty}^{\infty} V(R)\,dZ\right\} \tag{B.1}$$

in which $R^2 = R_0^2 + Z^2$, where for distant collisions $R_0 \simeq b$. Similarly, in Chapter 3, Eq. (3.52), the quantum impulse approximation or Born approximation gives the semiclassical phase shift as

$$\eta^{SC}(b) \simeq -(m/\hbar p_0)\int_{-\infty}^{\infty} V(R)\,dZ \tag{B.2}$$

for the straight-line trajectory, $Z = p_0 t/m$. Together, Eqs. (B.2) and (B.1) confirm the semiclassical relationship, $\chi(b) = (2\hbar/p_0)\partial\eta^{SC}/\partial b$, of Eq. (3.47). The integral in Eqs. (B.1) and (B.2) evaluated for the screened potentials, $V(R) = C_n\exp(-\beta R)/R^n$ of Eq. (3.63), becomes

$$\int_{-\infty}^{\infty} V\,dZ = C_n(-1)^{1-n}\cdot 2\frac{\partial^{1-n}}{\partial\beta^{1-n}}K_0(\beta b) \qquad \text{for } n \leq 1, \ n \text{ integer} \tag{B.3}$$

where $K_0(z)$ is a modified Bessel function of zero order [see Eq. (D.8)]. For large impact parameters $(b \gg \beta^{-1})$ and arbitrary n, we have

$$\int_{-\infty}^{\infty} V\,dZ \to \frac{2C_n}{b^{n-1}}\left(\frac{\pi}{2\beta b}\right)^{1/2} e^{-\beta b} \tag{B.4}$$

In the limit $\beta \to 0$, the simple power-law potentials yield

$$\int_{-\infty}^{\infty} V \, dZ = \frac{C_n}{b^{n-1}} \cdot 2 \int_0^{\infty} dx/(1 + x^2)^{n/2}$$

$$= \frac{C_n}{b^{n-1}} \left[\Gamma(1/2) \frac{\Gamma[(n-1)/2]}{\Gamma(n/2)} \right], \qquad n > 1 \qquad \text{(B.5)}$$

which is used to obtain Eq. (2.58), with $b \sim R_0$ and Eq. (5.1). Here $\Gamma(x)$ is defined as $\Gamma(x) = \int_0^{\infty} t^{x-1} e^{-t} \, dt$ and has the properties $\Gamma(x + 1) = x\Gamma(x)$, $\Gamma(x + 1) = x!$ for x integer, and $\Gamma(1/2) = (\pi)^{1/2}$. For other values, $\Gamma(x)$ is tabulated.

The exact classical deflection function is most easily evaluated using the Gauss–Mehler quadrature

$$\int_{-1}^{1} dx \, f(x)/(1 - x^2)^{1/2} \simeq \sum_{i=1}^{n} w_i \, f(x_i) \qquad \text{(B.6)}$$

where $x_i = \cos[(2i - 1)\pi/2n]$ and $w_i = \pi/n$. With the change of variable $x = R_0/R$, Eq. (2.55) becomes

$$\chi(b) = \pi - 2\beta \int_0^{\infty} dx/[1 - (x\beta)^2 - V/E]^{1/2}, \qquad \beta = b/R_0 \qquad \text{(B.7)}$$

which, when V/E is small, is close in form to Eq. (B.6). Using Eq. (B.6) with $f(x) = \beta\{(1 - x^2)/[1 - (x\beta)^2 - V/E]\}^{1/2}$ and $n = 2m$, as $f(x)$ is an even function, we have

$$\chi(b) \simeq \pi \left[1 - 1/m \sum_{i=1}^{m} f(x_i) \right] \qquad \text{(B.8)}$$

To evaluate the semiclassical action in Eq. (3.50a) or the radial phase shift in Eq. (3.50b), the integration procedure above may be used or a related integration expression

$$\int_{-1}^{1} f(x)(1 - x^2)^{1/2} \, dx \simeq \sum_{i=1}^{n} w_i \, f(x_i)$$

with $w_i = [\pi/(n + 1)] \sin^2[i\pi/(n + 1)]$, $x_i = \cos[i\pi/(n + 1)]$, and $f(x)$ determined from Eqs. (3.50).

For further information on the modified Bessel functions, the gamma functions, or these integration techniques, see, for instance, M. Abramowitz and I. A. Stegun, eds., *Handbook of Mathematical Functions*, National Bureau of Standards, U.S. Government Printing Office, Washington, D.C. (1964).

C. Collision with an Oscillator

The equation of motion for a harmonically bound charge of mass M_i in the field of a passing atom or ion is

$$M_i \ddot{\mathbf{r}} = -M_i \Gamma \dot{\mathbf{r}} - k\mathbf{r} + \mathbf{F}_i(t) \tag{C.1}$$

in which the terms on the right are a damping force, the binding force approximated as in Eq. (3.73), and the outside force on i, $\mathbf{F}_i(t)$, due to the passing particle, A, where $\mathbf{r}$ is the position of the bound charge, i, *in* the target B. The total work done on the charge by A is

$$Q_i = \int_{-\infty}^{\infty} \dot{\mathbf{r}} \cdot \mathbf{F}_i(t)\, dt \tag{C.2}$$

which can be evaluated using Fourier transforms. Taking the Fourier transform of Eq. (C.1) and solving for $\mathbf{r}(\omega)$ [the transform of $\mathbf{r}(t)$] in terms of $\mathbf{F}(\omega)$ [the transform of $\mathbf{F}(t)$] and substituting these into Eq. (C.2) we obtain

$$Q_i = (\pi/M_i) \int_{-\infty}^{\infty} |\mathbf{F}(\omega)|^2 g(\omega)\, d\omega \tag{C.3}$$

in which $g(\omega) = (1/\pi)\{\omega^2 \Gamma/[(\omega^2 - \omega_0^2)^2 + \omega^2 \Gamma^2]\}$, with $\omega_0^2 = k/M_i$. In the limit that the damping disappears (i.e., $\Gamma \to 0$), $g(\omega)$ becomes a delta function (Appendix A), $\delta(\omega - \omega_0)$. Hence, as $\Gamma \to 0$, using Eq. (A.1), we obtain

$$Q_i = (\pi/M_i)|\mathbf{F}(\omega_0)|^2 \tag{C.4}$$

The force between a particle A of charge Z_A and a bound electron on B is the coulomb force $\mathbf{F} = -Z_A e^2(\mathbf{R} - \mathbf{r})/|\mathbf{R} - \mathbf{r}|^3$ [cf. Eq. (4.4)]. If, in addition, A is assumed to be moving in a straight line, $R^2 = v^2 t^2 + b^2$ and $r \ll R$ [as in Eq. (4.10a)], then

$$F_\perp(\omega) \simeq (Z_A e^2/bv)(2/\pi)^{1/2}[(\omega b/v)K_1(\omega b/v)]$$

and

$$F_\parallel(\omega) \simeq -i(Z_A e^2/bv)(2/\pi)^{1/2}[(\omega b/v)K_0(\omega b/v)].$$

As in Appendix B, the K_v are modified Bessel functions and we used

$$\int_{-\infty}^{\infty} \frac{e^{i\alpha x}\, dx}{(1 + x^2)^{3/2}} = 2\alpha K_1(\alpha) \quad \text{and} \quad \int_{-\infty}^{\infty} \frac{x e^{i\alpha x}\, dx}{(1 + x^2)^{3/2}} = -2i\alpha K_0(\alpha)$$

Substituting these expressions in Eq. (C.4), we obtain $(M_i \to m_e)$

$$Q_e \simeq \frac{2Z_A^2 e^4}{m_e v^2} \frac{1}{b^2} [\omega b/v]^2 [K_1^2(\omega b/v) + K_0^2(\omega b/v)] \tag{C.5a}$$

The asymptotic expressions for K_v (cf. Appendix B),

$$K_v \xrightarrow{z \to 0} \tfrac{1}{2}\Gamma(v)(Z/2)^{-v}, \qquad v \neq 0$$

$$K_0(z) \xrightarrow{z \to 0} -\ln(1.123/z)$$

and

$$K_v(z) \xrightarrow{z \to \infty} (\pi/2z)^{1/2} e^{-z}$$

when used in Eq. (C.5a), yield

$$Q_e(b, \omega) \to \frac{2Z_A^2 e^4}{m_e v^2} \frac{1}{b^2} \begin{cases} 1, & \omega b/v \ll 1 \\ \pi(\omega b/v)e^{-2\omega b/v}, & \omega b/v \gg 1 \end{cases} \qquad \text{(C.5b)}$$

Defining b/v to be the collision time, we see that $\omega b/v \ll 1$ is the classical BEA limit of Chapter 2 in which the electron's bound motion is slow compared to the incident-particle motion, and $\omega b/v \gg 1$ is the adiabatic limit of Chapter 4.

The energy transfer to the electrons can be used to calculate the stopping power $(dE/dx)_e$. That is, using Eq. (2.71), we have

$$-(dE/dx)_e = \sum_i f_i n_B 2\pi \int_{b_{\min}}^{b_{\max}} Q_e(b, \omega_i)b \, db \qquad \text{(C.6)}$$

where f_i is the oscillator strength of Eqs. (4.19) and (4.20). The integral in Eq. (C.6), on substituting Eq. (C.5), yields

$$-(dE/dx)_e = n_B (4\pi Z_A^2 e^4/m_e v^2) \sum_i f_i [B_{\min} - B_{\max}] \qquad \text{(C.7)}$$

where $B = (\omega b/v)K_1(\omega b/v)K_0(\omega b/v)$. Estimates for $b_{\min}$ and $b_{\max}$ can be made using the arguments in Chapters 2 and 5. Letting $b_{\max} \to \infty$ (i.e., $v/\omega \ll d$, the distance to the neighboring target atom), and $b_{\min} \ll v/\omega$, we have

$$-(dE/dx)_e \simeq n_B (4\pi Z_A^2 e^4/m_e v^2)\ln(1.123v/\bar{\omega}b_{\min}) \qquad \text{(C.8)}$$

where $n_B \ln(\hbar\bar{\omega}) = \sum_i f_i \ln(\hbar\omega_i) \simeq \sum_j f_j \ln(I_{B_j})$. Comparing this to earlier discussions, we see that $\hbar\bar{\omega} = \bar{I}_B$ is the mean ionization energy of Eq. (5.22). The quantity $b_{\min}$ in Eq. (C.8) is the larger of $Z_A e^2/mv^2$ and $\hbar/mv$, the classical and quantum limits in impact parameter.

If A is an incident neutral atom, which interacts with the electrons on B via a screened coulomb potential $V = (Z_A e^2/R)e^{-\beta R}$, then a similar expression for Q_e is obtained. Using the integral

$$\int_{-\infty}^{\infty} \frac{e^{-\beta R}}{R} e^{i\omega t} dt = \frac{2K_0(b\beta')}{v} \qquad \text{(C.9)}$$

where $\beta'^2 = \beta^2 + (\omega/v)^2$, we find that the energy transfer becomes

$$Q_e = \frac{2Z_A^2 e^4}{m_e v^2} \frac{1}{b^2} [(\beta'b)^2 K_1^2(\beta'b) + (\omega b/v)^2 K_0^2(\beta'b)] \tag{C.10}$$

where we used $(\partial/\partial z)K_0(z) = -K_1(z)$ [see Eq. (D.8)]. This reduces to Eq. (C.5) when $\beta \to 0$. The stopping power now becomes, using $b_{max} \to \infty$,

$$\left| \frac{dE}{dx} \right|_e = n_B \frac{4\pi Z_A^2 e^4}{m_e v^2} \sum f_i \{(\beta'b_{min})K_1(\beta'b_{min})K_0(\beta'b_{min})$$

$$- \tfrac{1}{2}(\beta b_{min})^2 [K_1^2(\beta'b_{min}) - K_0^2(\beta'b_{min})]\} \tag{C.11}$$

In the limit $\beta'b_{min} \ll 1$,

$$\left| \frac{dE}{dx} \right|_e \sim n_B \frac{4\pi Z_A^2 e}{m_e v^2} \sum f_i \left[\ln \frac{1.123}{\beta'b_{min}} - \frac{1}{2} \left(\frac{\beta}{\beta'} \right)^2 \right] \tag{C.12}$$

For very fast collisions, $\beta = \beta'$, the term in brackets becomes a constant and $(dE/dx)_e$ goes as $1/E$ (e.g., BEA). Therefore, the screening length acts like a large b (small energy transfer) cutoff on the interaction as stated at the end of Chapter 2. Compare these expressions to Eqs. (5.18), (5.22), (5.29), and (2.72).

Additional information on these topics is contained in J. D. Jackson, *Classical Electrodynamics*, Chapter 13, Wiley, New York (1963).

D. Born-Approximation Cross Section and Transition Probabilities

The first Born approximation to the elastic scattering amplitude, $f(\chi)$, was given in Eq. (3.59) as

$$f^{(1)}(\chi) = \frac{-m}{2\pi\hbar^2} \int d^3R \exp[i(\mathbf{K}_0 - \mathbf{K}_f) \cdot \mathbf{R}]V(\mathbf{R}) \tag{D.1}$$

If the potential depends only on R, then the angular integrations can be performed, giving

$$f^{(1)}(\chi) = -\frac{2m}{\hbar^2} \int \frac{\sin(\Delta pR/\hbar)}{(\Delta pR/\hbar)} V(R)R^2 \, dR \tag{D.2}$$

For the screened power-law potentials, $V(R) = (C_n/R^n)e^{-\beta R}$, the scattering amplitude scales as

$$f^{(1)}(\chi) = -\frac{2m}{\hbar^2} \left(\frac{\hbar}{\Delta p} \right)^{3-n} C_n \int_0^\infty \frac{\sin x}{x^{n-1}} e^{-\gamma x} \, dx \tag{D.3}$$

where $\gamma = \beta\hbar/\Delta p$. In the limit $\gamma \to 0$ the expression for the power laws in Eq. (3.61) is obtained. The integrals in Eqs. (D.2) and (D.3) are fourier-sine integrals which are given in A. Erdelyi, *Tables of Integral Transforms*, Volume 1, McGraw-Hill, New York (1954). For the screened potential, for $n < 3$ and $\gamma < 0$, we have

$$f^{(1)}(\chi) = \frac{-2m}{\hbar^2} \left(\frac{\hbar}{\Delta p} \right)^{3-n} C_n g_n(\gamma),$$

$$g_n(\gamma) = \Gamma(2 - n)] \sin\left[(2 - n)\cot^{-1}(\gamma) \right]/(1 + \gamma^2)^{1 - n/2} \qquad (D.4)$$

This expression is valid for power laws (i.e., $\gamma = 0$) only for $1 \le n < 3$. The function $\Gamma(x)$ is the tabulated gamma function defined, for $x > 0$, as

$$\Gamma(x) = \int_0^\infty t^{x-1} e^{-t} \, dt \qquad (D.5)$$

For $x < 0$, see p. 255.

For $x < 0$, see p. 255.

The first-order transition amplitudes in the impact parameter method are given in Eq. (4.51) as

$$C_{0f}^1(\infty) = \frac{1}{i\hbar} \int_{-\infty}^\infty V_{f0}(R) \exp(i\omega_{f0} t) \, dt \qquad (D.6)$$

with $R^2 = b^2 + v^2 t^2$. Using the wave functions for the $H^+ + H(1s) \rightarrow H^+ + H(2s)$ transition, the reader can verify that $V_{1s, 2s} = -2^{3/2} \cdot (2 + 3R)e^{-3/2R}/27$. Therefore, we are generally interested in potentials of the form $V_{f0} = A_n R^n e^{-\beta R}$. Using Eq. (C.9), we see that the coefficients have a form like that in Eq. (B.3),

$$C_{0f}^1(\infty) = \frac{2A_n}{i\hbar v}(-1)^{n+1} \frac{\partial^{n+1}}{\partial \beta^{n+1}} K_0(b\beta') \qquad (D.7)$$

where $\beta'^2 = \beta^2 + (\omega_{f0}/v)^2$. The recursion relations for the modified Bessel functions are

$$\frac{\partial K_0(z)}{\partial z} = -K_1(z) \qquad (D.8)$$

$$2n K_n(z) = -z(K_{n-1} - K_{n+1}) \qquad 2\frac{\partial K_n(z)}{\partial z} = -(K_{n-1} + K_{n+1})$$

E. Transport Equations

In Chapters 1 and 2 various transport equations are considered either with regard to the discussion of collision cascades or to chemical kinetics and diffusion. These equations are simply conservation equations that account for all loses and gains in a small volume of the material. The local rate of change of the density of a species, $\partial n_i / \partial t$, in a given volume element is due to the difference in the flow into and out of the volume element and any production rates (sources), P_i, or loss rates (sinks), L_i, within this volume:

$$\frac{\partial n_i}{\partial t} = \text{Change due to flow} + P_i - L_i \tag{E.1}$$

If the flow is uniform, the change in density due to the flow is zero. However, if there is a gradient (e.g., more "in" one side than "out" the other), then n_i is affected. Writing the flow as $n_i \mathbf{w}_i$, Eq. (E.1) can be written

$$\frac{\partial n_i}{\partial t} = -\nabla n_i \mathbf{w}_i + P_i - L_i \tag{E.2}$$

where the reader should confirm the negative sign. This is referred to, generally, as the continuity equation.

The time-independent radiation cascade equation can be developed from the expression in Eq. (1.1). If the radiation intensity $I_i(\mathbf{p}, z)$ defined in the text is used, then $I_i(\mathbf{p}, z)(\hat{p} \cdot \hat{z}) d^3p$ is the number of particles per unit area per unit time of type i with momentum between $\mathbf{p}$ and $\mathbf{p} + d\mathbf{p}$ crossing a surface at depth z. Here $\hat{p} \cdot \hat{z}$ is the cosine between the incident direction $\hat{p}$ and the surface normal. As I_i is assumed uniform across the surface in our model, the total number of particles crossing the surface is

$$A \int I_i(\mathbf{p}, z)(p \cdot \hat{z}) d^3p$$

where A is the surface area. If $\omega(\mathbf{p}, \mathbf{p}') d^3p'$ is the probability per unit path length of a collision occurring in which $\mathbf{p} \to \mathbf{p}'$, then $\int \omega_i(\mathbf{p}, \mathbf{p}')\Delta z/(\hat{p} \cdot \hat{z}) d^3p'$ is the total probability of particles $\mathbf{p}$ changing their momentum when crossing a slab Δz thick. The path length through the slab is, of course, $\Delta z/(\hat{p} \cdot \hat{z})$. Using these definitions, we see that the change in flux across a slab Δz thick, as defined in Eq. (1.1), becomes

$$[I_i(\mathbf{p}, z + \Delta z) - I(\mathbf{p}, z)](\hat{p} \cdot \hat{z}) dp^3$$
$$= -I_i(\mathbf{p}, z)(\hat{p} \cdot \hat{z}) d^3p \int \omega_i(\mathbf{p}, \mathbf{p}') \left(\frac{\Delta z}{\hat{p} \cdot \hat{z}}\right) d^3p'$$
$$+ \left[\int I_i(\mathbf{p}', z)(\mathbf{p}' \cdot z)\omega_i(\mathbf{p}', \mathbf{p})\left(\frac{\Delta z}{\hat{p}' \cdot \hat{z}}\right) d^3p'\right] d^3p$$
$$+ \mathscr{S}_i(\mathbf{p}, z)\Delta z \, d^3p \tag{E.3}$$

The first term on the right is the loss due to scatterings of i type particles of momentum $\mathbf{p}$, where $(\Delta z/\hat{p} \cdot \hat{z})$ is their path length through the slab. The second term is the addition to the flux of particles of type $(i, \mathbf{p}')$ by scattering from $\mathbf{p}'$ into the interval $\mathbf{p}$ to $\mathbf{p} + d\mathbf{p}$, where the path length for the $\mathbf{p}'$

particles through the slab is $\Delta z/\hat{p}' \cdot \hat{z}$. The last term accounts for all other sources and sinks for particles of type $(i, \mathbf{p})$, that is, the probability of an absorption, ionization, nuclear reaction, etc. in the slab. This last term obviously contains a considerable amount of physics of the processes occurring in materials. The integral equation (E.3) can be converted into an integral-differential equation by writing

$$I_i(\mathbf{p}, z + \Delta z) \simeq I_i(\mathbf{p}, z) + \Delta z \frac{\partial I_i}{\partial z} (\mathbf{p}, z) \tag{E.4}$$

Using Eq. (E.4) and canceling the Δz and angular factors $(\hat{p} \cdot \hat{z})$, we see that Eq. (E.3) becomes

$$-\cos\theta \frac{\partial I_i}{\partial z} (\mathbf{p}, z) = \int [I_i(\mathbf{p}, z)\omega_i(\mathbf{p}, \mathbf{p}')$$

$$-I_i(\mathbf{p}', z)\omega_i(\mathbf{p}', \mathbf{p})] \, d^3p' - \mathscr{S}_i(\mathbf{p}, z) \tag{E.5}$$

where we have defined $\cos\theta = \hat{p} \cdot \hat{z}$. This is the expression used in Eq. (1.2). This result is related to that in Eq. (E.2) if the momenta are randomly oriented, so $\int I_i(\mathbf{p}, z) \, d^3p = 0$, and the net flow $(nw)_i = \int I_i(\mathbf{p}, z) \cos\theta \, d^3p$. Integrating Eq. (E.5) over all momenta, we see that the collision terms cancel (as many are scattered into any direction as out of) and the time-dependent continuity equation in one dimension is recovered: $\partial(nw)_i/\partial z = \mathscr{S}_i$. Here $\mathscr{S}_i = \int \mathscr{S}_i(\mathbf{p}, z) \, d^3p = P_i - L_i$.

In Chapter 2 we discuss the collision-cross-section differential in angle rather than the $\omega(\mathbf{p}, \mathbf{p}')$. These are related by $(d\sigma/d\Omega) \, d\Omega \, \delta[\mathbf{p}' - \mathbf{p}'(\Omega, \mathbf{p})]$ $d^3p' = (1/n_T)\omega(\mathbf{p}, \mathbf{p}') \, d^3p'$ where n_T is the target number density, generally written n_B in Chapter 2, where B identifies the target atoms or molecules. The quantity Ω indicates the solid angle associated with the scattered momentum direction $\hat{p}'$, as measured from the incident direction, here $\hat{p}$ (viz. Figure 2.6), and $\mathbf{p}'(\Omega, \mathbf{p})$ is the final momentum calculated by classical kinematics. The delta function is discussed in Chapter 2 and in Appendix A. The reader is also referred to Eq. (6.17) and the following discussion.

F. The Stationary-Phase Approximation

The stationary-phase approximation is used extensively in Chapters 3 and 4 to evaluate integrals of oscillatory functions having the general form

$$I = \int_{-\infty}^{\infty} A(x) \exp\left[i\alpha(x)\right] dx \tag{F.1}$$

where $A(x)$ is a slowly varying quantity. If the phase, $\alpha(x)$, changes rapidly (or randomly) with x, then the contributions to I tend to cancel. On the other hand, if $\alpha(x)$ has a stationary point, x_0, i.e., $\alpha' = d\alpha/dx|_{x=x_0} = 0$, then α will be slowly varying over a region of x about x_0, and a significant contribution to I may acrue, as indicated in Figure 3.6. Writing $\alpha(x) \simeq \alpha(x_0) + [(x - x_0)^2/2]\alpha''$ in a region about x_0 having extent Δx, we can approximate the integral by zero outside this region, and

$$I \approx \exp[i\alpha(x_0)] \int_{x_0 - \Delta x/2}^{x_0 + \Delta x/2} A(x) \exp\left[\frac{i\alpha''}{2}(x - x_0)^2\right] dx \qquad (F.2)$$

The slowly varying function $A(x)$ is replaced by its value $A(x_0)$ in this region. The remaining integral is now evaluated by extending the limits of integration to infinity,

$$I \approx \exp[i\alpha(x_0)] A(x_0) \int_{-\infty}^{\infty} \exp\left[\frac{i\alpha''}{2}(x - x_0)^2\right] dx \qquad (F.3)$$

This extension is valid since, when $|\Delta x| \gg |x_0|$, the integral in Eq. (F.3) oscillates rapidly as x changes, producing cancellation, and the primary contribution again comes from a small region about x_0. The correct procedure for evaluating the integral in Eq. (D.3) is to use complex integration methods. However, noting that $\int_{-\infty}^{\infty} \exp[-\beta(x - x_0)^2] dx = (\pi/\beta)^{1/2}$ and writing $\beta = \varepsilon - i\alpha''/2$, we see that the integral becomes, in the limit $\varepsilon \to 0$,

$$I \approx A(x_0) \exp[i\alpha(x_0)][\pi/(-i\alpha''/2)]^{1/2}$$

or

$$I \approx (2\pi/\alpha''(x_0))^{1/2} A(x_0) \exp\{i[\alpha(x_0) + \pi/4]\} \qquad (F.4)$$

This is the approximation used in Chapters 3 and 4.

G. Atomic State Labels

The wave function for the one-electron atom in Eq. (3.71) has the form $\psi_{nlm_l} = \mathcal{R}_{nl} Y_{lm_l}$, where the label n is an energy index, l ($l = 0, 1, 2, \ldots$) is the angular momentum index, and $m_l (m_l = l, l - 1, \ldots, -l + 1, -l)$, is an index for the angular momentum along the arbitrary axis through the atom. The latter is important only when the electron is in an applied field or the field of another electron or nucleus. Each electron also has a spin, which is another angular momentum quantity. As there are two possible spin orientations, the angular momentum labeling is $s = \frac{1}{2}$ and $m_s = \pm\frac{1}{2}$, by analogy with l and m_l. For a multielectron atom, the total angular momentum of a given atomic state is the only observable quantity. However, for light atoms, which we will concentrate on here, it is useful to imagine the atomic state as being built up from single-electron states. Each such state has a given angular momentum and spin, with the orbital and spin angular momentum combining separately. This assumes that the electrons in the atom orbit independently in the field of the nucleus and the average field of the other electrons.

The atomic-orbital method, based on the Pauli principle, assigns labels n and l for each electron orbital and allows each orbit type to be filled with only that number of electrons determined by the m_l values for each l and the two possible spin orientations. Each such orbital or shell, therefore, can contain at most $2(2l + 1)$ electrons. Using the spectroscopic notation (i.e., s, p, d, f, g, ... corresponds to $l = 0, 1, 2, 3, 4, ...$), one often labels a ground state hydrogen atom $(1s)$, where one is the energy index n, and s is the angular momentum index, $l = 0$. Helium becomes $(1s)^2$, implying that the two electrons have different spin orientations. Lithium, with three electrons, becomes $(1s)^2(2s)$; carbon, with six electrons, becomes $(1s)^2(2s)^2(2p)^2$; and neon completes the $n = 2$ shell with 10 electrons, $(1s)^2(2s)^2(2p)^6$.

The individual angular momenta combine to give a total orbital angular momentum, L, and spin angular momentum, S. Again the spectroscopic notation S, P, D, for sharp, principal, and diffuse lines, which were found to correspond to $L = 0, 1, 2$, is generally used to label the states. As the lines were often observed to be split slightly, each state was found to have a multiplicity related to the total spin, $(2S + 1)$. Hence hydrogen is labeled $(1s)$ ^2S or just ^2S, where the superscript is the spin multiplicity.

Helium is $(1s)^2$ ^1S and lithium $(1s)^2(2s)$ ^2S. The carbon atom has a number of angular-momentum states corresponding to $(1s)^2(2s)^2(2p)^2$ because of the vector addition of the angular momenta associated with the $(2p)$ electrons. By the so-called Hund's rules for angular-momentum coupling, the state lowest in energy has been found to be the one with the highest multiplicity and the lowest L, here ^3P. Neon has only one total angular-momentum state corresponding to the closed-shell configuration, $(1s)^2(2s)^2(2p)^6$ ^1S. The reader is referred to any one of the many modern physics or quantum-chemistry text for further information on the labeling of atomic states. The material presented above is adequate for understanding any references to atomic states in this text.

H. Thomas–Fermi Interaction: Results

Bohr, Firsov, Lindhard, and others have used the Thomas–Fermi model for calculating charge distributions to determine "universal" interaction potentials between atoms. The potential is written in the form

$$V_{AB}(R) = \frac{Z_A Z_B e^2}{R} \Phi \frac{R}{(a_{AB})} \tag{H.1}$$

where the function Φ does not depend on Z_A and Z_B explicitly. The variable is expressed as the internuclear separation scaled by the screening constant, a_{AB} [viz. Eq. (4.8)]. The Thomas–Fermi model is a statistical model for describing the electronic charge distribution. In this model the electrons are treated as a gas in which the uncertainty principle restricts the number of electrons having a given momentum in any volume of space. Lindhard and co-workers obtained an extensively used numerical solution for $\Phi(R/a_{AB})$, in which the screening length is $a_{AB}^L = 0.8853a_0(Z_A^{2/3} + Z_B^{2/3})^{-1/2}$. Lenz and Jensen earlier had obtained a variational solution to a similar interaction model,

$$\Phi_{LJ} = (1 + b_1 y + b_2 y^2 + b_3 y^3 + b_4 y^4)e^{-y} \tag{H.2}$$

where $y = (9.67R/a_{AB}^L)^{1/2}$, and $b_1 = 1$, $b_2 = 0.3344$, $b_3 = 0.0485$, and $b_4 = 0.002647$. These were employed in the comparison in Figure 5.6. Recent experimental evidence* suggests $b_1 = 0.9839$, $b_2 = 0.4272$, $b_3 = 0.01150$, and $b_4 = 0.01288$ for certain ion–atom collisions. Lindhard and co-workers used the Thomas–Fermi screening function to obtain a general form for the differential cross section to describe collisions involving many-electron atoms. In scaled units we have

$$\frac{d\sigma_{AB}}{dt} = \frac{\pi(a_{AB}^L)^2 f(t^{1/2})}{(2t^{3/2})} \tag{H.3}$$

where $t = \varepsilon^2 \, T/T_{max}$, with $\varepsilon = Ea_{AB}^L/Z_A Z_B e^2$, E is the CM energy, and $T_{max} = \gamma E_A$. The function $f(t^{1/2})$ is well approximated by $f(t^{1/2}) \simeq \lambda t^{1/6}/[1 + (2\lambda t^{2/3})^{2/3}]^{3/2}$, where $\lambda = 1.309$. Using this expression, we see that the corresponding nuclear component of the stopping cross section, S_n, in reduced units $[S_n = \pi(a_{AB}^L)^2(T_{max}/\varepsilon)S_n(\varepsilon)]$ is

$$s_n(\varepsilon) \simeq \tfrac{9}{8\varepsilon}\{\ln[A + (1 + A^2)^{1/2}] - A/(1 + A^2)^{1/2}\} \tag{H.4}$$

with $A = (2\lambda)^{1/3}\varepsilon^{4/9}$.

For background on this material see the following: W. Lenz, Z. Phys. **77**, 713 (1932); H. Jensen, Z. Phys. **77**, 722 (1932); O. B. Firsov, Sov. Phys. JETP **5**, 1192 (1957); J. Lindhard, V. Nielsen, and M. Scharff, Dan. Vidensk. Selsk. Mat. Fys. Medd. **36**, No. 10 (1968); P. Sigmund, in Physics of Ionized Gases, ed. M. Kurepa, (1972), p. 137; and J. Lindhard and M. Scharff, Phys. Rev. **124**, 128 (1961). For a discussion of the Thomas–Fermi interaction see I. M. Torrens, Interatomic Potentials, Academic Press, New York (1972), Chapter 3, and P. Gombás, Handbuch der Physik, XXXVI (1956).

I. Low Energy, Inelastic Cross Sections

Those inelastic cross sections which can be described by two states at low velocities often have simple analytic forms. From Eq. (2.24) and (4.61a),

* P. Lottager, F. Besenbacher, O. S. Jensen, and V. S. Sørensen, Phys. Rev. A, **20**, 1443 (1979).

(4.59), or (4.62), the cross sections can be written as

$$\sigma_{0\to f}^{(v)} \approx 2\pi \int_0^\infty \bar{P}_{0f} \cdot 2\sin^2(\Delta\varphi_{0f})b\,db \tag{I.1}$$

where $\Delta\varphi_{0f}$ is the net phase change. This integral requires numerical evaluation unless the random phase approximation on the oscillatory part can be used and $2\sin^2(\Delta\varphi_{0f})$ can be replaced by one. Firsov [O. Firsov, *Zh. Eksp. Teor. Fiz.* **21**, 1001 (1951)] makes this replacement for impact parameters less than that (b^*) at which the change in phase first becomes $\sim 1/\pi$. For an exponential interaction such that

$$(\Delta\varphi_{0f}) \sim \left(\frac{1}{\hbar}\int_{-\infty}^\infty V_{0f}\,dt\right),$$

and using Eq. (B.4) then b^* is estimated from

$$V_{0f}(b^*)(2\pi b^*/\beta)^{1/2}/v \approx 1/\pi. \tag{I.2}$$

For symmetric resonant, charge exchange [Eq. (4.62)] (Eq. I.1) becomes

$$\sigma_{c.t.}(v) \approx \pi b^{*2}/2. \tag{I.3}$$

For inelastic collisions the cross section in (Eq. I.1) becomes

$$\sigma_{0\to f} \approx \pi R^{*2} P_{0\to f}(v) \tag{I.4}$$

as in Eq. (2.11), where for the LZS approximation [Eq. (4.59)] $R^* = R_x$ and for the Demkov approximation [Eq. (4.61c)] R^* is the lesser of b^* and R_x. For the LZS case the reader can verify that

$$P_{0\to f} \approx 4\left(1 - \frac{V_{00}(R_x)}{E}\right)[E_3(\xi) - E_3(2\xi)] \tag{I.5}$$

where $E_3(\xi)$, the exponential integral $\int_1^\infty e^{-\xi x}dx/x^3$, is tabulated (Reference in appendix B) and $\xi = \tau_x/\tau_{0f}$ of Eq. (4.57) evaluated at $b = 0$. The Demkov (Eq. 4.61c) expression has a similar form

$$P_{0\to f} \approx 4(1 - V_{00}(R_x)/E)\int_1^\infty dx\, e^{-2\xi'x}/(1 + e^{-2\xi'x})^2 \tag{I.6}$$

where $\xi' = (\xi/2\hbar\beta) \cdot \Delta E_{0f}/v'$, where $v' = v[1 - V_{00}(R_x)/E]^{1/2}$ and $\Delta E_{0f} = \varepsilon_f + V_{ff} - \varepsilon_0 - V_{00})|_{R_x}$ the energy splitting at R_x, defined by $V_{0f}(R_x) = (1/2)\Delta E_{0f}$. Evaluating $P_{0\to f}$, with exponential coupling and constant ΔE_{0f} by numerically integrating the two state equations Olson [R. E. Olson, *Phys. Rev. A* **6**, 1822 (1972)] gives a numerical result which is more accurate than that in Eq. (I.6). Writing $P_{0\to f}(v) = (1 - V_{00}(R_x)/E)\tilde{P}$, values of $\tilde{P}$ are given in Table I.1.

When $b^* < R_x$, the Demkov (or Rozen-Zener) cross section has the

Table I.1. Values of **p**

$(\xi')^{-1}$	$\tilde{P}$	$(\xi')^{-1}$	$\tilde{P}$
0.5	0.015	3.0	0.54
1.0	0.15	3.5	0.53
1.5	0.33	4.0	0.52
2.0	0.48	4.5	0.51
2.5	0.52	5.0	0.50

form of Eq. (I.3), i.e., $\bar{P}_{0f} \rightarrow \frac{1}{2}$ in Eq. (I.1). For all these expressions the semiempirical coupling potential $V_{0f} = (I_0 I_f)^{1/2} (\beta R) e^{-.86\beta R}$ gives reasonable results where $\beta = 0.5 [(2I_0)^{1/2} + (2I_f)^{1/2}]$ with I_0 and I_f the binding energy of the electron in atomic units [R. E. Olson, F. T. Smith, and E. Bauer, *App. Optics* **10**, 1848 (1971)]. Smirnov [B. M. Smirnov, *Sov. Phys.-JETP* **20**, 345 (1965)] gives a more accurate expression for the symmetric resonant case.

When the initial state potential curve crosses a continuum or near continuum of final states [e.g., $A^{+n} + B \rightarrow A^{+(n-1)} + B^+$ or the Langerin process, Eq. (5.35)] then the total reaction cross section has the form of Eq. (I.4) with $P_{0 \rightarrow f} \rightarrow 1$. This is referred to as an absorbing sphere model and R^* is determined by that distance at which the reaction probability first becomes "significant": $\tau_x/\tau_{0f} \sim 0.15$ [R. E. Olson and A. Salop, *Phys. Rev. A.*, **14**, 579 (1976)]. For determining R^* as a tunneling problem see T. P. Grozdanov and R. Janer, *Phys. Rev. A.*, **17**, 880 (1978).

J. Constants and Units

Table A1. Physical Constants[a]

Designation of quantity	Symbol	Value	Units		
			SI	cgs	Other
Avogadro's number	N_A	6.0222	10^{26} kmol^{-1}	10^{23} mol^{-1}	
Atomic mass unit (^{12}C = 12)	amu	1.6605	10^{-27} kg	10^{-24} g	
Electron charge/mass ratio	e/m_e	1.7588	10^{11} C kg^{-1}	10^7 emu g^{-1}	
Electron charge	e	1.6022	10^{-19} C	10^{-20} emu	4.8033×10^{-10} esu
Electron mass	m_e	9.1096	10^{-31} kg	10^{-28} g	0.5110 MeV/c^2
Bohr radius (a.u.)	a_0	5.2918	10^{-11} m	10^{-9} cm	
Rydberg constant	R_∞	1.097373	10^7 m^{-1}	10^5 cm^{-1}	
Speed of light in a vacuum	c	2.99792	10^8 m s^{-1}	10^{10} cm s^{-1}	
Planck's constant	h	6.6262	10^{-34} J s	10^{-27} erg s	
Dirac's h ($h/2\pi$)	h	1.0546	10^{-34} J s	10^{-27} erg s	6.5819×10^{-16} eV s
Hartree (a.u.)	a.u.	4.3594	10^{-18} J	10^{-11} erg	27.21 eV
Gas constant	R	8.314	10^3 J kmol^{-1} K^{-1}	10^7 erg mol^{-1} K^{-1}	
Boltzmann's constant (R/N_A)	k	1.3806	10^{-23} J K^{-1}	10^{-16} erg K^{-1}	8.617×10^{-5} eV/°K

[a] From R. D. Levine and R. B. Bernstein, *Molecular Reaction Dynamic*, Oxford University Press, Oxford (1974).

Table A2. Atomic Units (a.u.)

Mass: $m_e = 1$, 1 amu = 1823 (based on ^{12}C = 12 amu)

Charge: $e = 1$

Angular momentum: $h = 1$

1 a.u. of length: $h^2/(m_e e^2) = 0.529 \times 10^{-8}$ cm = a_0 = Bohr radius of H

1 a.u. (energy, $m_e e^4/h^2$) = 27.2 eV = 1 hartree = (2 × ground-state energy of H)

1 a.u. (velocity, e^2/h) = 2.19×10^8 cm/s = $c/137.0$ = (speed of light × fine-structure constant)

1 a.u. (time, $h^3/m_e e^4$) = 0.242×10^{-16} s = (period of electron in ground state of H/2π)

Table A3. Approximate) Energy Conversion Factors[a]

	erg	J	cal	eV	a.u.	cm⁻¹	Hz	K	kJ mol⁻¹	kcal mol⁻¹
1 erg =	1	1.000(−7)	2.39(−8)	6.24(+11)	2.29(+10)	5.03(+15)	1.509(+26)	7.24(+15)	6.02(+13)	1.440(+13)
1 joule (J) =	1.000(+7)	1	2.39(−1)	6.24(+18)	2.29(+17)	5.03(+22)	1.509(+33)	7.24(+22)	6.02(+20)	1.440(+20)
1 cal =	4.184(+7)	4.184	1	2.61(+19)	9.58(+17)	2.10(+23)	6.31(+33)	3.03(+23)	2.52(+21)	6.02(+20)
1 eV =	1.602(−12)	1.602(−19)	3.83(−20)	1	3.68(−2)	8.07(+3)	2.42(+14)	1.161(+4)	9.65(+1)	2.31(+1)
1 hartree (au) =	4.36(−11)	4.36(−18)	1.042(−18)	2.72(+1)	1	2.19(+5)	6.58(+15)	3.16(+5)	2.63(+3)	6.28(+2)
1 cm⁻¹ =	1.986(−16)	1.986(−23)	4.75(−24)	1.240(−4)	4.56(−6)	1	3.00(+10)	1.439	1.200(−2)	2.86(−3)
1 Hz =	6.63(−27)	6.63(−34)	1.585(−34)	4.14(−15)	1.520(−16)	3.34(−11)	1	4.80(−11)	3.99(−13)	9.54(−14)
1°K (K) =	1.380(−16)	1.380(−23)	3.30(−24)	8.62(−5)	3.17(−6)	6.95(−1)	2.08(+10)	1	8.32(−3)	1.988(−3)
1 kJ mol⁻¹ =	1.659(−14)	1.659(−21)	3.97(−22)	1.035(−2)	3.81(−4)	8.36(+1)	2.50(+12)	1.202(+2)	1	2.39(−1)
1 kcal mol⁻¹ =	6.94(−14)	6.94(−21)	1.661(−21)	4.33(−2)	1.593(−3)	3.50(+2)	1.048(+13)	5.03(+2)	4.184	1

[a] Numbers in parentheses denote powers of ten by which the entry is to be multiplied. From R. D. Levine and R. B. Bernstein, *Molecular Reaction Dynamics*, Oxford University Press, Oxford (1974).

Table J.4

Elements	Z	M (a.m.u)	I (eV)	Term	Elements	Z	M (a.m.u)	I (eV)	Term
H	1	1.01	13.60	^2S	Cu	29	63.55	7.72	^2S
He	2	4.00	24.58	^1S	Zn	30	65.38	9.39	^1S
Li	3	6.94	5.39	^2S	Ga	31	69.72	6.00	^2P
Be	4	9.01	9.32	^1S	Ge	32	72.60	7.88	^3P
B	5	10.81	8.30	^2P	As	33	74.92	9.81	^4S
C	6	12.01	11.26	^3P	Se	34	78.96	9.75	^3P
N	7	14.01	14.54	^4S	Br	35	79.90	11.84	^2P
O	8	16.00	13.61	^3P	Kr	36	83.80	14.00	^1S
F	9	19.00	17.42	^2P	Rb	37	85.47	4.18	^2S
Ne	10	20.18	21.56	^1S	Sr	38	87.62	5.69	^1S
Na	11	22.99	5.14	^2S	Y	39	88.91	6.38	^2D
Mg	12	24.31	7.64	^1S	Zr	40	91.22	6.84	^3F
Al	13	26.98	5.98	^2P	Nb	41	92.91	6.88	^6D
Si	14	28.09	8.14	^3P	Mo	42	95.94	7.10	^7S
P	15	30.98	10.55	^4S	Tc	43	98.91	7.28	^6S
S	16	32.06	10.36	^3P	Ru	44	101.1	7.36	^5F
Cl	17	35.45	13.01	^2P	Rh	45	102.91	7.28	^6S
Ar	18	39.95	15.76	^1S	Pd	46	106.4	8.33	^1S
K	19	39.10	4.34	^2S	Ag	47	107.87	7.57	^2S
Ca	20	40.08	6.11	^1S	Cd	48	112.40	8.99	^1S
Sc	21	44.96	6.56	^2D	In	49	114.82	5.78	^2P
Ti	22	47.90	6.83	^3F	Sn	50	118.69	7.33	^3P
V	23	50.94	6.74	^4F	Sb	51	121.75	8.64	^4S
Cr	24	52.00	6.76	^7S	Te	52	127.60	9.01	^3P
Mn	25	54.94	7.43	^6S	I	53	126.90	10.44	^2P
Fe	26	55.85	7.90	^5D	Xe	54	131.30	12.13	^1S
Co	27	58.93	7.86	^4F	Cs	55	132.91	3.89	^2S
Ni	28	58.71	7.63	^3F	Ba	56	137.34	5.21	^1S

Subject Index

Author Reference Index

Interactions, Reactions and Collisions: References and Results*

$(A^+) + O_2^-$, 252
(A^+) + RNase, 238
(A^+) + Xe, 167
$Ar^+ + Ar$, 177, 212
$Ar^+ + Ne$, 177
(B) + T.E., 231
$Be^{+2} + H$, 187
(C) + T.E., 231
$CO_2 + N_2$, 11
$CO_2^+ + O$, 14, 208, 251
$Cs^+ + Cs$, 180
$Cs^+ + Rb$, 180
$Cs^+ + Xe$, 173
$D^+ + He$, 186
e + Ar, 169
$e + CO_2^+$, 251
e + H, 22
e + He, 162, 169
$e + He_2^+$, 210, 211
e + Kr, 169
$e + N_2$, 11
e + Ne, 169
$e + NO^+$, 253
$e + O_2^+$, 211, 247, 251
$e + O_3$, 211
$e + O_2 + (X)$, 211, 213, 252
e + Xe, 169
$F + H_2$, 203, 207

H^+ + Air, 232
$H^+ + Ar$, 126, 129, 133, 140, 144, 164, 232
$H^+ + CH_4$, 232
$H^+ + CO_2$, 232
$H^+ + H$, 112, 121, 127, 137, 145, 153, 154, 155, 185, 248
H + H, 129, 153, 154
$H^+ + H^-$, 129
$H^+ + H_2$, 185
$H^+ + He$, 185, 194, 232
H + He, 154
$(H^+) + H_2O$, 244
$H^+ + Kr$, 164, 232
$H^+ + N_2$, 232
$H^+ + Ne$, 194, 232
$H^+ + O$, 248
H^+ + T.E., 232
$H^+ + (X)$, 192
H + (X), 192
$H^+ + Xe$, 232
He^{++} + Air, 232
$He^+ + Ar$, 52, 184
$He^{++} + Ar$, 181, 232
$He^{++} + CH_4$, 232
$He^+ + CO$, 207, 208
$He^{++} + CO_2$, 232
$He^{++} + H$, 123, 129, 137

$(He^+) + H$, 194
$He^{++} + He$, 124, 186, 232
$He^+ + He$, 148
He + He, 22, 129
$(He^+) + H_2O$, 244
$He^{++} + Kr$, 232
$He^{++} + Ne$, 47, 48, 154, 181, 220, 232
$He^+ + Ne$, 160, 220, 232
$He_2^+ + Ne$, 199, 208, 210
$He^{++} + N_2$, 232
He^{++} + T.E., 232
$(He^+) + (X)$, 196
$He^{++} + Xe$, 232
$h\nu + CO_2$ 250
$h\nu + O$, 250
$K^+ + Ar$, 173
$K + Br_2$ 202
$K + I_2$, 205
K + Kr, 170
K + Xe, 163
KCl + NaBr, 204
$Li^+ + He$, 173
$N^{++} + He$, 177
$N_2^+ + N_2$, 179
$N_2^+ + O$, 14, 208
$N_2^+ + O_2$, 14
Na + Xe, 164

*The brackets ( ) imply arbitrary initial charge state and/or arbitrary molecular species. T.E. implies a gas mixture which is equivalent to tissue in having an appropriate mix of C, H, N, and O atoms.